Lecture Notes in Mathematics 1779

Editors:
J.-M. Morel, Cachan
F. Takens, Groningen
B. Teissier, Paris

Springer
Berlin
Heidelberg
New York
Barcelona
Hong Kong
London
Milan
Paris
Tokyo

Igor Chueshov

Monotone Random Systems Theory and Applications

 Springer

Author

Igor Chueshov
Dept. of Mechanics and Mathematics
Kharkov National University
4 Svobody Square
61077 Khrakov, Ukraine
e-mail: chueshov@ilt.kharkov.ua

Cataloging-in-Publication Data applied for

Die Deutsche Bibliothek - CIP-Einheitsaufnahme

Čuešov, Igor' D.:
Monotone random systems theory and applications / Igor Chueshov. –
Berlin ; Heidelberg ; New York ; Barcelona ; Hong Kong ; London ; Milan ; Paris ; Tokyo :
Springer, 2002
(Lecture notes in mathematics ; 1779)
ISBN 3-540-43246-9

Mathematics Subject Classification (2000):
PRIMARY: 37H10; SECONDARY 34C12, 34F05, 60H10, 37G35

ISSN 0075-8434
ISBN 3-540-4326-9 Springer-Verlag Berlin Heidelberg New York

Springer-Verlag Berlin Heidelberg New York a member of BertelsmannSpringer
Science + Business Media GmbH

http://www.springer.de

© Springer-Verlag Berlin Heidelberg 2002
Printed in Germany

Typesetting: Camera-ready TEX output by the author

SPIN: 10866602   41/3142/DU - 543210 - Printed on acid-free paper

Preface

There is a long history of the application of monotone methods and comparison arguments in deterministic dynamical systems (see, e.g., SMITH [102] and the literature quoted there). Monotonicity methods are now fully integrated within the framework of deterministic dynamical systems theory. The situation is quite different for random systems and stochastic differential equations. Monotonicity arguments have mainly been used for one-dimensional random or stochastic differential equations, relying on well-known comparison theorems for solutions of one-dimensional ordinary random (see, e.g., LADDE/LAKSHMIKANTHAM [75]) or stochastic (see, e.g., IKEDA/WATANABE [57]) equations. In particular, these theorems and also the analysis of some explicitly solvable models make it possible to give a complete description of random attractors and bifurcation scenarios for several rather complicated situations (see, e.g., ARNOLD [3, Chap. 9]).

Let us also mention that products of positive random matrices have been the subject of numerous studies (comprising, in particular, a random version of Perron-Frobenius theory) with applications notably in economics and biology (for a survey, see ARNOLD/DEMETRIUS/GUNDLACH [8]). KELLERER [65] found that independent identically distributed iterations of monotone random mappings on $\mathbb{R}_+$ are a model ideally suited for extending discrete Markov chain theory to uncountable state spaces.

Our main goal in this book is to present the basic ideas and methods for order-preserving (or monotone) random dynamical systems that have been developed over the past few years. We focus on the qualitative behaviour of these systems and our main objects are equilibria and attractors.

There is a deep analogy between the theory of random dynamical systems and the classical theory of dynamical systems. This analogy makes it possible to develop qualitative theory for stochastic systems relying on ideas of classical dynamical systems. In this book we try to expose this analogy in a clear and transparent way. We hope it makes the book accessible not only to experts in stochastic analysis but also to people working in the field of deterministic dynamical systems. It provides a bridge from classical theory to stochastic dynamics and it can be also used as an introductory textbook on random dynamical systems at the graduate level.

Our main application is to the so-called cooperative random and stochastic ordinary differential equations. These systems arise naturally from mathematical models in the field of ecology, epidemiology, economics and biochemistry (see, e.g., the literature quoted in SMITH [102]). Deterministic cooperative differential equations have been studied by many authors (see, e.g., SMITH [102] and the references therein). The books by KRASNOSELSKII [68, 69] and the series of papers by HIRSCH [52, 53, 54] (see also the references in SMITH [102]) lay the groundwork for the qualitative theory of deterministic cooperative systems. Monotone methods and comparison arguments are of prime importance in the study of these systems.

The results presented in this book rely on ideas and methods developed in collaboration with Ludwig Arnold (see ARNOLD/CHUESHOV [5], [6] and [7]). The author is extremely grateful to him for very stimulating and fruitful discussions on the subject. Warmest thanks are also due to Gunter Ochs, James Robinson and Björn Schmalfuss for their comments and suggestions, all of which improved the book.

The book was written while the author was spending the 2000/2001 academic year at the Institut für Dynamische Systeme, Universität Bremen. He would like to thank the people at that institution for their very kind hospitality during this period. He also gratefully acknowledges the financial support of the Deutsche Forschungsgemeinschaft.

September 2001   *Igor Chueshov*

Contents

Introduction 1

1. General Facts about Random Dynamical Systems 9
    1.1 Metric Dynamical Systems 9
    1.2 Concept of RDS 13
    1.3 Random Sets 18
    1.4 Dissipative, Compact and Asymptotically Compact RDS 24
    1.5 Trajectories 32
    1.6 Omega-limit Sets 34
    1.7 Equilibria 38
    1.8 Random Attractors 41
    1.9 Dissipative Linear and Affine RDS 45
    1.10 Connection Between Attractors and Invariant Measures 49

2. Generation of Random Dynamical Systems 55
    2.1 RDS Generated by Random Differential Equations 55
    2.2 Deterministic Invariant Sets 61
    2.3 The Itô and Stratonovich Stochastic Integrals 65
    2.4 RDS Generated by Stochastic Differential Equations 70
    2.5 Relations Between RDE and SDE 76

3. Order-Preserving Random Dynamical Systems 83
    3.1 Partially Ordered Banach Spaces 83
    3.2 Random Sets in Partially Ordered Spaces 88
    3.3 Definition of Order-Preserving RDS 93
    3.4 Sub-Equilibria and Super-Equilibria 95
    3.5 Equilibria 100
    3.6 Properties of Invariant Sets of Order-Preserving RDS 105
    3.7 Comparison Principle 109

4. Sublinear Random Dynamical Systems 113
    4.1 Sublinear and Concave RDS 113
    4.2 Equilibria and Semi-Equilibria for Sublinear RDS 116
    4.3 Almost Equilibria 122

4.4   Limit Set Trichotomy for Sublinear RDS 125
4.5   Random Mappings 132
4.6   Positive Affine RDS.................................... 138

5.   **Cooperative Random Differential Equations** 143
5.1   Basic Assumptions and the Existence Theorem 143
5.2   Generation of RDS 145
5.3   Random Comparison Principle 150
5.4   Equilibria, Semi-Equilibria and Attractors 156
5.5   Random Equations with Concavity Properties.............. 160
5.6   One-Dimensional Explicitly Solvable Random Equations..... 166
5.7   Applications 171
      5.7.1   Random Biochemical Control Circuit 171
      5.7.2   Random Gonorrhea Model........................ 175
      5.7.3   Random Model of Symbiotic Interaction............ 176
      5.7.4   Random Gross-Substitute System 178
5.8   Order-Preserving RDE with Non-Standard Cone 180

6.   **Cooperative Stochastic Differential Equations** 185
6.1   Main Assumptions...................................... 185
6.2   Generation of Order-Preserving RDS..................... 186
6.3   Conjugacy with Random Differential Equations 188
6.4   Stochastic Comparison Principle 192
6.5   Equilibria and Attractors 194
6.6   One-Dimensional Stochastic Equations 199
      6.6.1   Stochastic Equations on $\mathbb{R}_+$ 199
      6.6.2   Stochastic Equations on a Bounded Interval 206
6.7   Stochastic Equations with Concavity Properties 214
6.8   Applications 219
      6.8.1   Stochastic Biochemical Control Circuit............. 219
      6.8.2   Stochastic Gonorrhea Model 221
      6.8.3   Stochastic Model of Symbiotic Interaction 222
      6.8.4   Lattice Models of Statistical Mechanics 223

References 227

Introduction

The state of many physical, chemical and biological systems can be described by a single time-dependent variable $x(t)$ which satisfies the ordinary differential equation

$$\dot{x}(t) = f(\lambda, x(t)) . \tag{1}$$

This equation depends on parameters $\lambda = (\lambda_1, \ldots \lambda_m)$ which characterize the properties of the environment and are usually called external parameters. For example, the equation

$$\dot{x} = \alpha x - x^3 \tag{2}$$

can be used to describe the growth of a biological population. It contains the parameter $\alpha \in \mathbb{R}$ which takes into account the properties of the environment.

If there is an existence and uniqueness theorem for (1), then we can define an evolution operator S_t in $\mathbb{R}$ by the formula $S_t x_0 = x(t; x_0)$, where $x(t; x_0)$ is the solution to (1) with $x(0; x_0) = x_0$. The uniqueness theorem for (1) and the fact that $\mathbb{R}$ is a totally ordered set imply that one-dimensional equations generate monotone (or order-preserving) dynamical systems, i.e. $S_t x_1 \geq S_t x_2$ provided $x_1 \geq x_2$. This property drastically simplifies the dynamics. For example, for equation (2) we have either one or three equilibrium points depending on the parameter α and every solution is attracted by an equilibrium in a monotone way. Indeed, it is easy to see that any solution to (2) with initial data x_0 has the form

$$x(t) = \frac{x_0 e^{\alpha t}}{\left(1 + x_0^2 \cdot \alpha^{-1} \cdot (e^{2\alpha t} - 1)\right)^{1/2}} .$$

Therefore we have that (i) if $\alpha < 0$, then $x(t) \to 0$ as $t \to +\infty$ for any initial data x_0; (ii) if $\alpha > 0$ and $x_0 > 0$, then $x(t) \to \sqrt{\alpha}$ as $t \to +\infty$ and (iii) if $\alpha > 0$ and $x_0 < 0$, then $x(t) \to -\sqrt{\alpha}$ as $t \to +\infty$. Thus for the case $\alpha < 0$ we have a unique globally asymptotically stable equilibrium and in the case $\alpha > 0$ we have two stable equilibria and one unstable. In the latter case the global attractor is the interval $[-\sqrt{\alpha}, \sqrt{\alpha}]$ (by definition, the global attractor is a strictly invariant set which uniformly attracts every bounded set). Thus equilibria and their stability properties completely determine the long-time dynamics of the system.

Similar behaviour is observed for one-dimensional systems with discrete time which are generated from a nondecreasing continuous mapping $f : \mathbb{R} \mapsto \mathbb{R}$ via the formula

$$x_{n+1} = f(x_n), \quad n = 0, 1, \ldots .$$

The situation becomes more complicated in d-dimensional case. The phase space $\mathbb{R}^d$ is a partially ordered set with respect to the natural order relation ($x = (x_1, \ldots, x_d) \geq 0$ if and only if $x_i \geq 0$ for all i) and there is no monotonicity in general. For example, it is easy to see that the linear system

$$\dot{x}_1 = a_{11}x_1 + a_{12}x_2 ,$$
$$\dot{x}_2 = a_{21}x_1 + a_{22}x_2 ,$$

produces solutions which are monotone with respect to initial data if and only if $a_{12} \geq 0$ and $a_{21} \geq 0$. Nevertheless monotone multi-dimensional ordinary differential equations cover important classes of mathematical models arising in modern natural science (see discussion in SMITH [102]). The mathematical theory of deterministic monotone (order-preserving) systems is presently well-developed due to the efforts of many authors (see, e.g., KRASNOSELSKII [69], HIRSCH [52, 53, 54] and also SMITH [102] and the references therein). A well-posed autonomous system of ordinary differential equations

$$\dot{x}_i = f_i(x_1, \ldots, x_d), \quad i = 1, \ldots, d ,$$

generates an order-preserving (with respect to the natural order relation) dynamical system in $\mathbb{R}^d$ if and only if the mapping $x \mapsto (f_1(x), \ldots, f_d(x))$ from $\mathbb{R}^d$ into itself is cooperative (quasi-monotone), i.e.

$$f_i(x_1, \ldots, x_d) \leq f_i(y_1, \ldots, y_d)$$

for all $(x_1, \ldots, x_d)$ and $(y_1, \ldots, y_d)$ from $\mathbb{R}^d$ such that $x_i = y_i$ and $x_j \leq y_j$ for $j \neq i$, where $i = 1, \ldots, d$. For example, this relation holds for the following system of differential equations

$$\dot{x}_1(t) = g(x_d(t)) - \alpha_1 x_1(t) ,$$
$$\dot{x}_j(t) = x_{j-1}(t) - \alpha_j x_j(t), \quad j = 2, \ldots, d ,$$

where $\alpha_j > 0$ for $j = 1, \ldots, d$ and $g(x_d)$ is a nondecreasing function. A system of this type provides a simple model for positive feedback in biochemical control circuit (see, e.g., SELGRADE [96] and SMITH [102] and the references therein). The variables $x_j, j = 1, \ldots, d-1$, could represent the concentrations of a sequence of enzymes and x_d, the concentrations of their substrate.

It was shown by HIRSCH [52] that generic solutions to some classes of monotone systems converge to the set of equilibria. Thus, as in the one-dimensional case, we observe some simplification in the long-time dynamics. However an important construction due to Smale (see, e.g., SMITH [102,

Chap. 4]) shows that any (complicated) dynamics can occur on unstable invariant sets for monotone systems of sufficiently large dimension.

If the system is coupled to a fluctuating environment, then external parameters can become stochastic quantities. In many cases these quantities can be presented as stationary random processes. We refer to HORSTHEMKE/LEFEVER [55, Chap. 1] for a detailed discussion on the nature and sources of randomness in dynamical systems.

Thus taking into account random fluctuations of the environment for the system described by (1) leads to the equation

$$\dot{x}(t) = f(\lambda_0 + \xi(t, \omega), x(t)) \ ,$$

where λ_0 corresponds to the mean state of the environment and the stationary process $\xi(t, \omega)$ with zero expectation on some probability space $(\Omega, \mathcal{F}, \mathbb{P})$ describes environmental fluctuations around this mean state. For example, equation (2) turns into

$$\dot{x} = (\alpha + \xi(t, \omega)) \cdot x - x^3 \ . \tag{3}$$

As above we can show that the process $x(t; \omega, x_0)$ which solves (3) with initial data x_0 has the form

$$x(t; \omega, x_0) = \frac{x_0 \exp\{\alpha t + \eta(t, \omega)\}}{\left(1 + 2x_0^2 \cdot \int_0^t \exp\{2\alpha s + 2\eta(s, \omega)\} ds\right)^{1/2}} \ ,$$

where $\eta(t, \omega) = \int_0^t \xi(\tau, \omega) d\tau$. It is clear that the solutions $x(t; \omega, x_0)$ depend on x_0 in a monotone way, i.e. the relation $x_0 \geq x_0^*$ implies that $x(t; \omega, x_0) \geq x(t; \omega, x_0^*)$.

Assume that the strong law of large numbers is valid for the process $\xi(t, \omega)$, i.e. $t^{-1}\eta(t, \omega) \to 0$ almost surely as $t \to +\infty$. Then it is easy to see that all solutions $x(t; \omega, x_0)$ tend to 0 almost surely as $t \to +\infty$ in the case $\alpha < 0$. In the case $\alpha > 0$ the situation is a bit more complicated. However it is possible to prove (see ARNOLD [3, Chap. 9]) that there exists a stationary process $\zeta(t, \omega) > 0$ which solves equation (3) and such that the interval $[-\zeta(t, \omega), \zeta(t, \omega)]$ is a globally attracting set in some sense. Thus stationary solutions to equation (3) play a role of equilibria and the interval $[-\zeta(t, \omega), \zeta(t, \omega)]$ should be treated as a global attractor. As we will see in Chap.3, a similar picture is inherent in some classes of multi-dimensional monotone systems with both continuous and discrete time. However it is well to bear in mind that random monotone systems may display the long-time behaviour which is impossible in deterministic (autonomous or periodic) order-preserving systems. As an example we can consider the following differential equation

$$\dot{x} = \xi(t, \omega) \cdot x(1 - x)$$

in the interval $[0, 1] \subset \mathbb{R}$. Under some conditions concerning the stationary random process $\xi(t, \omega)$ the omega-limit set for any point from the open interval $(0, 1)$ is a non-trivial completely ordered set. We refer to Example 3.6.1 and Sects. 5.6 and 6.6 for details. This phenomenon does not take place in deterministic strongly order-preserving systems (see SMITH [102]) and this is one of obstacles which prevent the direct expansion of the results available for deterministic monotone systems.

To make the analogy with the deterministic case more precise it is convenient to involve the modern concept (see ARNOLD [3] and also Sect. 1.2 below) of a random dynamical system. This concept covers the most important families of dynamical systems with randomness, including random and stochastic ordinary and partial differential equations and random difference equations, and makes it possible to study randomness in the framework of classical dynamical systems theory with all its powerful machinery. Randomness could describe environmental or parametric perturbations, internal fluctuations, measurement errors, or just lack of knowledge. The theory of random dynamical systems has been developed intensively in recent years and contains a lot of interesting and deep results. From a probabilistic point of view this theory offers a new approach to the study of qualitative properties of stochastic differential equations. It became possible due to important results on two-parameter flows generated by stochastic equations (see, e.g., BELOPOLSKAYA/DALECKY [15], ELWORTHY [43], KUNITA [74] and the literature quoted there). For a detailed discussion of the theory and applications of random dynamical systems we refer to the monograph ARNOLD [3].

To present a clear explanation of the general concept of a random dynamical system (see Sect. 1.2 for the formal definition) we consider the following simple discrete dynamical system.

Assume that f_0 and f_1 are continuous mappings of a metric space X into itself. Let us consider X as the state space of some system that evolves as follows: if x is the state of the system at time k then its state at time $k + 1$ is either $f_0(x)$ or $f_1(x)$ with probability $1/2$ and the choice of f_0 or f_1 does not depend on time and the previous states. We can find the state of the system after a number of steps in time if we flip a coin and write down the sequence of events from right to left using 0 and 1. Assume, for example, that after 7 flips we get a following set of outcomes: 1001101. Here 1 corresponds to the head falling and 0 corresponds to the tail falling. Therewith the state of the system at time 7 will be written in the form

$$y = (f_1 \circ f_0 \circ f_0 \circ f_1 \circ f_1 \circ f_0 \circ f_1)(x).$$

This construction can be formalized as follows. Let Ω be the set of two-sided sequences $\omega = \{\omega_i \,|\, i \in \mathbb{Z}\}$ consisting of zeros and ones. On the set Ω there is a probability measure $\mathbb{P}$ such that

$$\mathbb{P}(C_{i_1 \ldots i_m}) = \mathbb{P}_0(C_1) \cdot \ldots \cdot \mathbb{P}_0(C_m)$$

for any "cylindrical" set

$$C_{i_1\ldots i_m} = \{\omega \,|\, \omega_{i_k} \in C_k,\ k = 1,\ldots,m\}\,,$$

where C_k is one of the sets $\emptyset$, $\{0\}$, $\{1\}$, $\{0,1\}$ and $\mathbb{P}_0(\emptyset) = 0$, $\mathbb{P}_0(\{0\}) = \mathbb{P}_0(\{1\}) = 1/2$, $\mathbb{P}_0(\{0,1\}) = 1$. Here $\{i_1,\ldots,i_m\}$ is an arbitrary m-tuple of integers. For every $n \in \mathbb{Z}$ we denote by θ_n the left shift operator in Ω, i.e.

$$\theta_n\{\omega_i \,|\, i \in \mathbb{Z}\} = \{\omega_{i+n} \,|\, i \in \mathbb{Z}\}, \quad n \in \mathbb{Z}\,.$$

It is clear that the shift operator preserves probabilities of sets from Ω.

For each $n \in \mathbb{Z}_+$ and $\omega \in \Omega$ we define the mapping $\pi_n(t,\omega)$ of $\Omega \times X$ into itself by the formula

$$\pi_n = \pi_1 \circ \pi_{n-1},\ n \in \mathbb{N},\quad \pi_0 = \mathrm{id}\,,$$

where $\pi_1(\omega, x) = (\theta_1\omega, f_{\omega_0}(x))$. This mapping π_n can be written in the form

$$\pi_n(\omega, x) = (\theta_n\omega, \varphi(n,\omega)x)\,, \tag{4}$$

where $\varphi(n,\omega)$ is defined by the formula

$$\varphi(n,\omega) = f_{\omega_{n-1}} \circ f_{\omega_{n-2}} \circ \ldots \circ f_{\omega_1} \circ f_{\omega_0}, \quad \omega = \{\omega_i \,|\, i \in \mathbb{Z}\}, \quad n \in \mathbb{N}\,,$$

and satisfies the *cocycle property*

$$\varphi(0,\omega) = \mathrm{id}, \quad \varphi(n+m,\omega) = \varphi(n,\theta_m\omega) \circ \varphi(m,\omega)$$

for all $n, m \in \mathbb{Z}_+$ and $\omega \in \Omega$. The pair $(\theta_n, \varphi(n,\omega))$ is called a random dynamical system with discrete time. The mapping θ_n models the evolution of some random environment and $\varphi(\omega, n)$ describes the dynamics of the system. If $X = \mathbb{R}$ and f_0 and f_1 are nondecreasing functions, then the mappings $\varphi(n,\omega)$ are order preserving, i.e. the relation $x_1 \geq x_2$ implies that $\varphi(n,\omega)x_1 \geq \varphi(n,\omega)x_2$ for all $n \in \mathbb{Z}_+$ and $\omega \in \Omega$.

It is easy to see that φ satisfies the cocycle property if and only if π_n given by (4) is a semigroup, i.e. $\pi_n \circ \pi_m = \pi_{n+m}$ for $n, m \in \mathbb{Z}_+$. Thus we obtain a dynamical system in the classical sense (i.e. a semiflow of mappings from some space into itself). We note that semiflows of a similar structure (see (4)) arise in the theory of nonautonomous (deterministic) differential equations and they are known as skew-product flows (see, e.g., CHICONE/LATUSHKIN [19] and the references therein). This observation is important in the study of the long-time behaviour of random dynamical systems.

The aim of this book is to present a recently developed approach which is suitable for investigating a variety of qualitative aspects of order-preserving random dynamical systems and to give the backgrounds for further development of the theory. We try to demonstrate the effectiveness of this approach by analyzing the long-time behaviour of some classes of random and stochastic ordinary differential equations which arise in many applications.

Although the most of general results in this book are proved for infinite-dimensional phase spaces, our examples and applications deal with finite-dimensional systems only. The book does not treat order-preserving random dynamical systems generated by random and stochastic partial differential equations. We refer to the papers CHUESHOV [21] and CHUESHOV/VUILLER-MOT [24, 25, 26], which are devoted to the application of monotone methods and comparison arguments to the study of long-time behaviour of random and stochastic parabolic PDEs (see also CHUESHOV [22] and SHEN/YI [99], where similar approaches are used for nonautonomous parabolic equations).

Now we describe the structure of the book.

We start with the preliminary Chapter 1 devoted to a description of some background material from the general theory of random dynamical systems and to a discussion of the simplest examples. Some results presented in this chapter are given without proofs. However for the sake of completeness we prove the theorem on the existence of a random (pull back) attractor. We also prove here several auxiliary facts which are important in our subsequent considerations. They are mainly concerned with measurability of trajectories and invariant sets. For a more detailed presentation on random dynamical systems we refer to the book ARNOLD [3].

In Chapter 2 we describe results on the generation of random dynamical systems by random and stochastic ordinary differential equations. We mainly follow the presentation given in ARNOLD [3, Chap. 2] and invoke some classical results on stochastic equations (see, e.g., IKEDA/WATANABE [57] and KUNITA [74]). We also prove a theorem on the existence of invariant deterministic domains for these equations and consider relations between random and stochastic differential equations. The reader who is primarily interested in the general theory of order-preserving random dynamical systems can omit this chapter on first reading.

Chapter 3 is central to the book. We develop here the general theory of order-preserving random dynamical systems. We first consider properties of partially ordered vector spaces and prove some auxiliary results concerning random sets in these spaces. After that we introduce the concept of an order-preserving random dynamical system and study properties of sub- and super-equilibria for these systems. We prove a theorem on the existence of equilibria between two ordered sub- and super-equilibria. These semi-equilibria are also proved to be very useful in the description of random attractors for the systems considered.

In Chapter 4 we study the asymptotic behavior of order-preserving random systems which have an additional concavity property called sublinearity (or subhomogeneity), frequently encountered in applications. Sublinear random systems are contractive with respect to some metric which is defined on parts of the cone. This implies that random equilibria are unique and asymptotically stable in each part of the cone. Our main result here is a random limit set trichotomy, stating that in a given part either (i) all orbits are unbounded, (ii) all orbits are bounded but their closure reaches out to the boundary of the part, or (iii) there exists a unique, globally attracting equilibrium. Several examples, including Markov chains and affine systems, are given.

In Chapters 5 and 6 we apply the results of Chapters 3 and 4 to study the qualitative behaviour of random and stochastic perturbations of cooperative ordinary differential equations. These applications are the main motivations for the development of the general theory presented in Chapters 3 and 4 and we believe that random and stochastic cooperative differential equations merit a detailed study of its own.

In Chapter 5 we consider random cooperative differential equations in $\mathbb{R}_+^d$ (real noise case). We first give conditions under which these equations generate order-preserving random dynamical systems in $\mathbb{R}_+^d$ and then study monotonicity properties of these systems. We prove several theorems on the existence of equilibria and random attractors. Systems with concavity properties are also considered. We apply general results from Chapters 3 and 4 to study the long-time behaviour of these systems and to obtain the limit set trichotomy theorem for random cooperative differential equations. We conclude Chapter 5 with a series of examples including a class of one-dimensional explicitly solvable equations to show possible scenarios of the long-time behaviour in monotone systems.

Chapter 6 is devoted to stochastic cooperative differential equations (white noise case). The hypotheses that guarantee order-preserving properties for this case lead to a special structure of the diffusion terms. In fact we consider here some class of stochastic perturbations of deterministic autonomous cooperative differential equations. We prove several assertions on the long-time behaviour, investigate properties of systems that possess concavity properties and establish a stochastic version of the limit set trichotomy theorem. We study the long-time dynamics in one-dimensional equations with details. We also discuss the stochastic versions of certain examples considered in the previous chapter. Although the results for the stochastic case are similar to the random case, Chapter 6 is not at all a duplication of Chapter 5 because the methods of proof are quite different.

1. General Facts about Random Dynamical Systems

In this chapter we recall some basic definitions and facts about random dynamical systems. For a more detailed discussion of the theory and applications of random dynamical systems we refer to the monograph ARNOLD [3]. We pay particular attention to dissipative systems and their random (pull back) attractors. These attractors were studied by many authors (see, e.g., ARNOLD [3], CRAUEL/DEBUSSCHE/FLANDOLI [35], CRAUEL/FLANDOLI [36], SCHENK-HOPPÉ [89], SCHMALFUSS [92, 93] and the references therein). The ideas that lead to the concept of a random attractor have their roots in the theory of deterministic dissipative systems which has been successfully developed in the last few decades (see, e.g., the monographs BABIN/VISHIK [13], CHUESHOV [20], HALE [50], TEMAM [104] and the literature quoted therein). The proof of the existence of random attractors given below follows almost step-by-step the corresponding deterministic argument (see, e.g., CHUESHOV [20], TEMAM [104]).

Throughout this book we will be concerned with a *probability space* by which we mean a triple $(\Omega, \mathcal{F}, \mathbb{P})$, where Ω is a space, $\mathcal{F}$ is a σ-algebra of sets in Ω, and $\mathbb{P}$ is a nonnegative σ-additive measure on $\mathcal{F}$ with $\mathbb{P}(\Omega) = 1$. We do not assume in general that the σ-algebra is complete. Below we will also use the symbol $\mathbb{T}$ for either $\mathbb{R}$ or $\mathbb{Z}$ and we will denote by $\mathbb{T}_+$ all nonnegative elements of $\mathbb{T}$. We will denote by $\mathcal{B}(X)$ the Borel σ-algebra of sets in a topological space X. By definition $\mathcal{B}(X)$ is the σ-algebra generated by the collection of open subsets of X. If $(X_1, \mathcal{F}_1)$ and $(X_2, \mathcal{F}_2)$ are measurable spaces, we denote by $\mathcal{F}_1 \times \mathcal{F}_2$ the product σ-algebra of subsets in $X_1 \times X_2$ which is defined as the σ-algebra generated by the cylinder sets $A = A_1 \times A_2$, $A_i \in \mathcal{F}_i$. We refer to COHN [30] for basic definitions and facts from the measure theory.

1.1 Metric Dynamical Systems

The random dynamical system is an object consisting of a metric dynamical system and a cocycle over this system. We need a metric dynamical system for modeling of random perturbations.

Definition 1.1.1. *A metric dynamical system (MDS)* $\theta \equiv (\Omega, \mathcal{F}, \mathbb{P}, \{\theta_t, t \in \mathbb{T}\})$ *with (two-sided) time* $\mathbb{T}$ *is a probability space* $(\Omega, \mathcal{F}, \mathbb{P})$ *with a family of transformations* $\{\theta_t : \Omega \mapsto \Omega, t \in \mathbb{T}\}$ *such that*

1. it is one-parameter group, i.e.

$$\theta_0 = \mathrm{id}, \quad \theta_t \circ \theta_s = \theta_{t+s} \quad \text{for all} \quad t, s \in \mathbb{T} \, ;$$

2. $(t, \omega) \mapsto \theta_t \omega$ *is measurable;*
3. $\theta_t \mathbb{P} = \mathbb{P}$ *for all* $t \in \mathbb{T}$, *i.e.* $\mathbb{P}(\theta_t B) = \mathbb{P}(B)$ *for all* $B \in \mathcal{F}$ *and all* $t \in \mathbb{T}$.

A set $B \in \mathcal{F}$ *is called* θ-*invariant if* $\theta_t B = B$ *for all* $t \in \mathbb{T}$. *A metric dynamical system* θ *is said to be* ergodic *under* $\mathbb{P}$ *if for any* θ-*invariant set* $B \in \mathcal{F}$ *we have either* $\mathbb{P}(B) = 0$ *or* $\mathbb{P}(B) = 1$.

We refer to CORNFELD/FOMIN/SINAI [29], MAÑÉ [79], RUDOLPH [88], SINAI [100] and WALTERS [106] for the references and presentation of MDS and ergodic theory.

From an applied point of view the use of metric dynamical systems to model external perturbations assumes implicitly that the external influence is stationary in some sense (see examples below). This means that we do not consider possible transient (random) process in the environment, i.e. we assume that all these processes are finished before we start to observe the dynamics of our system. This is also the reason why we consider MDS with two-sided time. We note that any one-sided MDS (with time $\mathbb{T}_+$) possesses a natural two-sided extension (see, e.g., CORNFELD/FOMIN/SINAI [29, Sect. 10.4] or ARNOLD [3, Appendix A]).

Now we give several important examples of metric dynamical systems. They show what kind of time dependence we can allow in the equations considered in Chaps. 5 and 6.

Example 1.1.1 (Periodic Case). Consider the probability space $(\Omega, \mathcal{F}, \mathbb{P})$, where Ω is a circle of unit circumference, $\mathcal{F}$ is its σ-algebra of Borel sets and $\mathbb{P}$ is the Lebesgue measure on Ω. Let $\{\theta_t, t \in \mathbb{R}\}$ be the group of rotations of the circle. It is easy to see that we obtain an ergodic MDS $(\Omega, \mathcal{F}, \mathbb{P}, \{\theta_t, t \in \mathbb{R}\})$ with continuous time.

Example 1.1.2 (Quasi-Periodic Case). Let Ω be d-dimensional torus, $\Omega = \mathrm{Tor}^d$. Assume that its points are written as $x = (x_1, \dots, x_d)$ with $x_i \in [0, 1)$. Let $\mathcal{F}$ be the σ-algebra of Borel sets of Tor^d and $\mathbb{P}$ be the Lebesgue measure on Tor^d. We define transformations $\{\theta_t, t \in \mathbb{T}\}$ by the formula

$$\theta_t x = (x_1 + t \cdot a_1 (\mathrm{mod}\, 1), \dots, x_d + t \cdot a_d (\mathrm{mod}\, 1)), \quad t \in \mathbb{T} \, ,$$

for a given $a = (a_1, \dots, a_d)$. Thus we obtain an MDS. If the numbers $a_1, \dots, a_d, 1$ are rationally independent, then this MDS is ergodic (see, e.g., RUDOLPH [88]).

Example 1.1.3 (Almost Periodic Case). Let $f(x)$ be a Bohr almost periodic function on $\mathbb{R}$. We define the hull $H(f)$ of the function f as the closure of the set $\{f(x+t),\, t \in \mathbb{R}\}$ in the norm $\|f\| = \sup_{x \in \mathbb{R}} |f(x)|$. The hull $H(f)$ is a compact metric space, and it has a natural commutative group structure. Therefore it possesses a Haar measure which, if normalized to unity, makes $H(f)$ into probability space. If we define transformations $\{\theta_t, t \in \mathbb{T}\}$ as shifts: $(\theta_t g)(x) = g(x+t)$, $g \in H(f)$, we obtain an ergodic MDS with continuous time. For details we refer to ELLIS [42] and LEVITAN/ZHIKOV [77].

Example 1.1.4 (Ordinary Differential Equations). MDS can be also generated by ordinary differential equations (ODE). Let us consider a system of ODEs in $\mathbb{R}^n$:

$$\frac{dx_i}{dt} = f_i(x_1, \ldots x_n), \quad i = 1, \ldots, n \,. \tag{1.1}$$

Assume that the Cauchy problem for this system is well-posed. We define $\{\theta_t, t \in \mathbb{R}\}$ by the formula $\theta_t x = x(t)$, where $x(t)$ is the solution of (1.1) with $x(0) = x$. Assume that a nonnegative smooth function $\rho(x_1, \ldots, x_n)$ satisfies the stationary Liouville equation

$$\sum_{i=1}^{n} \frac{\partial}{\partial x_i}\left(\rho(x_1, \ldots x_n) \cdot f_i(x_1, \ldots x_n)\right) = 0 \tag{1.2}$$

and possesses the property $\int_{\mathbb{R}^n} \rho(x)\,dx = 1$. Then $\rho(x)$ is a density of a probability measure on $\mathbb{R}^n$. By Liouville's theorem

$$\int_{\mathbb{R}^n} f(\theta_t x)\rho(x)\,dx = \int_{\mathbb{R}^n} f(x)\rho(x)\,dx$$

for any bounded continuous function $f(x)$ on $\mathbb{R}^n$ and therefore in this situation an MDS arises with $\Omega = \mathbb{R}^n$, $\mathcal{F} = \mathcal{B}(\mathbb{R}^n)$ and $\mathbb{P}(dx) = \rho(x)dx$. Here $\mathcal{B}(\mathbb{R}^n)$ is the Borel σ-algebra of sets in $\mathbb{R}^n$. Sometimes it is also possible to construct an MDS connected with the system (1.1), when the solution ρ to (1.2) is not integrable but the problem (1.1) possesses a first integral (e.g., if (1.1) is a Hamiltonian system) with appropriate properties (see, e.g., MAÑÉ [79] or SINAI [100] for details).

Example 1.1.5 (Bernoulli Shifts). Let $(\Omega_0, \mathcal{F}_0, \mathbb{P}_0)$ be a probability space and $(\Omega, \mathcal{F}, \mathbb{P})$ be the probability space of infinite sequences $\omega = \{\omega_i\}$, where $\omega_i \in \Omega_0$, $i \in \mathbb{Z}$. Here $\mathcal{F}$ is the σ-algebra generated by finite-dimensional cylinders

$$C_{i_1 \ldots i_m} = \{\omega \,|\, \omega_{i_k} \in C_k,\, k = 1, \ldots, m\}\,,$$

where $C_k \in \mathcal{F}_0$ and $\{i_1, \ldots, i_m\}$ is an arbitrary m-tuple of integers. The probability measure $\mathbb{P}$ is defined such that $\mathbb{P}(C_{i_1 \ldots i_m}) = \mathbb{P}_0(C_1) \cdot \ldots \cdot \mathbb{P}_0(C_m)$.

We define transformations $\{\theta_t, t \in \mathbb{Z}\}$ by the formula $\theta_t \omega = \omega^*$, where $\omega = \{\omega_i\}$ and $\omega^* = \{\omega_{i+t}\}$. Since

$$\theta_t C_{i_1 \ldots i_m} = \{\omega \mid \omega_{i_k - t} \in C_k, \ k = 1, \ldots, m\} ,$$

the probability measure $\mathbb{P}$ is invariant under θ_t. Thus we obtain an MDS. In the particular case when $\Omega_0 = \{0, 1\}$ is a two-point set and $\mathbb{P}_0(\{0\}) = \mathbb{P}_0(\{1\}) = 1/2$, we have the standard Bernoulli shift. In the general case we can interpret this MDS as one generated by an infinite sequence of independent identically distributed random variables.

Example 1.1.6 (Stationary Random Process). Let $\xi = \{\xi(t), t \in \mathbb{T}\}$ be a stationary random process on a probability space $(\Omega, \mathcal{F}, \mathbb{P})$, where $\mathcal{F}$ is the σ-algebra generated by ξ. Assume that in the continuous case ($\mathbb{T} = \mathbb{R}$) the process ξ possesses the *càdlàg* property: all trajectories are right-continuous and have limits from the left. Then the shifts $\xi(t) \mapsto (\theta_\tau \xi)(t) = \xi(t + \tau)$ generate an MDS. See ARNOLD [3] and the references therein for details.

In the framework of stochastic equations the following example of an MDS is of importance.

Example 1.1.7 (Wiener Process). Let $W_t = (W_t^1, \ldots, W_t^d)$ be a Wiener process with values in $\mathbb{R}^d$ and two-sided time $\mathbb{R}$. Let $(\Omega, \mathcal{F}, \mathbb{P})$ be the corresponding canonical Wiener space. More precisely, let $C_0(\mathbb{R}, \mathbb{R}^d)$ be the space of continuous functions ω from $\mathbb{R}$ into $\mathbb{R}^d$ such that $\omega(0) = 0$ endowed with the compact-open topology, i.e. with the topology generated by the metric

$$\varrho(\omega, \omega^*) := \sum_{n=1}^{\infty} \frac{1}{2^n} \frac{\varrho_n(\omega, \omega^*)}{1 + \varrho_n(\omega, \omega^*)} , \quad \varrho_n(\omega, \omega*) = \max_{t \in [-n,n]} |\omega(t) - \omega^*(t)| .$$

Let $\tilde{\mathcal{F}}$ be the corresponding Borel σ-algebra of $C_0(\mathbb{R}, \mathbb{R}^d)$, and let $\mathbb{P}$ be the Wiener measure on $\tilde{\mathcal{F}}$. We suppose Ω is the subset in $C_0(\mathbb{R}, \mathbb{R}^d)$ consisting of the functions that have a growth rate less than linear for $t \to \pm\infty$ and $\mathcal{F}$ is the restriction of $\tilde{\mathcal{F}}$ to Ω. In this realization $W_t(\omega) = \omega(t)$, where $\omega(\cdot) \in \Omega$, i.e. the elements of Ω are identified with the paths of the Wiener process. We define a metric dynamical system θ by $\theta_t \omega(\cdot) := \omega(t + \cdot) - \omega(t)$. These transformations preserve the Wiener measure and are ergodic. Thus we have an ergodic MDS. The flow $\{\theta_t\}$ is called the Wiener shift. We note that the σ-algebra $\mathcal{F}$ is not complete with respect to $\mathbb{P}$ and we cannot use its completion $\bar{\mathcal{F}}^{\mathbb{P}}$ to construct MDS because $(t, \omega) \mapsto \theta_t \omega$ is not a measurable mapping from $(\mathbb{R} \times \Omega, \mathcal{B}(\mathbb{R}) \times \bar{\mathcal{F}}^{\mathbb{P}})$ into $(\Omega, \bar{\mathcal{F}}^{\mathbb{P}})$. This is one of the reasons why the completeness of $\mathcal{F}$ is not assumed in the basic definitions. See ARNOLD [3] for details. We also note that this realization of a Wiener process makes it possible to introduce the white noise process as the derivative $\dot{W}_t$ of W_t with respect to t in the sense of generalized functions. From an applied point of view white noise processes correspond to an extremely short memory of the environment in comparison with the memory of the system (see the discussion in HORSTHEMKE/LEFEVER [55], for instance).

1.2 Concept of RDS

Let X be a Polish space, i.e. a separable complete metric space. We equip X with the Borel σ-algebra $\mathcal{B} = \mathcal{B}(X)$ generated by open sets of X. We need the following concept of a (continuous) random dynamical system (cf. ARNOLD [3]).

Definition 1.2.1 (Random Dynamical System). *A random dynamical system (RDS) with (one-sided) time* $\mathbb{T}_+$ *and state (phase) space* X *is a pair* (θ, φ) *consisting of a metric dynamical system* $\theta \equiv (\Omega, \mathcal{F}, \mathbb{P}, \{\theta_t, t \in \mathbb{T}\})$ *and a cocycle* φ *over* θ *of continuous mappings of* X *with time* $\mathbb{T}_+$, *i.e. a measurable mapping*

$$\varphi : \mathbb{T}_+ \times \Omega \times X \mapsto X, \quad (t, \omega, x) \mapsto \varphi(t, \omega, x),$$

such that

(i) *the mapping* $x \mapsto \varphi(t, \omega, x) \equiv \varphi(t, \omega)x$ *is continuous for every* $t \geq 0$ *and* $\omega \in \Omega$,

(ii) *the mappings* $\varphi(t, \omega) := \varphi(t, \omega, \cdot)$ *satisfy the* cocycle property:

$$\varphi(0, \omega) = \text{id}, \quad \varphi(t + s, \omega) = \varphi(t, \theta_s \omega) \circ \varphi(s, \omega)$$

for all $t, s \in \mathbb{T}_+$ *and* $\omega \in \Omega$. *Here* $\circ$ *means composition of mappings.*

We emphasize the following peculiarities of this definition.

Remark 1.2.1. (i) While the metric dynamical system (modeling the random perturbations) is assumed to have two-sided time $\mathbb{T} = \mathbb{R}$ or $\mathbb{Z}$, the cocycle is only required to have one-sided time $\mathbb{T}_+ = \mathbb{R}_+$ or $\mathbb{Z}_+$. This reflects the fact that evolution operators are often non-invertible. However this set-up allows us to consider $\varphi(t, \theta_s \omega)$ for $t \in \mathbb{T}_+$, but starting at an arbitrary (possibly negative) time $s \in \mathbb{T}$ which will be crucial for the construction of equilibria and attractors. In the case of continuous time ($\mathbb{T} = \mathbb{R}$) the standard definition of a continuous RDS requires the continuity of the mappings $(t, x) \mapsto \varphi(t, \omega)x$ for all $\omega \in \Omega$ (see ARNOLD [3, Sect. 1.1]). This property is usually true for the RDS generated by finite-dimensional random and stochastic equations. However, as we will see, many general results on the long-time behaviour can be proved under a weaker assumption of the continuity of the mapping $x \mapsto \varphi(t, \omega)x$ for each $t \geq 0$ and $\omega \in \Omega$. We also note that the cocycle property reduces to the classical semiflow property if φ is independent of ω. Hence deterministic dynamical systems are particular cases of RDS.

(ii) If in Definition 1.2.1 the cocycle is defined on a θ-invariant set Ω^* of full measure, then we can extend it to the whole Ω by the formula

$$\tilde{\varphi}(t, \omega) := \begin{cases} \varphi(t, \omega) & \text{if } \omega \in \Omega^*, \\ \text{id} & \text{if } \omega \notin \Omega^*. \end{cases} \tag{1.3}$$

Thus we obtain the cocycle $\tilde{\varphi}(t,\omega)$ which is indistinguishable from $\varphi(t,\omega)$. We recall that by definition the indistinguishability of $\varphi(t,\omega)$ and $\tilde{\varphi}(t,\omega)$ means that there exists a set $\mathcal{N} \in \mathcal{F}$ such that $\mathbb{P}(\mathcal{N}) = 0$ and

$$\{\omega \ : \ \varphi(t,\omega) \neq \tilde{\varphi}(t,\omega) \quad \text{for some} \quad t \in \mathbb{R}_+\} \subset \mathcal{N} \ .$$

In our case the cocycles coincide on the θ-invariant set Ω^* and we can set $\mathcal{N} = \Omega \setminus \Omega^*$. In further considerations we do not distinguish cocycles which coincide on θ-invariant sets of full measure.

(iii) In the definition of an RDS we require some properties to be valid for *all* $\omega \in \Omega$. However the stochastic analysis deals usually with *almost* all elementary events ω. Solutions to stochastic differential equations are defined almost surely, for example. Therefore to construct RDS connected with stochastic equations we need extend the corresponding evolution operator to all $\omega \in \Omega$ and prove the cocycle property for this extension. This can be done for many cases which are important from the point of view of applications. This procedure is usually referred to as *perfection*. Roughly speaking the perfection of cocycles (or other objects) can be done in the following way. First we prove a property for some θ-invariant set Ω^* of full measure. After that we define the cocycle on $\Omega \setminus \Omega^*$ in an appropriate way (cf. (1.3)). Perfection theorems have been shown in various different cases, see, e.g., ARNOLD/SCHEUTZOW [10], SCHEUTZOW [90], KAGER/SCHEUTZOW [61], SHARPE [98] and also the discussion in ARNOLD [3].

We also recall the following definitions ARNOLD [3].

Definition 1.2.2 (Smooth RDS). *Let X be an open subset of a Banach space. A random dynamical system (θ, φ) is said to be a smooth RDS of class C^k or a C^k RDS, where $1 \leq k \leq \infty$, if it satisfies the following property: for each $(t,\omega) \in \mathbb{T}_+ \times \Omega$ the mapping $x \mapsto \varphi(t,\omega)x$ from X into itself is k times Frechet differentiable with respect to x and the derivatives are continuous with respect to x.*

Definition 1.2.3 (Affine RDS). *Let X be a linear Polish space. The RDS (θ, φ) is said to be affine if the cocycle φ is of the form*

$$\varphi(t,\omega)x = \Phi(t,\omega)x + \psi(t,\omega) \ , \tag{1.4}$$

where $\Phi(t,\omega)$ is a cocycle over θ consisting of bounded linear operators of X, and $\psi : \mathbb{T}_+ \times \Omega \to X$ is a measurable function. If $\psi(t,\omega) \equiv 0$ then the affine RDS is said to be linear.

If (θ, Φ) is a linear RDS, then the cocycle property for the mapping φ defined by (1.4) is equivalent to the relation

$$\psi(t+s,\omega) = \Phi(t,\theta_s\omega)\psi(s,\omega) + \psi(t,\theta_s\omega), \quad t,s \geq 0 \ . \tag{1.5}$$

A thorough treatment of affine RDS in $\mathbb{R}^d$ can be found in Sect. 5.6 of ARNOLD [3].

Any RDS (θ, φ) generates a *skew-product semiflow* $\{\pi_t, t \in \mathbb{T}_+\}$ on $\Omega \times X$ by the formula

$$\pi_t(\omega, x) = (\theta_t \omega, \varphi(t, \omega)x), \quad t \in \mathbb{T}_+ . \qquad (1.6)$$

Since $(\omega, x) \mapsto \pi_t(\omega, x)$ is an $(\mathcal{F} \times \mathcal{B})$-measurable mapping from $\Omega \times X$ into itself, we obtain a *measurable dynamical system* on $(\Omega \times X, \mathcal{F} \times \mathcal{B})$. Here $\mathcal{B}$ is the σ-algebra of Borel sets in X. The cocycle property for φ is equivalent to the semigroup property for π. We note that the standard theory of skew-product flows (see, e.g., SHEN/YI [99], CHICONE/LATUSHKIN [19] and the references therein) usually requires that both Ω and X are topological spaces and $\{\theta_t\}$ are continuous mappings. In the RDS case we have no topology on Ω in general.

The simplest examples of RDS are described below.

Example 1.2.1 (Markov Chain). This is a generalization of the example considered in the Introduction. Let $(\Omega_0, \mathcal{F}_0, \mathbb{P}_0)$ be a probability space and X be a Polish space. Assume that $f(\alpha, x)$ is a measurable mapping from $\Omega_0 \times X$ into X which is continuous with respect to x for every fixed $\alpha \in \Omega_0$. Let $(\Omega, \mathcal{F}, \mathbb{P})$ be the probability space of infinite sequences $\omega = \{\omega_i\}$, where $\omega_i \in \Omega_0$, $i \in \mathbb{Z}$, and $\theta = (\Omega, \mathcal{F}, \mathbb{P}, \{\theta_t, t \in \mathbb{Z}\})$ be the metric dynamical system constructed in Example 1.1.5. For every $\omega = \{\omega_i : i \in \mathbb{Z}\} \in \Omega$ we introduce the function $f_\omega : X \mapsto X$ by the formula $f_\omega(x) = f(\omega_0, x)$ and for each $n \in \mathbb{Z}_+$ and $\omega \in \Omega$ we define the mapping $\varphi(n, \omega)$ by the formula

$$\varphi(n, \omega) = f_{\theta_{n-1}\omega} \circ f_{\theta_{n-2}\omega} \circ \ldots \circ f_{\theta_1 \omega} \circ f_\omega, \quad \omega = \Omega, \, n \in \mathbb{N} . \qquad (1.7)$$

We also suppose $\varphi(0, \omega) = \mathrm{id}$. It is easy to see that the sequence $\varphi(n, \omega)x$ solves the difference equation

$$x_{n+1} = f_{\theta_n \omega}(x_n), \, n \in \mathbb{Z}_+, \quad x_0 = x ,$$

and the mappings $\varphi(n, \omega)$ possess the cocycle property. Thus we obtain a discrete RDS. It is a C^k-RDS, if $X \subset \mathbb{R}^d$ and $f(\alpha, \cdot) \in C^k(X, X)$. If X is a linear Polish space and $f(\alpha, \cdot)$ are affine mappings, i.e. $f(\alpha, x) = K_\alpha x + h_\alpha$, where K_α are continuous linear operators in X and h_α are elements from X, then the RDS constructed above is affine. It is a linear RDS when $h_\alpha = 0$ for $\alpha \in \Omega_0$.

Since all random mappings $f_{\theta_n \omega}$, $n \in \mathbb{Z}$, are independent and identically distributed (i.i.d.), the RDS constructed above generates (see ARNOLD [3, p. 53]) the homogeneous Markov chain

$$\{\Phi_n^x := \varphi(n, \omega)x \, : \, n \in \mathbb{Z}_+, \, x \in X\}$$

with state space X and transition probability

$$P(x, B) := \mathbb{P}\{\Phi_{n+1} \in B \mid \Phi_n = x\}$$

$$= \mathbb{P}\{\omega \ : \ f_\omega(x) \in B\} \equiv \mathbb{P}_0\{\alpha \ : \ f(\alpha, x) \in B\}, \quad B \in \mathcal{B}(X) \ .$$

For a detailed presentation of the theory of Markov chains we refer to GIH-MAN/SKOROHOD [48, Chap. 2], for example. We note that the inverse problem of constructing an RDS of i.i.d. mappings with a prescribed transition probability is not unique in general and so far largely unsolved. We refer to ARNOLD [3] and KIFER [66] for discussions of this problem.

Example 1.2.2 (Kick Model). Let $\{\xi_k \ : \ k \in \mathbb{Z}\}$ be a stationary random process (chain) in X on a probability space $(\Omega, \mathcal{F}, \mathbb{P})$ and θ be the corresponding metric dynamical system such that $\xi_k(\omega) = \xi_0(\theta_k \omega)$ for all $k \in \mathbb{Z}$ (cf. Example 1.1.6). Suppose that mappings $f_\omega : X \mapsto X$ have the form

$$f_\omega(x) = g(x, \xi_1(\omega)), \quad \omega \in \Omega \ ,$$

where g is a continuous function from $X \times X$ into X. In this case the cocycle φ defined by (1.7) generates the sequence $x_n = \varphi(n, \omega)x$ which solves the difference equation

$$x_{n+1} = g(x_n, \xi_{n+1}(\omega)), \quad n \in \mathbb{Z}_+, \ x_0 = x \ .$$

If X is a Banach space and $g(x, \xi) = g(x) + \xi$, then this equation has the form

$$x_{n+1} = g(x_n) + \xi_{n+1}(\omega), \quad n \in \mathbb{Z}_+, \ x_0 = x \ . \tag{1.8}$$

A *kick force model* corresponds to the case when the mapping $g : X \mapsto X$ has the form $g(x) = y(T; x)$, where $T > 0$ is a fixed number and $y(t) := y(t; x)$ solves the equation

$$\dot{y}(t) = h(y(t)), \ t > 0, \quad y(0) = x \ . \tag{1.9}$$

Here h is a mapping from X into itself such that equation (1.9) generates a (deterministic) continuous dynamical system. In this case

$$\varphi(n, \omega)x = z(n \cdot T + 0, \omega; x), \quad n \in \mathbb{Z}_+ \ .$$

Here $z(t) := z(t, \omega; x)$ is a generalized solution to the problem

$$\dot{z}(t) = h(z(t)) + \sum_{k \in \mathbb{Z}} \xi_k(\omega) \cdot \delta(t - k \cdot T), \quad z(+0) = x \ ,$$

where $\delta(t)$ is a Dirac δ-function of time. Thus the kick model describes the situation when the deterministic system (1.9) gets random kicks with some period T and evolves freely between kicks. We note that kick models are sufficiently popular in the study of turbulence phenomena.

The next examples present the simplest versions of RDS considered in Chaps. 2, 5 and 6 with details.

Example 1.2.3 (1D Random Equation). Let $\theta = (\Omega, \mathcal{F}, \mathbb{P}, \{\theta_t, t \in \mathbb{R}\})$ be a metric dynamical system. Consider the pathwise ordinary differential equation

$$\dot{x}(t) = f(\theta_t \omega, x(t)) . \qquad (1.10)$$

Under some natural conditions (see Sect. 2.1 below) on the function f : $\Omega \times \mathbb{R} \mapsto \mathbb{R}$ this equation generates an RDS with state space $\mathbb{R}$ and with the cocycle given by the formula $\varphi(t, \omega)x = x(t)$, where $x(t)$ is the solution to (1.10) with $x(0) = x$. This RDS is affine if $f(\omega, x) = a(\omega) \cdot x + b(\omega)$ for some random variables $a(\omega)$ and $b(\omega)$.

Example 1.2.4 (Binary Biochemical Model). Consider the system of ordinary differential equations

$$\begin{aligned}
\dot{x}_1 &= g(x_2) - \alpha_1(\theta_t \omega)x_1 , \\
\dot{x}_2 &= x_1 - \alpha_2(\theta_t \omega)x_2 ,
\end{aligned} \qquad (1.11)$$

over a metric dynamical system θ. This is a two-dimensional version of the deterministic model considered in the Introduction. If we assume that $g(x)$ is a globally Lipschitz function and $\alpha_i(\omega)$ is a random variable such that $\alpha_i(\theta_t \omega) \in L_{loc}^1(\mathbb{R})$ for $i = 1, 2$ and $\omega \in \Omega$, then equations (1.11) generate an RDS in $\mathbb{R}^2$ with $\varphi(t, \omega)x = x(t)$, where $x(t) = (x_1(t), x_2(t))$ is the solution to (1.11) with $x(0) = x$.

Example 1.2.5 (1D Stochastic Equation). Let $\{W_t\}$ be the one-dimensional Wiener process (see Example 1.1.7). Then the Itô stochastic differential equation in $\mathbb{R}$

$$dx(t) = b(x(t))dt + \sigma(x(t))dW_t , \qquad (1.12)$$

where the scalar functions $b(x)$ and $\sigma(x)$ possess some regularity properties (see Sect. 2.4 below), also generates an RDS. Of course, the same conclusion remains true, if we understand the stochastic equation (1.12) in the Stratonovich sense. We note that formally equation (1.12) can be written in the form

$$\dot{x}(t) = b(x(t)) + \sigma(x(t))\dot{W}_t$$

and the corresponding RDS can be interpreted as a system in a white noise environment.

More detailed presentation of the last three examples and their generalizations can be found in Chaps. 5 and 6. We also refer to Sects. 2.1 and 2.4 in

Chap. 2 for a description of the basic properties of random and stochastic differential equations.

As in the deterministic case the following concept of topological equivalence (or conjugacy) of two random dynamical systems is of importance in our study. In particular below we will use equivalence between some classes of random and stochastic differential equations.

Definition 1.2.4 (Equivalence of RDS). *Let (θ, φ_1) and (θ, φ_2) be two RDS over the same MDS θ with phase spaces X_1 and X_2 resp. These RDS (θ, φ_1) and (θ, φ_2) are said to be (topologically) equivalent (or conjugate) if there exists a mapping $T : \Omega \times X_1 \mapsto X_2$ with the properties:*

(i)   the mapping $x \mapsto T(\omega, x)$ is a homeomorphism from X_1 onto X_2 for every $\omega \in \Omega$;
(ii) the mappings $\omega \mapsto T(\omega, x_1)$ and $\omega \mapsto T^{-1}(\omega, x_2)$ are measurable for every $x_1 \in X_1$ and $x_2 \in X_2$;
(iii) the cocycles φ_1 and φ_2 are cohomologous, i.e.

$$\varphi_2(t, \omega, T(\omega, x)) = T(\theta_t \omega, \varphi_1(t, \omega, x)) \quad \text{for any} \quad x \in X_1. \quad (1.13)$$

We refer to ARNOLD [3], KELLER/SCHMALFUSS [63] and also to the recent papers IMKELLER/LEDERER [58] and IMKELLER/SCHMALFUSS [59] for more details concerning equivalence of RDS.

1.3 Random Sets

One of the goals in this book is to describe the long-time behaviour of RDS and the limit regimes of these systems. These limit regimes typically depend on an event ω and therefore to characterize their attractivity properties we should at least be able to calculate the distance between (random) trajectories and (random) limit objects and treat this distance as a random variable. It is also crucial to decide whether the limit regimes contain a random variable representing the different states of the system. These circumstances lead to a notion of a random set which is stronger than simply a collection of sets depending on ω. We introduce this notion of a random set following to CASTAING/VALADIER [18] and HU/PAPAGEORGIOU [56] (see also CRAUEL [32] and ARNOLD [3]).

 Below any mapping from Ω into the collection of all subsets of X is said to be *a multifunction* (or a set valued mapping) from Ω into X.

Definition 1.3.1 (Random Set). *Let X be a metric space with a metric ϱ. The multifunction $\omega \mapsto D(\omega) \neq \emptyset$ is said to be a random set if the mapping $\omega \mapsto \text{dist}_X(x, D(\omega))$ is measurable for any $x \in X$, where $\text{dist}_X(x, B)$ is the distance in X between the element x and the set $B \subset X$. If $D(\omega)$ is closed for each $\omega \in \Omega$ then D is called a random closed set. If $D(\omega)$ are compact sets* *for all $\omega \in \Omega$ then D is called a* random compact set. *A random set $\{D(\omega)\}$ is said to be* bounded *if there exist $x_0 \in X$ and a random variable $r(\omega) > 0$ such that*

$$D(\omega) \subset \{x \in X \ : \ \varrho(x, x_0) \leq r(\omega)\} \quad \text{for all} \quad \omega \in \Omega \ .$$

For ease of notation we denote the random set $\omega \mapsto D(\omega)$ by D or $\{D(\omega)\}$.

Remark 1.3.1. (i) The property of D being a random closed set is slightly stronger than

$$\text{graph}(D) = \{(\omega, x) \in \Omega \times X \ : \ x \in D(\omega)\}$$

being $\mathcal{F} \times \mathcal{B}(X)$-measurable and $D(\omega)$ being closed; the two properties are equivalent if $\mathcal{F}$ is $\mathbb{P}$-complete, i.e. if for any set $A \in \mathcal{F}$ with zero probability all subsets of A also belong to $\mathcal{F}$ (see CASTAING/VALADIER [18]).
   (ii) For any $x \in X$ and bounded sets A and B from X we have the relation

$$|\text{dist}_X(x, A) - \text{dist}_X(x, B)| \leq h(A|B) \ ,$$

where $h(A|B)$ is the Hausdorff distance defined by the formula

$$h(A|B) = \sup_{a \in A} \text{dist}_X(a, B) + \sup_{b \in B} \text{dist}_X(b, A) \ .$$

Therefore, if for a multifunction $\omega \mapsto D(\omega)$ there exists a sequence $\{D_n\}$ of random bounded sets such that

$$\lim_{n \to \infty} h(D_n(\omega)|D(\omega)) = 0 \quad \text{for all} \quad \omega \in \Omega \ ,$$

then $\overline{D(\omega)} = \cap_{n \geq 0} \overline{\cup_{k \geq n} D_k(\omega)}$ for every $\omega \in \Omega$ and $\omega \mapsto D(\omega)$ is a random bounded set ($\overline{D}$ denotes the closure of D in X).

Example 1.3.1 (Random Ball). Let $X = \mathbb{R}^d$. Suppose that $r(\omega) \geq 0$ is a random variable and $a(\omega)$ is a random vector from $\mathbb{R}^d$. Then the multifunction

$$\omega \mapsto B(\omega) = \{x \ : \ |x - a(\omega)| \leq r(\omega)\}$$

is a random compact set. Here $|\cdot|$ is the Euclidean distance in $\mathbb{R}^d$. This fact follows from the formula

$$\text{dist}_X(y, B(\omega)) = \begin{cases} 0 & \text{if } y \in B(\omega) \ , \\ |y - a(\omega)| - r(\omega) & \text{if } y \notin B(\omega) \ , \end{cases}$$

which implies that $\text{dist}_X(y, B(\omega)) = \max\{0, |y - a(\omega)| - r(\omega)\}$. It is also clear that $\text{int} B(\omega) = \{x \ : \ |x - a(\omega)| < r(\omega)\}$ is a random (open) set.

More general examples are described in Proposition 1.3.1(vi) and in Proposition 1.3.6.

We need the following properties of random sets (for the proofs we refer to HU/PAPAGEORGIOU [56, Chap. 2], see also CASTAING/VALADIER [18], CRAUEL [32] and ARNOLD [3]).

Proposition 1.3.1. *Let X be a Polish space. The following assertions hold:*

(i)   *D is a random set in X if and only if the set $\{\omega : D(\omega) \cap U \neq \emptyset\}$ is measurable for any open set $U \subset X$;*

(ii) *D is a random set in X if and only if $\{\overline{D(\omega)}\}$ is a random closed set $(\overline{D(\omega)}$ denotes the closure of $D(\omega)$ in $X)$;*

(iii) *D is a random compact set in X if and only if $D(\omega)$ is compact for every $\omega \in \Omega$ and the set $\{\omega : D(\omega) \cap C \neq \emptyset\}$ is measurable for any closed set $C \subset X$;*

(iv) *if $\{D_n, n \in \mathbb{N}\}$ is a sequence of random closed sets with non-void intersection and there exists $n_0 \in \mathbb{N}$ such that D_{n_0} is a random compact set, then $\cap_{n \in \mathbb{N}} D_n$ is a random compact set in X;*

(v)   *if $\{D_n, n \in \mathbb{N}\}$ is a sequence of random sets, then $D = \cup_{n \in \mathbb{N}} D_n$ is also a random set in X;*

(vi) *if $f : \Omega \times X \mapsto X$ is a mapping such that $f(\omega, \cdot)$ is continuous for all ω and $f(\cdot, x)$ is measurable for all x, then $\omega \mapsto f(\omega, D(\omega))$ is a random set in X provided D is a random set in X; similarly, $\omega \mapsto f(\omega, D(\omega))$ is a random compact set in X provided D is a random compact set.*

The following representation theorem (see IOFFE [60]) provides us with a convenient description of random closed sets.

Theorem 1.3.1. *Let D be a random closed set in a Polish space X. Then there exist a Polish space Y and a mapping $g(\omega, y) : \Omega \times Y \mapsto X$ such that*

(i)   *$g(\omega, \cdot)$ is continuous for all $\omega \in \Omega$ and $g(\cdot, y)$ is measurable for all $y \in Y$;*

(ii) *for all $\omega \in \Omega$ and $y_1, y_2 \in Y$ one has*

$$\varrho(g(\omega, y_1), g(\omega, y_2)) \leq (1 + \varrho(g(\omega, y_1), g(\omega, y_2))) \cdot r(y_1, y_2) ,$$

where $\varrho(\cdot, \cdot)$ and $r(\cdot, \cdot)$ are distances in X and Y;

(iii) *for all $\omega \in \Omega$ one has $D(\omega) = g(\omega, Y)$, the range of $g(\omega, \cdot)$.*

This theorem immediately implies the following assertion.

Proposition 1.3.2 (Measurable Selection Theorem). *Let a multifunction $\omega \mapsto D(\omega)$ take values in the subspace of closed non-void subsets of a Polish space X. Then $\{D(\omega)\}$ is a random closed set if and only if there exists a sequence $\{v_n : n \in \mathbb{N}\}$ of measurable maps $v_n : \Omega \mapsto X$ such that*

$$v_n(\omega) \in D(\omega) \quad and \quad D(\omega) = \overline{\{v_n(\omega), n \in \mathbb{N}\}} \quad for\ all \quad \omega \in \Omega .$$

In particular if $\{D(\omega)\}$ is a random closed set, then there exists a measurable selection, i.e. a measurable map $v : \Omega \mapsto X$ such that $v(\omega) \in D(\omega)$ for all $\omega \in \Omega$.

Below we also need the following assertion on the measurability of projections (see, e.g., CASTAING/VALADIER [18, p. 75]). It deals with the σ-algebra $\mathcal{F}^u$ of universally measurable sets associated with the measurable space $(\Omega, \mathcal{F})$ which is defined by the formula

$$\mathcal{F}^u = \bigcap_{\nu} \bar{\mathcal{F}}^{\nu} \,,$$

where the intersection taken over all probability measures ν on $(\Omega, \mathcal{F})$ and $\bar{\mathcal{F}}^{\nu}$ denotes the completion of the σ-algebra $\mathcal{F}$ with respect to the measure ν. We call $\mathcal{F}^u$ the universal σ-algebra and $\bar{\mathcal{F}}^{\nu}$ the ν-completion of $\mathcal{F}$ for shortness. Recall that the $\mathbb{P}$-completion $\bar{\mathcal{F}}^{\mathbb{P}}$ is the σ-algebra consisting of all subsets A of Ω for which there are sets U and V in $\mathcal{F}$ such that $U \subset A \subset V$ and $\mathbb{P}(U) = \mathbb{P}(V)$. The probability measure $\mathbb{P}$ can be extended from $\mathcal{F}$ to $\bar{\mathcal{F}}^{\mathbb{P}}$ such that $\bar{\mathcal{F}}^{\mathbb{P}}$ is a complete σ-algebra with respect to the extended probability measure. For details we refer to COHN [30], for instance. We also note that $\theta_t \bar{\mathcal{F}}^{\mathbb{P}} = \bar{\mathcal{F}}^{\mathbb{P}}$ for any fixed $t \in \mathbb{R}$. This property follows from the relation $\mathbb{P}(\theta_t U) = \mathbb{P}(U)$ for any $U \in \mathcal{F}$ and $t \in \mathbb{R}$.

Proposition 1.3.3 (Projection Theorem). *Let X be a Polish space and $M \subset \Omega \times X$ be a set which is measurable with respect to the product σ-algebra $\mathcal{F} \times \mathcal{B}(X)$. Then the set*

$$\mathrm{proj}_{\Omega} M = \{\omega \in \Omega \ : \ (\omega, x) \in M \text{ for some } x \in X\}$$

is universally measurable, i.e. belongs to $\mathcal{F}^u$. In particular it is measurable with respect to the $\mathbb{P}$-completion $\bar{\mathcal{F}}^{\mathbb{P}}$ of $\mathcal{F}$.

Now we introduce the following set valued analog of a separable process (cf. GIHMAN/SKOROHOD [48, p. 165]).

Definition 1.3.2. *Let I be a set in $\mathbb{R}$. A collection $\{C_t : t \in I\}$ of random sets is said to be* separable *if there exists an everywhere dense countable set Q in I such that*

$$C_t(\omega) \subset \bigcap_{n \in \mathbb{N}} \overline{\bigcup \{C_\tau(\omega) \ : \ \tau \in [t - n^{-1}, t + n^{-1}] \cap Q\}} \qquad (1.14)$$

for all $t \in I$ and $\omega \in \Omega$. The set Q is called the separability set *of the collection $\{C_t\}$. A process $\{v(t, \omega) \ : \ t \in I\}$ is said to be* separable *if the collection of random sets $C_t(\omega) = \{v(t, \omega)\}$ is separable.*

It is easy to see that $\{C_t : t \in I\}$ is a separable collection with a separability set Q if and only if for any $t \in I$ and $x \in C_t(\omega)$ there exist sequences $\{t_n\} \subset Q$ and $\{x_n\} \subset X$ such that $t_n \to t$ and $x_n \to x$ as $n \to \infty$ and $x_n \in C_{t_n}(\omega)$.

The following proposition gives examples of separable collections of random closed sets.

Proposition 1.3.4. *Let D be a random closed set and $I = (\alpha, \beta) \subset \mathbb{R}$. Assume that the function $h(t, \omega, x) : I \times \Omega \times X$ satisfies*

(i) for each $t \in I$ the function $h(t, \omega, \cdot)$ is continuous for all $\omega \in \Omega$ and $h(t, \cdot, x)$ is measurable for all $x \in X$;

(ii) $h(\cdot, \omega, x)$ is a right continuous function for all $\omega \in \Omega$ and $x \in X$.

Then $\omega \mapsto \overline{h(t, \omega, D(\omega))}$ is a separable collection of random closed sets whose separability set Q is an arbitrary everywhere dense countable set from (α, β). The same conclusion holds if $h(\cdot, \omega, x)$ is a left continuous function.

Proof. Proposition 1.3.1(vi) implies that $\omega \mapsto \overline{h(t, \omega, D(\omega))}$ is a random closed set for every t. From Theorem 1.3.1 we have that

$$h(t, \omega, D(\omega)) = \{h(t, \omega, g(\omega, y)) : y \in Y\}$$

Thus by (ii) for any $t \in I$ there exists a sequence $\{t_k\} \subset Q$ such that $t_k > t$ and

$$h(t, \omega, g(\omega, y)) = \lim_{t_k \to t} h(t_k, \omega, g(\omega, y))$$

for every $y \in Y$ and $\omega \in \Omega$. This property easily implies

$$h(t, \omega, g(\omega, y)) \in \bigcap_{n \in \mathbb{N}} \overline{\bigcup \{C_\tau(\omega) : \tau \in [t - n^{-1}, t + n^{-1}] \cap Q\}}$$

for all $y \in Y$ and $\omega \in \Omega$, where $C_t(\omega) = \overline{h(t, \omega, D(\omega))}$. This relation gives the separability of $\{\overline{h(t, \omega, D(\omega))}\}$.   □

The main property of separable collections of random closed sets which is important in the considerations below is given in the following proposition.

Proposition 1.3.5. *Let $\{C_t : t \in I\}$ be a separable collection of random sets. Then the multifunction*

$$\omega \mapsto C(\omega) = \overline{\bigcup_{t \in I} C_t(\omega)}$$

is a random closed set.

Proof. It follows from (1.14) that $\overline{\cup_{t \in I} C_t(\omega)} = \overline{\cup_{t \in I \cap Q} C_t(\omega)}$. Therefore we can apply Proposition 1.3.1(v).   □

Below we also need the following assertion.

Proposition 1.3.6. *Let $V : X \mapsto \mathbb{R}$ be a continuous function on a Polish space X and $R(\omega)$ be a random variable. If the set $V_R(\omega) := \{x : V(x) \leq R(\omega)\}$ is non-empty for any $\omega \in \Omega$, then it is a random closed set.*

Proof. The idea of the proof is borrowed from SCHENK-HOPPÉ [89]. It is clear that $V_R(\omega)$ is closed for any $\omega \in \Omega$. Due to Proposition 1.3.1(i) it is sufficient to prove that $\{\omega : V_R(\omega) \cap U \neq \emptyset\}$ is measurable for every open set $U \subset X$. This is equivalent to measurability of the set

$$\{\omega : V_R(\omega) \cap U = \emptyset\} \equiv \{\omega : U \subset X \setminus V_R(\omega)\} .$$

This measurability follows from the relation

$$\{\omega : U \subset X \setminus V_R(\omega)\} = \{\omega : R(\omega) < s \text{ for any } s \in V(U)\} \qquad (1.15)$$

which we now prove. Since

$$X \setminus V_R(\omega) = V^{-1}(\mathbb{R}) \setminus V^{-1}((-\infty, R(\omega)]) = V^{-1}((R(\omega), +\infty)) ,$$

we have that $U \subset X \setminus V_R(\omega)$ if and only if $V(U) \subset (R(\omega), +\infty)$. This implies (1.15) and therefore

$$\{\omega : U \subset X \setminus V_R(\omega)\} = \bigcap_{n \in \mathbb{N}} \{\omega : R(\omega) < s_n\},$$

where $s_n \in V(U)$ and $s_n \to \inf V(U)$ as $n \to \infty$.   □

The following notions of random tempered sets and variables play an important role in applications of the general theory of RDS connected with random and stochastic equations (cf. Chaps. 4 and 5). Roughly speaking, that a random variable which describes an influence of the random environment is tempered means that this environment evolves in non-explosive way.

Definition 1.3.3 (Tempered Random Set). *A random set $\{D(\omega)\}$ is said to be* tempered *with respect to MDS $\theta = (\Omega, \mathcal{F}, \mathbb{P}, \{\theta_t, t \in \mathbb{T}\})$ if there exist a random variable $r(\omega)$ and an element $y \in X$ such that*

$$D(\omega) \subset \{x \mid \text{dist}_X(x, y) \leq r(\omega)\} \quad \text{for all} \quad \omega \in \Omega$$

and $r(\omega)$ is a tempered random variable with respect to θ, i.e

$$\sup_{t \in \mathbb{T}} \left\{ e^{-\gamma|t|} |r(\theta_t \omega)| \right\} < \infty \quad \text{for all} \quad \omega \in \Omega \quad \text{and} \quad \gamma > 0 . \qquad (1.16)$$

A random variable $v(\omega)$ with values in X is said to be tempered *if the one-point random set $\{v(\omega)\}$ is tempered.*

It is clear that every deterministic set is tempered. We note that non-tempered random variables exist on any standard probability space with ergodic and aperiodic θ (see ARNOLD/CONG/OSELEDETS [9]). Sometimes (see, e.g., ARNOLD [3, p. 164]) the definition of a tempered random variable is based on the relation

$$\lim_{|t|\to\infty} \frac{1}{|t|} \log\{1 + |r(\theta_t\omega)|\} = 0 \quad \text{for all} \quad \omega \in \Omega \, .$$

which is weaker than (1.16). However we prefer to use (1.16) because it allows us to simplify some calculations in the applications below. We also note that if θ is ergodic, the only alternative to property (1.16) is that

$$\lim_{|t|\to\infty} \frac{1}{|t|} \log\{1 + |r(\theta_t\omega)|\} = +\infty \quad \text{for almost all} \quad \omega \in \Omega \, ,$$

see ARNOLD [3, p. 165].

As in the deterministic case we need a notion of an invariant set for the description of qualitative properties of RDS. It is convenient to introduce this notion for multifunctions to cover all types of random sets.

Definition 1.3.4 (Invariance Property). *Let (θ, φ) be a random dynamical system. A multifunction $\omega \mapsto D(\omega)$ is said to be*

(i)   forward invariant with respect to (θ, φ) if $\varphi(t, \omega)D(\omega) \subseteq D(\theta_t\omega)$ for all $t > 0$ and $\omega \in \Omega$, i.e. if $x \in D(\omega)$ implies $\varphi(t, \omega)x \in D(\theta_t\omega)$ for all $t \geq 0$ and $\omega \in \Omega$;

(ii)   backward invariant with respect to (θ, φ) if $\varphi(t, \omega)D(\omega) \supseteq D(\theta_t\omega)$ for all $t > 0$ and $\omega \in \Omega$, i.e. for every $t > 0$, $\omega \in \Omega$ and $y \in D(\theta_t\omega)$ there exists $x \in D(\omega)$ such that $\varphi(t, \omega)x = y$;

(iii) invariant with respect to (θ, φ) if $\varphi(t, \omega)D(\omega) = D(\theta_t\omega)$ for all $t > 0$ and $\omega \in \Omega$, i.e. if it is both forward and backward invariant.

We note that the forward invariance of the multifunction $\omega \mapsto D(\omega)$ means that

$$\text{graph}(D) = \{(\omega, x) \in \Omega \times X : x \in D(\omega)\}$$

is a forward invariant set in $\Omega \times X$ with respect to the semiflow $\{\pi_t\}$ defined by (1.6), i.e. $\pi_t\text{graph}(D) \subset \text{graph}(D)$ for all $t > 0$. The same is true for the property of invariance.

1.4 Dissipative, Compact and Asymptotically Compact RDS

In this section we start to develop methods for studying the qualitative behaviour of random dynamical systems. Our main goal is to investigate the behaviour of expressions of the form $x(t) = \varphi(t, \theta_{-t}\omega)x$ when $t \to +\infty$. At first sight this object looks a bit strange. However there are at least three reasons to study the limiting structure of $\varphi(t, \theta_{-t}\omega)x$.

The first one is connected with the question of what limiting dynamics we want to observe. The point is that in many applications RDS are generated by equations whose coefficients depend on $\theta_t \omega$. These coefficients describe the internal evolution of the environment and $\theta_{-t} \omega$ represents the state of the environment at time $-t$ which transforms into the "real" state (ω) at the time of observation (time 0, after a time t has elapsed). Furthermore the two-parameter mapping $U(\tau, s) := \varphi(\tau - s, \theta_s \omega)$ describes the evolution of the system from moment s to time τ, $\tau > s$. Therefore the limiting structure of $U(0, -t)x = \varphi(t, \theta_{-t} \omega)x$ when $t \to +\infty$ can be interpreted as the state of our system which we observe now $(t = 0)$ provided it was in the state x in the infinitely distant past $(t = -\infty)$. Thus the union of all these limits provides us with the real picture of the present state of the system.

The second reason is that the asymptotic behaviour of $\varphi(t, \theta_{-t} \omega)x$ provides us with some information about the long-time future. Indeed, since $\{\theta_t\}$ are measure preserving, we have that

$$\mathbb{P}\{\omega \ : \ \varphi(t, \omega)x \in D\} = \mathbb{P}\{\omega \ : \ \varphi(t, \theta_{-t} \omega)x \in D\}$$

for any $x \in X$ and $D \in \mathcal{B}(X)$. Therefore

$$\lim_{t \to +\infty} \mathbb{P}\{\omega \ : \ \varphi(t, \omega)x \in D\} = \lim_{t \to +\infty} \mathbb{P}\{\omega \ : \ \varphi(t, \theta_{-t} \omega)x \in D\} \ ,$$

if the limit on the right hand side exists. Thus the limiting behaviour of $\varphi(t, \theta_{-t} \omega)x$ for all ω determines the long-time behaviour of $\varphi(t, \omega)x$ with respect to convergence in probability.

The third reason is purely mathematical. If on the set of random variables $a(\omega)$ with values in X we define the operators T_t by the formula

$$(T_t a)(\omega) = \varphi(t, \theta_{-t} \omega)a(\theta_{-t} \omega), \quad t \in \mathbb{R}_+ \ ,$$

then the family $\{T_t, t \in \mathbb{R}_+\}$ is a one-parameter semigroup. Indeed, using the cocycle property we have

$$(T_s[T_t a])(\omega) = \varphi(s, \theta_{-s} \omega)(T_t a)(\theta_{-s} \omega) = \varphi(s, \theta_{-s} \omega)\varphi(t, \theta_{-t-s} \omega)a(\theta_{-t-s} \omega)$$

$$= \varphi(t + s, \theta_{-t-s} \omega)a(\theta_{-t-s} \omega) = (T_{t+s} a)(\omega) \ .$$

Thus it becomes possible to use ideas from the theory of deterministic (autonomous) dynamical systems for which the semigroup structure of the evolution operator is crucial. Below we introduce several important dynamical notions and study the qualitative behaviour of RDS relying on this observation.

Let $\mathcal{D}$ be a family of random closed sets which is closed with respect to inclusions (i.e. if $D_1 \in \mathcal{D}$ and a random closed set $\{D_2(\omega)\}$ possesses the property $D_2(\omega) \subset D_1(\omega)$ for all $\omega \in \Omega$, then $D_2 \in \mathcal{D}$). Sometimes the collection $\mathcal{D}$ is called a *universe* of sets (see, e.g., SCHENK-HOPPÉ [89]) or an IC-system (see FLANDOLI/SCHMALFUSS [44]). The simplest example of a universe is the collection of all one-point subsets of X. However the concept of a universe allows us to include the consideration of local regimes of the system into the theory in a natural way. We refer to SCHENK-HOPPÉ [89]) for a further discussion of this concept. In the applications presented in Chaps.5 and 6 we deal with the universe of all tempered subsets of the phase space.

Definition 1.4.1 (Absorbing Set). *A random closed set $\{B(\omega)\}$ is said to be* absorbing *for the RDS (θ, φ) in the universe $\mathcal{D}$, if for any $D \in \mathcal{D}$ and for any ω there exists $t_0(\omega)$ such that*

$$\varphi(t, \theta_{-t}\omega)D(\theta_{-t}\omega) \subset B(\omega) \quad \text{for all} \quad t \geq t_0(\omega) \quad \text{and} \quad \omega \in \Omega .$$

Definition 1.4.2 (Dissipative RDS). *An RDS (θ, φ) is said to be* dissipative *in the universe $\mathcal{D}$, if there exists an absorbing set B for the RDS (θ, φ) in the universe $\mathcal{D}$ such that*

$$B(\omega) \subset B_{r(\omega)}(x_0) \equiv \{x \ : \ \mathrm{dist}_X(x, x_0) \leq r(\omega)\}, \tag{1.17}$$

for some $x_0 \in X$ and random variable $r(\omega)$ and for all $\omega \in \Omega$. If X is a linear space and $x_0 = 0$, then the variable $r(\omega)$ is said to be a radius of dissipativity *of the RDS (θ, φ) in the universe $\mathcal{D}$.*

The simplest examples of dissipative RDS are the following ones.

Example 1.4.1 (Discrete Dissipative RDS). Let us consider the RDS constructed in Example 1.2.1. Let $X = \mathbb{R}$ and $\Omega_0 = \{0, 1\}$ be a two-point set. Assume that the continuous functions f_0 and f_1 possess the property

$$|f_i(x)| \leq a|x| + b \quad \text{with some} \quad 0 \leq a < 1, \ b \geq 0 .$$

In this case Ω is the set of two-sided sequences $\omega = \{\omega_i \,|\, i \in \mathbb{Z}\}$ consisting of zeros and ones and

$$\varphi(n, \omega) = f_{\omega_{n-1}} \circ f_{\omega_{n-2}} \circ \ldots \circ f_{\omega_1} \circ f_{\omega_0}, \quad \omega = \{\omega_i \,|\, i \in \mathbb{Z}\}, \quad n \in \mathbb{N} .$$

Using the cocycle property it is easy to see that

$$|\varphi(n+1, \omega)x| \leq a \cdot |\varphi(n, \omega)x| + b, \quad n \in \mathbb{Z}_+ . \tag{1.18}$$

Therefore after n iterations we obtain

$$|\varphi(n, \omega)x| \leq a^n \cdot |x| + b \cdot (1-a)^{-1}, \quad n \in \mathbb{Z}_+ . \tag{1.19}$$

Let $\mathcal{D}$ be the family of all tempered (with respect to θ) random closed sets in $\mathbb{R}$. Let $D \in \mathcal{D}$ and $D(\omega) \subset \{x \ : \ |x| \leq r(\omega)\}$, where $r(\omega)$ possesses the property (1.16) (i.e. is a tempered random variable). Then (1.19) implies that

$$|\varphi(n, \theta_{-n}\omega)x(\theta_{-n}\omega)| \leq a^n r(\theta_{-n}\omega) + b \cdot (1-a)^{-1}, \quad \text{for all} \quad x(\omega) \in D(\omega) .$$

Since $0 \leq a < 1$, it follows from (1.16) that $a^n r(\theta_{-n}\omega) \to 0$ as $n \to +\infty$. Therefore for every $\omega \in \Omega$ there exists $n_0(\omega)$ such that $a^n r(\theta_{-n}\omega) \leq 1$ for $n \geq n_0(\omega)$. Consequently we have

$$\varphi(n, \theta_{-n}\omega)D(\theta_{-n}\omega) \subset B := [-1 - b \cdot (1-a)^{-1}, 1 + b \cdot (1-a)^{-1}]$$

for $n \geq n_0(\omega)$. Thus the RDS considered is dissipative in the universe $\mathcal{D}$ of all tempered random closed sets from $\mathbb{R}$. Using (1.18) with $n = 0$ one can easily see that B is a forward invariant set from $\mathcal{D}$.

Example 1.4.2 (Kick Model). Let X be a Banach space and $g : X \mapsto X$ be a continuous mapping such that

$$\|g(x)\| \leq a\|x\| + b, \quad 0 \leq a < 1, \; b \geq 0 \,. \tag{1.20}$$

Consider the RDS (θ, φ) generated by the difference equation

$$x_{n+1} = g(x_n) + \xi(\theta_{n+1}\omega), \quad n \in \mathbb{Z}_+ \,, \tag{1.21}$$

over a metric dynamical system $(\Omega, \mathcal{F}, \mathbb{P}, \{\theta_n, n \in \mathbb{Z}\})$, where $\xi(\omega)$ is a tempered random variable in X. Using (1.20) and (1.21) we have

$$\|\varphi(n, \omega)x\| \leq a^n\|x\| + R(\theta_n\omega), \quad n \in \mathbb{Z}_+ \,,$$

where

$$R(\omega) = b(1-a)^{-1} + \sum_{k=0}^{\infty} a^k \|\xi(\theta_{-k}\omega)\|$$

is a tempered random variable. It is easy to see that for every $\delta > 0$ the ball $B^\delta(\omega) = \{x : \|x\| \leq (1+\delta)R(\omega)\}$ is a forward invariant absorbing set for (θ, φ) in the universe $\mathcal{D}$ of all tempered random closed sets from X.

Example 1.4.3 (Continuous Dissipative RDS). Let (θ, φ) be the RDS considered in Example 1.2.3 from the random ODE $\dot{x} = f(\theta_t\omega, x)$. Assume additionally that the function $f(\omega, x)$ possesses the property

$$xf(\omega, x) \leq -\alpha|x|^2 + \beta, \quad \text{for all} \quad \omega \in \Omega \,,$$

where $\alpha > 0$ and $\beta \geq 0$ are nonrandom constants. Then it is easy to see that

$$\frac{1}{2} \cdot \frac{d}{dt}|x(t)|^2 \leq -\alpha|x(t)|^2 + \beta, \quad t > 0 \,,$$

for any solution to (1.10). Therefore, since $\varphi(t, \omega)x = x(t)$, we have

$$|\varphi(t, \omega)x|^2 \leq e^{-2\alpha t}|x|^2 + \frac{\beta}{\alpha} \cdot \left(1 - e^{-2\alpha t}\right), \quad t > 0 \,.$$

As in Example 1.4.1 this property implies that (θ, φ) is dissipative in the universe $\mathcal{D}$ of all tempered (with respect to θ) random closed sets from $\mathbb{R}$. Moreover the absorbing set $B = \{x : |x| \leq 1 + \beta/\alpha\}$ is a forward invariant set from $\mathcal{D}$.

The situation described in Example 1.4.3 admits the following generalization which can be also considered as an extension of well-known deterministic results (see, e.g., BABIN/VISHIK [13], CHUESHOV [20] or HALE [50]) to the random case.

Proposition 1.4.1. *Assume that the phase space X of RDS (θ, φ) is a separable Banach space with the norm $\|\cdot\|$ and there exists a continuous function $V : X \mapsto \mathbb{R}$ with the properties:*

(i)   *$V(\varphi(t,\omega)x)$ is absolutely continuous with respect to t for any $(\omega, x) \in \Omega \times X$;*

(ii) *there exists a constant $\alpha > 0$ and a tempered random variable $\beta(\omega) \geq 0$ such that for every $(\omega, x) \in \Omega \times X$ we have the inequality*

$$\frac{d}{dt}V(\varphi(t,\omega)x) + (\alpha + \varrho(\theta_t\omega)) \cdot V(\varphi(t,\omega)x) \leq \beta(\theta_t\omega) \qquad (1.22)$$

for almost all $t > 0$, where $\varrho(\omega)$ is a random variable such that $\varrho(\theta_t\omega)$ lies in $L^1_{loc}(\mathbb{R})$ for every $\omega \in \Omega$ and

$$\lim_{t \to +\infty} \frac{1}{t} \int_0^t \varrho(\theta_\tau\omega) \, d\tau = \lim_{t \to +\infty} \frac{1}{t} \int_{-t}^0 \varrho(\theta_\tau\omega) \, d\tau = 0 \qquad (1.23)$$

for all $\omega \in \Omega$;

(iii) *there exist positive constants $b_1, b_2, \delta_1, \delta_2$ and nonnegative numbers c_1 and c_2 such that*

$$b_1\|x\|^{\delta_1} - c_1 \leq V(x) \leq b_2\|x\|^{\delta_2} + c_2, \quad x \in X . \qquad (1.24)$$

Then the RDS (θ, φ) is dissipative in the universe $\mathcal{D}$ of all tempered random closed sets in X. Moreover there exists a tempered random variable $R(\omega) \geq 0$ such that for any positive ϵ the set

$$B_\epsilon(\omega) = \{x : V(x) \leq (1 + \epsilon)R(\omega)\} \qquad (1.25)$$

is a forward invariant absorbing tempered random closed set.

Proof. Let $D \in \mathcal{D}$ and $x(\omega) \in D(\omega)$ for all $\omega \in \Omega$. From (1.22) we have that

$$V(\varphi(t,\omega)x(\omega)) \leq V(x(\omega)) \cdot \exp\left\{-\alpha t - \int_0^t \varrho(\theta_\tau\omega) \, d\tau\right\}$$

$$+ \int_0^t \beta(\theta_s\omega) \cdot \exp\left\{-\alpha(t - s) - \int_s^t \varrho(\theta_\tau\omega) \, d\tau\right\} ds .$$

Therefore

$$V(\varphi(t, \theta_{-t}\omega)x(\theta_{-t}\omega)) \leq V(x(\theta_{-t}\omega)) \cdot \exp\left\{-\alpha t - \int_{-t}^{0} \varrho(\theta_\tau\omega)\,d\tau\right\}$$

(1.26)

$$+ \int_{-t}^{0} \beta(\theta_s\omega) \cdot \exp\left\{\alpha s - \int_{s}^{0} \varrho(\theta_\tau\omega)\,d\tau\right\}ds.$$

It follows from (1.23) that for any $\varepsilon > 0$ and $\omega \in \Omega$ there exists $c(\omega) > 0$ such that

$$\left|\int_{0}^{t} \varrho(\theta_\tau\omega)\,d\tau\right| \leq \varepsilon|t| + c(\omega), \quad t \in \mathbb{R}, \ \omega \in \Omega.$$

(1.27)

Therefore, since $\beta(\omega)$ is tempered, for all $\omega \in \Omega$ the integral

$$R(\omega) = \int_{-\infty}^{0} \beta(\theta_s\omega) \cdot \exp\left\{\alpha s - \int_{s}^{0} \varrho(\theta_\tau\omega)\,d\tau\right\}ds$$

(1.28)

exists. It follows from (1.27) that

$$R(\theta_t\omega) \leq C(\omega)e^{\varepsilon|t|} \int_{-\infty}^{0} e^{\alpha s}e^{(\gamma+\varepsilon)|t+s|}ds \cdot \sup_{\tau}\left\{e^{-\gamma|\tau|}\beta(\theta_\tau\omega)\right\}$$

$$\leq \frac{1}{\alpha - \gamma - \varepsilon}e^{(\gamma+2\varepsilon)|t|}\sup_{\tau}\left\{e^{-\gamma|\tau|}\beta(\theta_\tau\omega)\right\}$$

for all $\varepsilon > 0$ and $\gamma > 0$ such that $\gamma + \varepsilon < \alpha$. This implies that $R(\omega)$ is a tempered random variable. Proposition 1.3.6 and relation (1.24) imply that $B_\epsilon(\omega)$ given by (1.25) is a tempered random closed set. Let

$$e(t, \omega) = \exp\left\{\alpha t - \int_{t}^{0} \varrho(\theta_\tau\omega)\,d\tau\right\}.$$

Then from (1.26) for any $x(\omega) \in B_\epsilon(\omega)$ we have that

$$V(\varphi(t, \theta_{-t}\omega)x(\theta_{-t}\omega)) \leq (1+\epsilon)R(\theta_{-t}\omega) \cdot e(-t, \omega) + \int_{-t}^{0} \beta(\theta_s\omega) \cdot e(s, \omega)ds.$$

Since $e(-t, \omega) \cdot e(s, \theta_{-t}\omega) = e(s - t, \omega)$, it follows from (1.28) that

$$R(\theta_{-t}\omega) \cdot e(-t, \omega) = \int_{-\infty}^{-t} \beta(\theta_s\omega) \cdot e(s, \omega)ds.$$

Therefore

$$V(\varphi(t, \theta_{-t}\omega)x(\theta_{-t}\omega)) \leq (1+\epsilon)R(\omega).$$

Thus $B_\epsilon(\omega)$ is forward invariant. It follows from (1.24) and (1.26) that

$$V(\varphi(t, \theta_{-t}\omega)x(\theta_{-t}\omega)) \le \left(b_2 \|x(\theta_{-t}\omega)\|^{\delta_2} + c_2\right) \cdot e^{-\alpha t} + R(\omega) .$$

This relation implies that $B_\epsilon(\omega)$ is absorbing in the universe $\mathcal{D}$.   □

Remark 1.4.1. If θ is an ergodic metric dynamical system, assumption (ii) in Proposition 1.4.1 can be replaced by the inequality

$$\frac{d}{dt} V(\varphi(t,\omega)x) + \tilde{\alpha}(\theta_t\omega) \cdot V(\varphi(t,\omega)x) \le \beta(\theta_t\omega) , \qquad (1.29)$$

there $\beta(\omega) \ge 0$ is a tempered random variable and $\tilde{\alpha}(\omega) \in L^1(\Omega, \mathcal{F}, \mathbb{P})$ is a random variable such that $\mathbb{E}\tilde{\alpha} > 0$. Indeed, it follows from the Birkhoff-Khintchin ergodic theorem (see, e.g., ARNOLD [3, Appendix]) that

$$\lim_{|t| \to \infty} \frac{1}{t} \int_0^t \tilde{\alpha}(\theta_\tau\omega)d\tau = \mathbb{E}\tilde{\alpha}, \quad \omega \in \Omega^*,$$

where $\Omega^* \subseteq \Omega$ is a θ-invariant set of full measure. Without loss of generality we can suppose that $\Omega^* = \Omega$ (see Remark 1.2.1(ii)). Therefore we can apply Proposition 1.4.1 with $\alpha = \mathbb{E}\tilde{\alpha}$ and $\varrho(\omega) = \tilde{\alpha}(\omega) - \mathbb{E}\tilde{\alpha}$.

Example 1.4.4 (Binary Biochemical Model). Consider the RDS (θ, φ) generated in $\mathbb{R}^2$ by equations (1.11) over an ergodic metric dynamical system θ. Let the hypotheses concerning g and α_i listed in Example 1.2.4 hold. We assume in addition that

$$\alpha_{\min}(\omega) = \min\{\alpha_1(\omega), \alpha_2(\omega)\} \in L^1(\Omega, \mathcal{F}, \mathbb{P}) \quad \text{and} \quad \alpha_0 = \mathbb{E}\alpha_{\min} > 0 .$$

If

$$x_1 \cdot (x_2 + g(x_2)) \le \frac{\alpha_0}{2} \cdot (x_1^2 + x_2^2) + \beta_0, \quad (x_1, x_2) \in \mathbb{R}_+^2 ,$$

where $\beta_0 \ge 0$ is a constant, then (1.29) holds with $V(x) = x_1^2 + x_2^2$, $\tilde{\alpha} = 2\alpha_{\min}(\omega) - \alpha_0$ and $\beta(\omega) \equiv 2\beta_0$. Thus the RDS (θ, φ) is dissipative in the universe of all tempered random closed sets from $\mathbb{R}^2$.

The following concepts are useful when the phase space X is infinite-dimensional.

Definition 1.4.3 (Compact RDS). *An RDS (θ, φ) is said to be compact in the universe $\mathcal{D}$, if it is dissipative in $\mathcal{D}$ and the absorbing set B is a random compact set.*

If the phase space X of an RDS (θ, φ) is compact, then (θ, φ) is a compact RDS. If X is a finite-dimensional space, then any dissipative RDS is compact.

Example 1.4.5 (Kick Model). Let (θ, φ) be the RDS considered in Example 1.4.2. Assume additionally that g is a compact mapping, i.e. $\overline{g(B)}$ is a compact set for every bounded set B from X. The set

$$C(\omega) = \overline{\varphi(1, \theta_{-1}\omega) B^{\delta}(\theta_{-1}\omega)} = \overline{g(B^{\delta}(\theta_{-1}\omega)) + \xi(\omega)}$$

is an absorbing forward invariant random compact set for (θ, φ) in the universe $\mathcal{D}$ of all tempered random closed sets from X.

Definition 1.4.4 (Asymptotically Compact RDS). *An RDS (θ, φ) is said to be* asymptotically compact *in the universe $\mathcal{D}$, if there exists an attracting random compact set $\{B_0(\omega)\}$, i.e. for any $D \in \mathcal{D}$ and for any $\omega \in \Omega$ we have*

$$\lim_{t \to +\infty} d_X \{\varphi(t, \theta_{-t}\omega) D(\theta_{-t}\omega) \,|\, B_0(\omega)\} = 0 \,, \qquad (1.30)$$

where $d_X \{A|B\} = \sup_{x \in A} \mathrm{dist}_X(x, B)$.

It is clear that any compact RDS is asymptotically compact. Deterministic examples of asymptotically compact systems which are not compact can be found in BABIN/VISHIK [13], CHUESHOV [20], HALE [50] and TEMAM [104].

The following assertion shows that every asymptotically compact RDS is dissipative.

Proposition 1.4.2. *Let (θ, φ) be an asymptotically compact RDS in $\mathcal{D}$ with an attracting random compact set $\{B_0(\omega)\}$. Then it is dissipative in $\mathcal{D}$.*

Proof. For any $x_0 \in X$ we can find a random variable $r(\omega) \in (0, +\infty)$ such that

$$B_0(\omega) \subset \{x \,:\, \mathrm{dist}_X(x, x_0) \leq r(\omega)\} \quad \text{for all} \quad \omega \in \Omega \,. \qquad (1.31)$$

To prove this we note that by Theorem 1.3.1

$$B_0(\omega) = \{g(\omega, y) \,:\, y \in Y\} \quad \text{for all} \quad \omega \in \Omega \,,$$

where Y is a Polish space and the mapping $g(\omega, y) \,:\, \Omega \times Y \mapsto X$ is such that $g(\omega, \cdot)$ is continuous for all $\omega \in \Omega$ and $g(\cdot, y)$ is measurable for all $y \in Y$. Since $B_0(\omega)$ is a compact set and Y is separable, $r(\omega)$ defined by

$$r(\omega) := \sup_{y \in Y} \mathrm{dist}_X(x_0, g(\omega, y)) \in (0, +\infty), \quad \omega \in \Omega \,,$$

is a random variable and (1.31) holds.

It follows from (1.30) that for any $D \in \mathcal{D}$ and for any ω there exists a $t_0(\omega)$ such that

$$\varphi(t, \theta_{-t}\omega) D(\theta_{-t}\omega) \subset B^*(\omega) := \{x \,:\, \mathrm{dist}_X(x, x_0) \leq 1 + r(\omega)\} \quad \text{for } t \geq t_0(\omega) \,.$$

Thus (θ, φ) is dissipative. $\qquad \qquad \square$

The notions of dissipative, compact and asymptotically compact random systems differ only in infinite-dimensional phase spaces.

1.5 Trajectories

In this section we describe some measurable properties of the trajectories of RDS.

Definition 1.5.1. *Let* $D : \omega \mapsto D(\omega)$ *be a multifunction. We call the multifunction*

$$\omega \mapsto \gamma_D^t(\omega) := \bigcup_{\tau \geq t} \varphi(\tau, \theta_{-\tau}\omega)D(\theta_{-\tau}\omega)$$

the tail *(from the moment t) of the pull back trajectories emanating from* D. *If* $D(\omega) = \{v(\omega)\}$ *is a single valued function, then* $\omega \mapsto \gamma_v(\omega) \equiv \gamma_D^0(\omega)$ *is said to be the* (pull back) *trajectory (or orbit) emanating from* v.

In the deterministic case Ω is a one-point set and $\varphi(t, \omega) = \varphi(t)$ is a semigroup of continuous mappings. Therefore in this case the tail γ_D^t has the form

$$\gamma_D^t = \bigcup_{\tau \geq t} \varphi(\tau)D = \bigcup_{\tau \geq 0} \varphi(\tau)(\varphi(t)D) = \gamma_{\varphi(t)D}^0 \ ,$$

i.e. γ_D^t is a collection of the "normal" trajectories emanating from $\varphi(t)D$.

We note that any tail is a forward invariant multifunction. It also follows from Proposition 1.3.1(v) that in the case of discrete time ($\mathbb{T} = \mathbb{Z}$) the closure $\overline{\gamma_D^t(\omega)}$ of any tail $\gamma_D^t(\omega)$ is a random closed set. For continuous time we have the following proposition.

Proposition 1.5.1. *For any random closed set* $\{D(\omega)\}$ *the closure* $\overline{\gamma_D^t(\omega)}$ *of any tail* $\gamma_D^t(\omega)$ *of the pull back trajectories emanating from* D *is a random closed set with respect to the* σ-algebra $\mathcal{F}^u$ *of universally measurable sets.*

Proof. The idea of the proof is borrowed from CRAUEL/FLANDOLI [36]. The Representation Theorem 1.3.1 gives that $D(\omega) = g(\omega, Y)$, where Y is a Polish space, $g(\omega, \cdot)$ is continuous for all $\omega \in \Omega$ and $g(\cdot, y)$ is measurable for all $y \in Y$. Therefore for every $x \in X$ we have

$$d(t, \omega) := \mathrm{dist}_X(x, \varphi(t, \theta_{-t}\omega)D(\theta_{-t}\omega)) = \inf_k \mathrm{dist}_X(x, \varphi(t, \theta_{-t}\omega)g(\theta_{-t}\omega, y_k)) \ ,$$

where $\{y_k\}$ is a dense sequence in Y. Since $(t, \omega) \mapsto (t, \theta_{-t}\omega)$ is a measurable mapping and $(t, \omega) \mapsto d_k(t, \omega) := \mathrm{dist}_X(x, \varphi(t, \omega)g(\omega, y_k))$ is a measurable function, the function $(t, \omega) \mapsto d_k(t, \theta_{-t}\omega)$ is also measurable. Consequently the function $(t, \omega) \mapsto d(t, \omega)$ is $\mathcal{B}(\mathbb{R}_+) \times \mathcal{F}$-measurable. It is also clear that

$$\mathrm{dist}(x, \overline{\gamma_D^t(\omega)}) = \mathrm{dist}(x, \gamma_D^t(\omega)) = \inf_{\tau \geq t} d(\tau, \omega) \ .$$

For any $a \in \mathbb{R}_+$ we have

$$\left\{\omega \; : \; \inf_{\tau \geq t} d(\tau, \omega) < a\right\} = \mathrm{proj}_{\Omega}\left\{(\tau, \omega) \; : \; d(\tau, \omega) < a, \; \tau \geq t\right\},$$

where proj_{Ω} is the canonical projection of $\mathbb{R}_+ \times \Omega$ on Ω defined by

$$\mathrm{proj}_{\Omega} M = \{\omega \in \Omega \; : \; (t, \omega) \in M \text{ for some } t \in \mathbb{R}_+\}.$$

Hence Proposition 1.3.3 implies that $\{\omega \; : \; \inf_{\tau \geq t} d(\tau, \omega) < a\}$ is a universally measurable set and therefore $\omega \mapsto \overline{\gamma_D^t(\omega)}$ is a random closed set with respect to $\mathcal{F}^u$. □

As a direct consequence of Proposition 1.3.5 we also have the following assertions.

Proposition 1.5.2. *Let $a(\omega)$ be a random variable in X. Assume that $t \mapsto \varphi(t, \theta_{-t}\omega)a(\theta_{-t}\omega)$ is a separable process, $t \in \mathbb{R}_+$. Then $\omega \mapsto \overline{\gamma_a^t(\omega)}$ is a forward invariant random closed set with respect to $\mathcal{F}$. In particular, if for some $x \in X$ the mapping $t \mapsto \varphi(t, \theta_{-t}\omega)x$ is a right continuous function for all $t > 0$ and $\omega \in \Omega$, then $\omega \mapsto \overline{\gamma_x^t(\omega)}$ is a forward invariant random closed set with respect to $\mathcal{F}$.*

Proof. It is clear that

$$\overline{\gamma_D^t(\omega)} = \overline{\bigcup \{\varphi(\tau, \theta_{-\tau}\omega)a(\theta_{-\tau}\omega) \; : \; \tau \geq t, \; \tau \in Q\}},$$

where Q is a separability set of the process $t \mapsto \varphi(t, \theta_{-t}\omega)a(\theta_{-t}\omega)$. Therefore we can apply Proposition 1.3.1(v). □

Proposition 1.5.3. *Let (θ, φ) be an RDS such that the function*

$$(t, x) \mapsto \varphi(t, \theta_{-t}\omega)x \text{ is a continuous mapping} \tag{1.32}$$

from $\mathbb{R}_+ \times X$ into X. Assume that D is a random closed set such that $\{D(\theta_t\omega) : t \leq 0\}$ is a separable collection. Then the closure $\overline{\gamma_D^t}$ of the tail γ_D^t is a forward invariant random closed set with respect to $\mathcal{F}$ for every $t \geq 0$. In particular, $\overline{\gamma_D^t}$ possesses this property for every deterministic D.

Proof. Since $\{D(\theta_t\omega) \; : \; t \leq 0\}$ is a separable collection, we can find an everywhere dense countable set Q such that for any $t \geq 0$ and $x \in D(\theta_{-t}\omega)$ there exist $t_n \in Q$ and $x_n \in D(\theta_{-t_n}\omega)$ such that $x_n \to x$ and $t_n \to t$ as $n \to \infty$. Property (1.32) implies that $\varphi(t_n, \theta_{-t_n}\omega)x_n \to \varphi(t, \theta_{-t}\omega)x$ as $n \to \infty$. Therefore $\{\varphi(t, \theta_{-t}\omega)D(\theta_{-t}\omega) \; : \; t \geq t_0\}$ is a separable collection for any $t_0 \geq 0$. Thus we can apply Propositions 1.3.5 and 1.3.1(v). □

Remark 1.5.1. Assume that the mappings $\varphi(t, \omega)$ are restrictions to $\mathbb{R}_+$ of mappings $\tilde{\varphi}(t, \omega)$ which satisfy the conditions listed in Definition 1.2.1 for all $t, s \in \mathbb{R}$ and such that $(t, x) \mapsto \tilde{\varphi}(t, \omega)x$ is a continuous mapping from $\mathbb{R} \times X$ into X for every $\omega \in \Omega$. This situation is typical for RDS generated by finite-dimensional random and stochastic differential equations (for instance, this is true for the RDS considered in Examples 1.2.4 and 1.4.4). The cocycle property for $\tilde{\varphi}$ implies that

$$\tilde{\varphi}(t, \theta_{-t}\omega) \circ \tilde{\varphi}(-t, \omega) = \tilde{\varphi}(-t, \omega) \circ \tilde{\varphi}(t, \theta_{-t}\omega) = \mathrm{id}, \quad t \in \mathbb{R}, \ \omega \in \Omega \ .$$

Hence $(t, x) \mapsto (t, \tilde{\varphi}(-t, \omega))$ is a bijective mapping from $\mathbb{R} \times X$ into itself and

$$\varphi(t, \theta_{-t}\omega) = \tilde{\varphi}(t, \theta_{-t}\omega) = [\tilde{\varphi}(-t, \omega)]^{-1}, \quad t \geq 0 \ .$$

Therefore by Proposition 1.1.6 (ARNOLD [3]) $(t, x) \mapsto \varphi(t, \theta_{-t}\omega)$ is a continuous mapping from $\mathbb{R} \times X$ into X for every $\omega \in \Omega$ provided that X is either a compact Hausdorff space or a finite-dimensional topological manifold. Therefore in this case by Proposition 1.5.2 $\{\overline{\gamma_a^t(\omega)}\}$ is a forward invariant random closed set with respect to $\mathcal{F}$ for every $a(\omega)$ such that the mapping $t \mapsto a(\theta_t\omega)$ is continuous for all $\omega \in \Omega$. By Proposition 1.5.3 the same is true for $\overline{\gamma_D^t}$, where D is a deterministic subset in X.

We note that if X is a separable Banach space, then the set of random variables $v(\omega)$ such that $t \mapsto v(\theta_t\omega)$ is a C^∞-function for every ω is dense in the set of all random variables with respect to convergence in probability (see the argument given in the proof of Proposition 8.3.8 ARNOLD [3]). We also note that in the case considered the function $t \mapsto \varphi(t, \theta_{-t}\omega)a(\theta_{-t}\omega)$ is a stochastically continuous process (i.e. it is continuous with respect to convergence in probability) for any random variable $a(\omega)$. This property follows from the stochastic continuity of the process $t \mapsto a(\theta_t\omega)$ (see ARNOLD [3, Appendix A.1]).

1.6 Omega-limit Sets

To describe the asymptotic behaviour of RDS as in the deterministic case (cf. HARTMAN [51] and also HALE [50], TEMAM [104], CHUESHOV [20], for example) we use the concept of an omega-limit set. As in CRAUEL/FLANDOLI [36] our definition concerns pull back trajectories.

Definition 1.6.1. *Let $D : \omega \mapsto D(\omega)$ be a multifunction. We call the multifunction*

$$\omega \mapsto \Gamma_D(\omega) := \bigcap_{t > 0} \overline{\gamma_D^t(\omega)} = \bigcap_{t > 0} \overline{\bigcup_{\tau \geq t} \varphi(\tau, \theta_{-\tau}\omega)D(\theta_{-\tau}\omega)}$$

the (pull back) omega-limit set of the trajectories emanating from D.

The following assertion gives another description of omega-limit sets.

Proposition 1.6.1. *Let $\Gamma_D(\omega)$ be the omega-limit set of the trajectories emanating from a multifunction D. Then $x \in \Gamma_D(\omega)$ if and only if there exist sequences $t_n \to +\infty$ and $y_n \in D(\theta_{-t_n}\omega)$ such that*

$$x = \lim_{n \to +\infty} \varphi(t_n, \theta_{-t_n}\omega)y_n . \tag{1.33}$$

Proof. Let $x \in \Gamma_D(\omega)$. Then we have

$$x \in \overline{\bigcup_{\tau \geq n} \varphi(\tau, \theta_{-\tau}\omega)D(\theta_{-\tau}\omega)} \quad \text{for all} \quad n = 1, 2, \ldots .$$

Therefore there exists an element b_n such that

$$b_n \in \bigcup_{\tau \geq n} \varphi(\tau, \theta_{-\tau}\omega)D(\theta_{-\tau}\omega) \tag{1.34}$$

and $\mathrm{dist}(x, b_n) \leq 1/n$, $n = 1, 2, \ldots$. It follows from (1.34) that there exist $t_n \geq n$ and $y_n \in D(\theta_{t_n}\omega)$ such that $b_n = \varphi(t_n, \theta_{-t_n}\omega)y_n$. It is clear that we have (1.33) for these t_n and y_n.

Vice versa, assume that an element x possesses property (1.33). It is obvious that for any $t > 0$ there exists t_n such that

$$\varphi(t_n, \theta_{-t_n}\omega)y_n \in \bigcup_{\tau \geq t} \varphi(\tau, \theta_{-\tau}\omega)D(\theta_{-\tau}\omega) \subset \overline{\bigcup_{\tau \geq t} \varphi(\tau, \theta_{-\tau}\omega)D(\theta_{-\tau}\omega)} .$$

Therefore

$$x \in \overline{\bigcup_{\tau \geq t} \varphi(\tau, \theta_{-\tau}\omega)D(\theta_{-\tau}\omega)} \quad \text{for all} \quad t > 0.$$

This implies that $x \in \Gamma_D(\omega)$. $\qquad\square$

We note that Proposition 1.6.1 provides us with a description of omega-limit sets. But it does not guarantee that they are nonempty. The following assertion gives us conditions under which $\Gamma_D(\omega)$ is nonempty.

Proposition 1.6.2. *Assume that the RDS (θ, ϕ) is asymptotically compact in a universe $\mathcal{D}$ with the attracting random compact set $\{B_0(\omega)\}$. Then for any $D \in \mathcal{D}$ and for all $\omega \in \Omega$ the omega-limit set $\Gamma_D(\omega)$ is a nonempty compact set and $\Gamma_D(\omega) \subset B_0(\omega)$. The multifunction $\omega \mapsto \Gamma_D(\omega)$ is invariant and it is a random compact set with respect to the σ-algebra $\mathfrak{F}^u$ of universally measurable sets (with respect to $\mathfrak{F}$, in the case of discrete time).*

Proof. Let $t_n \to \infty$ and $y_n \in D(\theta_{-t_n}\omega)$ be arbitrary sequences. From (1.30) we have that

$$\varphi(t_n, \theta_{-t_n}\omega)y_n \to B_0(\omega) \quad \text{when} \quad n \to +\infty ,$$

i.e. there exists a sequence $\{b_n\} \subset B_0(\omega)$ such that

$$\text{dist}_X(\varphi(t_n, \theta_{-t_n}\omega)y_n, b_n) \to 0 \quad \text{when} \quad n \to +\infty .$$

The compactness of $B_0(\omega)$ implies that for some subsequence $\{n_k\}$ and some $b \in B_0(\omega)$ we have that $b_{n_k} \to b$. This implies that

$$\varphi(t_{n_k}, \theta_{-t_{n_k}}\omega)y_{n_k} \to b \in B_0(\omega) \quad \text{when} \quad k \to +\infty .$$

Thus $\Gamma_D(\omega)$ is nonempty. It is clear from (1.30) that any element of the form (1.33) belongs to $B_0(\omega)$. Therefore we have $\Gamma_D(\omega) \subset B_0(\omega)$ and, since $\Gamma_D(\omega)$ is closed, $\Gamma_D(\omega)$ is a compact set.

Let us prove that $\omega \mapsto \Gamma_D(\omega)$ is invariant. Using the cocycle property we have

$$\varphi(t, \omega)x = \lim_{n \to \infty} \varphi(t, \omega) \circ \varphi(t_n, \theta_{-t_n}\omega)y_n = \lim_{n \to \infty} \varphi(t + t_n, \theta_{-t-t_n} \circ \theta_t\omega)y_n$$

for any $x \in \Gamma_D(\omega)$ of the form (1.33). Due to Proposition 1.6.1 this implies that $\varphi(t, \omega)x \in \Gamma_D(\theta_t\omega)$. Thus $\varphi(t, \omega)\Gamma_D(\omega) \subset \Gamma_D(\theta_t\omega)$ for all $t > 0$ and $\omega \in \Omega$.

Assume that $x \in \Gamma_D(\theta_t\omega)$ for some $t > 0$ and $\omega \in \Omega$. Proposition 1.6.1 implies that

$$x = \lim_{n \to \infty} \varphi(t_n, \theta_{-t_n} \circ \theta_t\omega)y_n , \tag{1.35}$$

where $y_n \in D(\theta_{-t_n} \circ \theta_t\omega)$ and $t_n \to \infty$. The cocycle property gives that

$$x = \lim_{n \to \infty} \varphi(t, \omega)z_n \quad \text{with} \quad z_n = \varphi(t_n - t, \theta_{-t_n+t}\omega)y_n . \tag{1.36}$$

From (1.30) we have that $z_n \to B_0(\omega)$ as $n \to \infty$. Since $B_0(\omega)$ is compact, there exist $\{n_k\}$ and $b \in B_0(\omega)$ such that $z_{n_k} \to b$ as $k \to \infty$. Moreover Proposition 1.6.1 implies that $b \in \Gamma_D(\omega)$. From (1.36) we obtain that $x = \varphi(t, \omega)b$. Therefore $\Gamma_D(\theta_t\omega) \subset \varphi(t, \omega)\Gamma_D(\omega)$ for all $t > 0$ and $\omega \in \Omega$. Thus $\{\Gamma_D(\omega)\}$ is invariant.

To prove that $\{\Gamma_D(\omega)\}$ is a random compact set with respect to $\overline{\mathcal{F}^u}$ we use Proposition 1.5.1 and the obvious formula $\Gamma_D(\omega) = \cap_{n \in \mathbb{Z}_+} \overline{\gamma_D^n(\omega)}$ which implies in our case that

$$\text{dist}(x, \Gamma_D(\omega)) = \lim_{n \to \infty} \text{dist}(x, \gamma_D^n(\omega)), \quad \omega \in \Omega . \tag{1.37}$$

Indeed, since $\Gamma_D(\omega) \subset \gamma_D^{n+1}(\omega) \subset \gamma_D^n(\omega)$, we have that

$$\text{dist}(x, \gamma_D^n(\omega)) \leq \text{dist}(x, \gamma_D^{n+1}(\omega)) \leq \text{dist}(x, \Gamma_D(\omega))$$

for any $x \in X$. Therefore the limit in (1.37) exists and

$$\text{dist}(x, \Gamma_D(\omega)) \geq \lim_{n \to \infty} \text{dist}(x, \gamma_D^n(\omega)), \quad \omega \in \Omega.$$

Let $x_n \in \gamma_D^n(\omega)$ be such that

$$\text{dist}(x, x_n) \leq \text{dist}(x, \gamma_D^n(\omega)) + \frac{1}{n}, \quad n = 1, 2, \ldots$$

Since $\gamma_D^n(\omega) \to B_0(\omega)$ as $n \to \infty$ for all $\omega \in \Omega$, there exist a subsequence $n_k = n_k(\omega)$ and $b \in B_0(\omega)$ such that $x_{n_k} \to b$. By Proposition 1.6.1 $b \in \Gamma_D(\omega)$. Therefore

$$\text{dist}(x, \Gamma_D(\omega)) \leq \text{dist}(x, b) = \lim_{k \to \infty} \text{dist}(x, x_{n_k}) \leq \lim_{n \to \infty} \text{dist}(x, \gamma_D^n(\omega)).$$

Thus we obtain (1.37). By Proposition 1.5.1 $\omega \mapsto \text{dist}(x, \gamma_D^n(\omega))$ is $\mathcal{F}^u$-measurable. Therefore $\omega \mapsto \text{dist}(x, \Gamma_D(\omega))$ is also $\mathcal{F}^u$-measurable. Hence Γ_D is a random set with respect to the universal σ-algebra $\mathcal{F}^u$. $\qquad\square$

Remark 1.6.1. The existence and measurability of omega-limit sets with respect to the universal σ-algebra can be proved under a weaker property than the asymptotic compactness of RDS (θ, φ). Assume that $\{D(\omega)\}$ is a random closed set and for every $\omega \in \Omega$ there exists a compact set $B_D(\omega) \subset X$ such that

$$\lim_{t \to +\infty} d_X\{\varphi(t, \theta_{-t}\omega)D(\theta_{-t}\omega) \mid B_D(\omega)\} = 0,$$

where $d_X\{A|B\} = \sup_{x \in A} \text{dist}_X(x, B)$. Then, as in the proof of Proposition 1.6.2, it follows from Proposition 1.5.1 that Γ_D exists and $\omega \mapsto \Gamma_D(\omega)$ is an invariant random compact set with respect to the universal σ-algebra $\mathcal{F}^u$. If we additionally assume that the closure $\overline{\gamma_D^t(\omega)}$ of the tail $\gamma_D^t(\omega)$ is a random closed set for every $t \geq 0$ (cf. Proposition 1.5.3 and Remark 1.5.1), then Γ_D is a random compact set with respect to $\mathcal{F}$. We refer to CRAUEL [33] for other results concerning the measurability of omega-limit sets.

The following two assertions provide us with conditions which guarantee that $\{\Gamma_D(\omega)\}$ is a random compact set with respect to the σ-algebra $\mathcal{F}$.

Proposition 1.6.3. *If $\{D(\omega)\}$ is a forward invariant random compact set for the RDS (θ, φ), then the multifunction $\omega \mapsto \Gamma_D(\omega)$ is an invariant random compact set with respect to $\mathcal{F}$ and $\Gamma_D(\omega) \subset D(\omega)$.*

Proof. Since $\{D(\omega)\}$ is a forward invariant set, we have

$$\Gamma_D(\omega) = \bigcap_{t>0} \varphi(t, \theta_{-t}\omega)D(\theta_{-t}\omega) = \bigcap_{n\in\mathbb{Z}_+} \varphi(n, \theta_{-n}\omega)D(\theta_{-n}\omega). \qquad (1.38)$$

Proposition 1.3.1(vi) implies that $\omega \mapsto D_n(\omega) := \varphi(n, \omega)D(\omega)$ is a random compact set. Therefore $\omega \mapsto D_n(\theta_{-n}\omega)$ is also a random compact set. Consequently it follows from Proposition 1.3.1(iv) that $\Gamma_D(\omega)$ is a random compact set. It is clear from (1.38) that $\Gamma_D(\omega)$ is a forward invariant set. Let us prove its backward invariance. Let $x \in \Gamma_D(\theta_t\omega)$ for some $t > 0$ and $\omega \in \Omega$. Then as above by Proposition 1.6.1 we have (1.35) and (1.36) with $z_n \in D(\omega)$. Since $D(\omega)$ is compact, we can choose a convergent subsequence $\{z_{n_k}\}$ and apply the same argument as in the proof of Proposition 1.6.2. $\square$

Proposition 1.6.4. *Let $a(\omega)$ be a random variable in X. Assume that the process $t \mapsto \varphi(t, \theta_{-t}\omega)a(\theta_{-t}\omega)$ is separable for $t \in \mathbb{R}_+$ and for each $\omega \in \Omega$ there exists $t^* = t^*(\omega)$ such that $\overline{\gamma_a^{t^*}(\omega)}$ is a compact set. Then the omega-limit set $\omega \mapsto \Gamma_a(\omega)$ is a random compact set with respect to $\mathcal{F}$.*

Proof. The compactness of $\overline{\gamma_a^{t^*}(\omega)}$ implies that $\Gamma_a(\omega)$ is a nonempty compact set for all $\omega \in \Omega$. Therefore we can use Proposition 1.5.2, the formula $\Gamma_D(\omega) = \cap_{n\in\mathbb{Z}_+} \overline{\gamma_D^n(\omega)}$ and the argument given in the proof of Proposition 1.6.2. $\square$

1.7 Equilibria

A special case of omega-limit sets are random equilibria. They are the random analog of deterministic fixed points and generate stationary stochastic orbits (cf. ARNOLD [3], ARNOLD/SCHMALFUSS [11] and SCHMALFUSS [94]).

Definition 1.7.1. *A random variable $u : \Omega \mapsto X$ is said to be an equilibrium (or fixed point, or stationary solution) of the RDS (θ, φ) if it is invariant under φ, i.e. if*

$$\varphi(t, \omega)u(\omega) = u(\theta_t\omega) \quad \text{for all} \quad t \geq 0 \quad \text{and all} \quad \omega \in \Omega.$$

It is clear that if $u = u(\omega)$ is an equilibrium, then $\Gamma_u(\omega) = u(\omega)$.

Example 1.7.1 (Kick Model). If in Example 1.4.2 we additionally assume that g is a linear mapping such that $\|g\| \leq a < 1$, then it is easy to see that

$$u(\omega) = \sum_{k=0}^{\infty} g^k(\xi(\theta_{-k}\omega))$$

is an equilibrium for the RDS generated by (1.21).

Remark 1.7.1. The problem of the construction of equilibria for general RDS is rather complicated. The following example demonstrates the difficulties in the construction of equilibria. Let us consider the RDS on $\mathbb{R}_+$ constructed in the Introduction (cf. also Example 1.4.1) with $f_0(x) = \frac{1}{2}x$ and $f_1(x) = \frac{1}{2} + f_0(x) = \frac{1}{2}(1 + x)$. Both functions $f_0(x)$ and $f_1(x)$ have a fixed point: $f_0(0) = 0$ and $f_1(1) = 1$. To obtain an equilibrium we should look for a solution to the equation $f_{\omega_0}(u(\omega)) = u(\theta_1 \omega)$, where $\omega = \{\omega_i \mid i \in \mathbb{Z}\}$ is a two-sided sequence consisting of zeros and ones and θ_1 is the left one-symbol shift operator. It is clear that an equilibrium $u(\omega)$ is not simply a random variable which takes as its values the fixed points 0 and 1 of the mappings $f_0(x)$ and $f_1(x)$. The variable $u(\omega)$ can really depend on the sequence $\omega = \{\omega_i \mid i \in \mathbb{Z}\}$ in a very complicated way. However we prove in Chap. 3 that this RDS possesses a unique globally asymptotically stable equilibrium in $\mathbb{R}_+$ with its values inside the interval $(0, 1)$.

We also note that the results by OCHS/OSELEDETS [87] and OCHS [85] show that it is impossible to generalize topological fixed point theorems to the case of random dynamical systems. However, as we will see in Chaps.3–6, there are more simple approaches which allow us to construct equilibria for monotone RDS.

The following simple assertion makes it possible to prove the uniqueness of equilibria, if they exist, in several important cases (see, e.g., Sect. 4.2 below).

Proposition 1.7.1. *Let $\omega \mapsto D(\omega)$ be a forward invariant multifunction for the RDS (θ, φ). Assume that on the set*

$$G = \{(\omega, u, v) \; : \; u, v \in D(\omega), \; \omega \in \Omega\} \subset \Omega \times X \times X$$

there exists a function $V : G \mapsto \mathbb{R}$ satisfying

(i) $V(\omega, u(\omega), v(\omega))$ is measurable for any random variables $u(\omega)$ and $v(\omega)$ from $D(\omega)$;

(ii) for any u and v from $D(\omega)$ we have

$$V(\theta_t \omega, \varphi(t, \omega)u, \varphi(t, \omega)v) \leq V(\omega, u, v) \quad \text{for all} \quad t > 0, \; \omega \in \Omega \; ; \quad (1.39)$$

(iii) we have strict inequality in (1.39), if $u \neq v$.

Then any two equilibria $u_1(\omega)$ and $u_2(\omega)$ with the property $u_1(\omega), u_2(\omega) \in D(\omega)$ for all $\omega \in \Omega$ are equal on the set of full measure which is invariant with respect to θ.

Proof. Assume that the RDS (θ, φ) has two equilibria u_1 and u_2 in D such that $u_1(\omega) \neq u_2(\omega)$ on a measurable set $U \subset \Omega$ with $\mathbb{P}(U) > 0$. It follows from condition (iii) that

$$V(\theta_t \omega, \varphi(t, \omega)u_1(\omega), \varphi(t, \omega)u_2(\omega)) < V(\omega, u_1(\omega), u_2(\omega)) < \infty \quad (1.40)$$

for all $\omega \in U$ and $t > 0$. Since u_1 and u_2 are equilibria, (1.40) is equivalent to

$$V(\theta_t \omega, u_1(\theta_t \omega), u_2(\theta_t \omega)) < V(\omega, u_1(\omega), u_2(\omega)) < \infty$$

for all $\omega \in U$ and $t > 0$. From (1.39) we also have

$$V(\theta_t \omega, u_1(\theta_t \omega), u_2(\theta_t \omega)) \leq V(\omega, u_1(\omega), u_2(\omega)) < \infty$$

for all $\omega \in \Omega$ and $t > 0$. However the functions

$$f_t(\omega) := V(\theta_t \omega, u_1(\theta_t \omega), u_2(\theta_t \omega)) \quad \text{and} \quad f(\omega) := V(\omega, u_1(\omega), u_2(\omega))$$

have the same probability distribution for every $t > 0$, but satisfy $f_t(\omega) \leq f(\omega)$ for $\omega \in \Omega$ and $f_t(\omega) < f(\omega)$ for $\omega \in U$. This contradicts the assumption $\mathbb{P}(U) > 0$. Thus for any fixed $t > 0$ we have $f(\theta_t \omega) = f(\omega)$ on a set of full measure. Let

$$\Omega_n = \{\omega : f(\theta_n \omega) = f(\omega)\}, \quad n \in \mathbb{Z}_+ .$$

The sets Ω_n are $\mathcal{F}$-measurable and $\mathbb{P}(\Omega_n) = 1$. Property (1.39) implies that $f(\theta_t \omega) = f(\omega)$ for all $t \in [0, n]$ and $\omega \in \Omega_n$. Therefore $f(\theta_{n-k}\theta_s \omega) = f(\theta_s \omega)$ for all $s \in [0, k]$ and $\omega \in \Omega_n$, where $k \leq n$. Thus

$$\theta_s \Omega_n \subset \Omega_{n-k} \quad \text{for all} \quad 0 \leq s \leq k \leq n . \tag{1.41}$$

Let $\Omega^* = \cap_{n \geq 1} \Omega_n$. It is clear that $\mathbb{P}(\Omega^*) = 1$ and $f(\theta_t \omega) = f(\omega)$ for all $t \geq 0$ and $\omega \in \Omega^*$. From (1.41) we also easily have that $\theta_s \Omega^* \subseteq \Omega^*$ for all $s \geq 0$. Therefore $\tilde{\Omega} = \cap_{s \geq 0} \theta_s \Omega^* = \cap_{n \in \mathbb{Z}_+} \theta_n \Omega^*$ is $\mathcal{F}$-measurable θ-invariant set such that $\mathbb{P}(\tilde{\Omega}) = 1$. Since $\tilde{\Omega} \subset \Omega^*$, we have that $u_1(\omega) = u_2(\omega)$ for all $\omega \in \tilde{\Omega}$. $\square$

We note that Proposition 1.7.1 is wrong without the assumption (iii). Indeed, the identical mapping $f(x) = x$ in $\mathbb{R}$ possesses the property $|f(x) - f(y)| = |x - y|$ and every point $x \in \mathbb{R}$ is an equilibrium for f. See also the example of an RDS given in Remark 4.2.1 in Chap.4.

Example 1.7.2. Consider the one-dimensional random differential equation

$$\dot{x}(t) = (g(x(t)) + \xi(\theta_t \omega)) \cdot h(x(t))$$

over some metric dynamical system θ. Here ξ is a random variable and $g, h : \mathbb{R} \mapsto \mathbb{R}$ are smooth functions. Assume that this equation generates RDS in some interval $(a, b) \subseteq \mathbb{R}$ and $h(x) > 0$ for all $x \in (a, b)$. If $g(x)$ is strictly decreasing on (a, b), then the function

$$V(u, v) = \left| \int_v^u \frac{ds}{h(s)} \right| , \quad u, v \in (a, b) ,$$

satisfies the hypotheses of Proposition 1.7.1. The same is true for $V^*(u, v) := -V(u, v)$ provided that $g(x)$ is strictly increasing.

1.8 Random Attractors

Below we also need the following concept of a random attractor of an RDS (see, e.g., ARNOLD [3], CRAUEL/DEBUSSCHE/FLANDOLI [35], CRAUEL/FLANDOLI [36], SCHENK-HOPPÉ [89], SCHMALFUSS [92, 93] and the references therein). The appearance of this concept is motivated by the corresponding definition of a global attractor (cf. BABIN/VISHIK [13], CHUESHOV [20], HALE [50], LADYZHENSKAYA [76], TEMAM [104], for example).

Definition 1.8.1. *Let $\mathcal{D}$ be a universe. A random closed set $\{A(\omega)\}$ from $\mathcal{D}$ is said to be a* random pull back attractor *of the RDS (θ, φ) in $\mathcal{D}$ if $A(\omega) \neq X$ for every $\omega \in \Omega$ and the following properties hold:*

(i) A is an invariant set, i.e. $\varphi(t, \omega)A(\omega) = A(\theta_t \omega)$ for $t \geq 0$ and $\omega \in \Omega$;
(ii) A is attracting in $\mathcal{D}$, i.e. for all $D \in \mathcal{D}$

$$\lim_{t \to +\infty} d_X \{\varphi(t, \theta_{-t}\omega)D(\theta_{-t}\omega) \,|\, A(\omega)\} = 0, \quad \omega \in \Omega, \qquad (1.42)$$

where $d_X\{A|B\} = \sup_{x \in A} \operatorname{dist}_X(x, B)$.

Below for brevity we sometimes say "random attractor" instead of "random pull back attractor".

Remark 1.8.1. (i) If A is a random attractor, then the convergence in (1.42) and the invariance of the measure $\mathbb{P}$ with respect to θ imply that

$$d_X \{\varphi(t, \omega)D(\omega) \,|\, A(\theta_t \omega)\} \to 0, \quad D \in \mathcal{D},$$

in probability as $t \to \infty$, i.e.

$$\lim_{t \to +\infty} \mathbb{P}\{\omega : d_X\{\varphi(t, \omega)D(\omega) \,|\, A(\theta_t\omega)\} > \delta\} = 0, \quad D \in \mathcal{D},$$

for any $\delta > 0$. Thus any pull back attractor is a forward attractor with respect to convergence in probability. We refer to OCHS [86] for some discussion of the theory of attractors based on convergence in probability. We note that an example given in ARNOLD [3] shows that pull back convergence (1.42) does not imply forward convergence, i.e. the closeness of $\varphi(t, \omega)D(\omega)$ and $A(\theta_t\omega)$ in the topology of the space X for every $\omega \in \Omega$. We also refer to SCHEUTZOW [91] for a short survey of other (non-equivalent) definitions of a random attractor.

(ii) An attractor depends crucially on a choice of universe $\mathcal{D}$. Indeed, the deterministic dynamical system in $\mathbb{R}$ generated by the equation

$$\dot{x} = x - x^3$$

has one-point attractor $A = \{1\}$ in the universe of all compact subsets of $\mathbb{R}_+ \setminus \{0\}$ (see the formula for solutions given in the Introduction). The same formula implies that the interval $[-1, 1]$ is the attractor in the universe of all bounded subsets of $\mathbb{R}$ and the set $\{-1, 0, 1\}$ is the attractor in the universe of all one-point subsets of $\mathbb{R}$. We also note that there exists some classification of random attractors (see, e.g., CRAUEL [34]) depending on the choice of families of sets which are attracted (set attractors, point attractors, etc.).

(iii) Sometimes it is convenient to consider random attractors which do not belong to the corresponding universe (see CRAUEL [33, 34], CRAUEL/DEBUS-SCHE/FLANDOLI [35], CRAUEL/FLANDOLI [36]).

Proposition 1.8.1. *If the RDS (θ, φ) possesses a random attractor in the universe $\mathcal{D}$, then this attractor is unique in $\mathcal{D}$.*

Proof. Assume that there exist two random attractors $A_1(\omega)$ and $A_2(\omega)$ in the universe $\mathcal{D}$. Since $\varphi(t, \omega)A_1(\omega) = A_1(\theta_t\omega)$, we have

$$d_X\{A_1(\omega) \,|\, A_2(\omega)\} = d_X\{\varphi(t, \theta_{-t}\omega)A_1(\theta_{-t}\omega) \,|\, A_2(\omega)\}$$

for all $t > 0$. Therefore the attraction property (1.42) implies that

$$d_X\{A_1(\omega) \,|\, A_2(\omega)\} = 0 \ .$$

Thus $A_1(\omega) \subset A_2(\omega)$. The same argument gives $A_2(\omega) \subset A_1(\omega)$.   □

In a similar way we can prove the following assertion.

Proposition 1.8.2. *If the RDS (θ, φ) possesses a random attractor in the universe $\mathcal{D}$, then any backward invariant random closed set from $\mathcal{D}$ lies in the attractor. In particular the attractor contains every equilibrium $u(\omega)$ with the property $\{u(\omega)\} \in \mathcal{D}$.*

Now we prove a theorem on the existence of random attractors.

Theorem 1.8.1. *Let (θ, φ) be an asymptotically compact RDS in the universe $\mathcal{D}$ with an attracting random compact set $B_0 \in \mathcal{D}$. Then this RDS possesses a unique random compact pull back attractor $\{A(\omega)\}$ in the universe $\mathcal{D}$, and $A(\omega) \subset B_0(\omega)$ for all $\omega \in \Omega$. This attractor has the form*

$$A(\omega) = \Gamma_{B_0}(\omega) \equiv \bigcap_{t>0} \overline{\bigcup_{\tau \geq t} \varphi(\tau, \theta_{-\tau}\omega) B_0(\theta_{-\tau}\omega)} \quad \text{for all} \ \ \omega \in \Omega \ . \quad (1.43)$$

We also have the relation

$$A(\omega) = \bigcap_{n \geq N} \varphi(n, \theta_{-n}\omega) B_0(\theta_{-n}\omega) \quad \text{for all} \ \ \omega \in \Omega, \ N \in \mathbb{Z}_+ \ . \quad (1.44)$$

Proof. We follow the line of arguments given for the deterministic case (see, e.g., TEMAM [104] or CHUESHOV [20]).

Let $A(\omega)$ be defined by (1.43). Proposition 1.6.2 implies that $A(\omega)$ is a nonempty invariant set and is a compact subset of $B_0(\omega)$ for all $\omega \in \Omega$.

Let us prove the attraction property (1.42). Let $D \in \mathcal{D}$. Proposition 1.6.2 shows that $\Gamma_D(\omega)$ is a nonempty compact set and $\Gamma_D(\omega) \subset B_0(\omega)$ for every $\omega \in \Omega$. It is also easy to see from the invariance of $\Gamma_D(\omega)$ that

$$\Gamma_D(\omega) \subset \Gamma_{B_0}(\omega) = A(\omega) \quad \text{for all} \quad \omega \in \Omega . \tag{1.45}$$

Assume now that property (1.42) is not true for some $D \in \mathcal{D}$. Then there exist $\epsilon > 0$ and sequences $t_n \to \infty$ and $y_n \in D(\theta_{-t_n}\omega)$ such that

$$\mathrm{dist}_X(\varphi(t_n, \theta_{-t_n}\omega)y_n, A(\omega)) \geq \epsilon, \quad n = 1, 2, \ldots , \tag{1.46}$$

for some $\omega \in \Omega$. It follows from (1.30) that there exists a sequence $\{b_n\} \subset B_0(\omega)$ such that

$$\lim_{t \to +\infty} \mathrm{dist}_X(\varphi(t_n, \theta_{-t_n}\omega)y_n, b_n) = 0 .$$

Therefore the compactness of $B_0(\omega)$ implies that the limit

$$z = \lim_{k \to +\infty} \varphi(t_{n_k}, \theta_{-t_{n_k}}\omega)y_{n_k}$$

exists for some subsequence $\{n_k\}$. Proposition 1.6.1 and relation (1.45) imply that $z \in \Gamma_D(\omega) \subset A(\omega)$. Thus we have

$$\lim_{k \to +\infty} \mathrm{dist}_X(\varphi(t_{n_k}, \theta_{-t_{n_k}}\omega)y_{n_k}, A(\omega)) = 0 ,$$

contradicting equation (1.46).

Now we prove (1.44). Let

$$\Gamma_N^*(\omega) = \bigcap_{n \geq N} \varphi(n, \theta_{-n}\omega)B_0(\theta_{-n}\omega) \quad \text{with} \quad N \in \mathbb{Z}_+ .$$

Since $\{A(\omega)\}$ is invariant and $A(\omega) \subset B_0(\omega)$ for all $\omega \in \Omega$, we have

$$A(\omega) = \varphi(n, \theta_{-n}\omega)A(\theta_{-n}\omega) \subset \varphi(n, \theta_{-n}\omega)B_0(\theta_{-n}\omega) \quad \text{for all} \quad n \in \mathbb{Z}_+ .$$

Therefore $A(\omega) \subset \Gamma_N^*(\omega)$ for any $\omega \in \Omega$ and $N \in \mathbb{Z}_+$. On the other hand, it is clear from (1.43) that $\Gamma_N^*(\omega) \subset A(\omega)$. Thus (1.44) is proved.

To prove that $\{A(\omega)\}$ is a random compact set we use Proposition 1.3.1(iv) and relation (1.44). $\qquad\square$

Remark 1.8.2. (i) It is clear that if the RDS (θ, φ) has a random compact attractor, then (θ, φ) is asymptotically compact. Thus Theorem 1.8.1 implies that (θ, φ) possesses a random compact attractor in $\mathcal{D}$ if and only if this RDS is asymptotically compact in $\mathcal{D}$ with an attracting set from $\mathcal{D}$.

(ii) Under the hypotheses of Theorem 1.8.1 similarly to the deterministic case (see, e.g., CHUESHOV [20, Sect. 1.5.2]) we can prove that

$$\lim_{t \to +\infty} h\left(A(\omega) \,|\, \varphi(t, \theta_{-t}\omega)B(\theta_{-t}\omega)\right) = 0, \quad \omega \in \Omega ,$$

for any absorbing set $B \in \mathcal{D}$ of the RDS (θ, φ), where $h(A|B)$ is the Hausdorff distance defined by the equality

$$h(A|B) = d_X\{A|B\} + d_X\{B|A\} \quad \text{with} \quad d_X\{A|B\} = \sup_{x \in A} \text{dist}_X(x, B) .$$

This property means that the set $A_t^B(\omega) := \varphi(t, \theta_{-t}\omega)B(\theta_{-t}\omega)$ provides us with an approximate image of the random attractor $A(\omega)$ for t large enough. We also refer to ARNOLD/SCHMALFUSS [12] for the study of stability properties of random attractors for finite-dimensional RDS.

Corollary 1.8.1. *Let (θ, φ) be a dissipative RDS in the universe $\mathcal{D}$ with an absorbing set from $\mathcal{D}$. Assume that the phase space X is locally compact. Then the RDS (θ, φ) possesses a unique global random attractor in the universe $\mathcal{D}$.*

Proof. In this case any closed bounded set is compact. Therefore (θ, φ) is a compact RDS and we can apply Theorem 1.8.1. $\qquad\square$

Corollary 1.8.2. *Assume that for the RDS (θ, φ) the hypotheses of Proposition 1.4.1 on the dissipativity of an RDS possessing a Lyapunov type function hold. Let the phase space X be finite-dimensional. Then the RDS (θ, φ) possesses a unique random attractor in the universe $\mathcal{D}$ of all tempered random closed sets in X.*

Proof. Since X is finite-dimensional, Proposition 1.4.1 implies that (θ, φ) is a compact RDS. Thus we can apply Theorem 1.8.1. $\qquad\square$

Theorem 1.8.1 and Corollaries 1.8.1 and 1.8.2 imply the existence of random attractors for the RDS considered in Examples 1.4.1, 1.4.3, 1.4.4 and 1.4.5.

Below we also need the following simple assertion concerning attractors of equivalent RDS (cf. KELLER/SCHMALFUSS [63] and IMKELLER/SCHMALFUSS [59]).

Proposition 1.8.3. *Let (θ, φ_1) and (θ, φ_2) be two RDS over the same MDS θ with phase spaces X_1 and X_2 resp. Assume that the systems (θ, φ_1) and (θ, φ_2) are conjugate by a random homeomorphism T from X_1 onto X_2 (see Definition 1.2.4) and there exists a compact random attractor A_1 for the RDS (θ, φ_1) in the universe $\mathcal{D}_1$. Then the RDS (θ, φ_2) possesses a random attractor A_2 in the universe*

$$\mathcal{D}_2 = \left\{ \{T(\omega, D(\omega))\} : \{D(\omega)\} \in \mathcal{D}_1 \right\} .$$

The attractors A_1 and A_1 are conjugated by the random homeomorphism T, i.e. $T(\omega, A_1(\omega)) = A_2(\omega)$ for all $\omega \in \Omega$.

Proof. Since T is a homeomorphism, Proposition 1.3.1(vi) implies that $A_2(\omega) := T(\omega, A_1(\omega))$ is an invariant random compact set. From (1.13) we also have that

$$d_2(\omega, t) := d_{X_2}\{\varphi_2(t, \theta_{-t}\omega)D_2(\theta_{-t}\omega) \mid A_2(\omega)\}$$

$$= d_{X_2}\{T(\omega, \varphi_1(t, \theta_{-t}\omega)D_1(\theta_{-t}\omega)) \mid T(\omega, A_1(\omega))\},$$

where $D_2(\omega) = T(\omega, D_1(\omega))$ and $d_X\{A|B\} = \sup_{x \in A} \operatorname{dist}_X(x, B)$. If $d_2(\omega, t)$ does not tend to 0 as $t \to \infty$ for some ω, then there exist $t_n \to \infty$ and $b_n \in D_1(\theta_{-t_n}\omega)$ such that

$$\operatorname{dist}_{X_2}(T(\omega, x_n(\omega)), T(\omega, A_1(\omega))) \geq \varepsilon, \quad n \in \mathbb{Z}_+, \tag{1.47}$$

for some $\varepsilon > 0$, where $x_n(\omega) = \varphi_1(t_n, \theta_{-t_n}\omega)b_n$. Since $A_1(\omega)$ is an attractor for (θ, φ_1), there exists a sequence $\{a_n\} \subset A_1(\omega)$ such that

$$\operatorname{dist}_{X_1}(x_n(\omega), a_n) \to 0 \quad \text{as} \quad n \to \infty.$$

The compactness of $A_1(\omega)$ implies that $x_{n_k}(\omega) \to a$ for some subsequence $\{n_k\}$ and $a \in A_1(\omega)$. Therefore $\operatorname{dist}_{X_2}(T(\omega, x_{n_k}(\omega)), T(\omega, a)) \to 0$. This contradicts (1.47). Thus A_2 is a random attractor for (θ, φ_2). □

1.9 Dissipative Linear and Affine RDS

In this section we prove several results on global attractors for dissipative linear and affine random dynamical systems in a real separable Banach space X. By Definition 1.2.3 the cocycle φ of an affine RDS has the form

$$\varphi(t, \omega)x = \Phi(t, \omega)x + \psi(t, \omega), \tag{1.48}$$

where $\Phi(t, \omega)$ is a cocycle over θ consisting of bounded linear operators of X, and $\psi : \mathbb{T}_+ \times \Omega \to X$ satisfies

$$\psi(t + s, \omega) = \Phi(t, \theta_s\omega)\psi(s, \omega) + \psi(t, \theta_s\omega), \quad t, s \geq 0. \tag{1.49}$$

If $\psi(t, \omega) \equiv 0$ we obtain a linear RDS (θ, Φ).

Our first result gives a criterion for dissipativity of linear RDS.

Proposition 1.9.1. *Assume that $\mathcal{D}$ is a universe of subsets of X such that for any $D \in \mathcal{D}$ and for any $\lambda > 0$ the set $\omega \mapsto \lambda D(\omega) := \{x : x\lambda^{-1} \in D(\omega)\}$ belongs to $\mathcal{D}$. Then the linear RDS (θ, Φ) is dissipative in $\mathcal{D}$ if and only if*

$$\lim_{t \to +\infty} \sup_{v \in D(\theta_{-t}\omega)} \|\Phi(t, \theta_{-t}\omega)v\| = 0 \tag{1.50}$$

for any $D \in \mathcal{D}$.

Proof. Let $r(\omega)$ be a radius of dissipativity of (θ, Φ). Then for any $D \in \mathcal{D}$ and for any $\lambda > 0$ there exists a time $t_{\lambda,D}(\omega) > 0$ such that

$$\|\Phi(t, \theta_{-t}\omega)v\| \le r(\omega), \quad v \in \lambda D(\theta_{-t}\omega), \quad t \ge t_{\lambda,D}(\omega) .$$

Therefore

$$\sup_{v \in D(\theta_{-t}\omega)} \|\Phi(t, \theta_{-t}\omega)v\| \le \frac{r(\omega)}{\lambda}, \quad t \ge t_{\lambda,D}(\omega) .$$

Hence

$$\limsup_{t \to +\infty} \sup_{v \in D(\theta_{-t}\omega)} \|\Phi(t, \theta_{-t}\omega)v\| \le \frac{r(\omega)}{\lambda}$$

for all $\lambda > 0$. This implies (1.50).

Vice versa, (1.50) implies that the deterministic ball $\{x : \|x\| \le 1\}$ is an absorbing set for (θ, Φ). $\square$

From Proposition 1.9.1 we easily have the following assertion.

Corollary 1.9.1. *Let $\mathcal{D}$ be the universe consisting of one-point subsets of X. Then*

$$\lim_{t \to +\infty} \|\Phi(t, \theta_{-t}\omega)x\| = 0 \quad \text{for any} \quad x \in X$$

if and only if the RDS (θ, Φ) is dissipative in $\mathcal{D}$.

Remark 1.9.1. Let $\mathcal{D}$ be a universe such that $\{0\} \in \mathcal{D}$. It is easy to see that the dissipativity of the affine RDS (θ, φ) implies the dissipativity of its linear part (θ, Φ).

Now we consider asymptotically compact affine RDS.

Proposition 1.9.2. *Assume that $\mathcal{D}$ is a universe of subsets of X such that $\{0\} \in \mathcal{D}$ and for any $D \in \mathcal{D}$ and $\lambda > 0$ the set $\omega \mapsto \lambda D(\omega) := \{x : x\lambda^{-1} \in D(\omega)\}$ belongs to $\mathcal{D}$. Let (θ, φ) be an asymptotically compact affine RDS with the cocycle given by (1.48) and with an attracting random compact set $B_0 \in \mathcal{D}$. Then the limit*

$$u(\omega) := \lim_{t \to +\infty} \psi(t, \theta_{-t}\omega) \tag{1.51}$$

exists for all $\omega \in \Omega$ and is an equilibrium for the RDS (θ, φ). This equilibrium is globally asymptotically (pull back) stable in $\mathcal{D}$, i.e.

$$\lim_{t \to +\infty} \sup_{v \in D(\theta_{-t}\omega)} \|\varphi(t, \theta_{-t}\omega)v - u(\omega)\| = 0 \tag{1.52}$$

for any $D \in \mathcal{D}$. Moreover $\{u(\omega)\} \in \mathcal{D}$ and the RDS (θ, φ) possesses a unique equilibrium with this property.

Proof. From (1.49) we get

$$\psi(\tau, \theta_{-\tau}\omega) = \Phi(t, \theta_{-t}\omega)\psi(\tau - t, \theta_{-\tau}\omega) + \psi(t, \theta_{-t}\omega), \quad \tau > t \geq 0 . \quad (1.53)$$

Since $\{0\} \in \mathcal{D}$, we have that

$$\psi(\tau, \theta_{-\tau}\omega) = \varphi(\tau, \theta_{-\tau}\omega)0 \to B_0(\omega) \quad \text{as} \quad \tau \to \infty. \quad (1.54)$$

Hence there exist $\tau_n = \tau_n(\omega) \to \infty$ and $b \in B_0(\omega)$ such that

$$\psi(\tau_n, \theta_{-\tau_n}\omega) \to b \quad \text{as} \quad n \to \infty .$$

Since

$$\psi(\tau - t, \theta_{-\tau}\omega) = \varphi(\tau - t, \theta_{-\tau}\omega)0 \to B_0(\theta_{-t}\omega) \quad \text{as} \quad \tau \to \infty ,$$

we can choose a subsequence $\{\tau_{n_k}\}$ and an element $b_1(t) \in B_0(\theta_{-t}\omega)$ such that $\psi(\tau_{n_k} - t, \theta_{-\tau_{n_k}}\omega) \to b_1(t)$ as $n \to \infty$. Consequently from (1.53) we have

$$b = \Phi(t, \theta_{-t}\omega)b_1(t) + \psi(t, \theta_{-t}\omega) . \quad (1.55)$$

Relation (1.54) implies that (θ, Φ) is asymptotically compact in $\mathcal{D}$. Therefore by Proposition 1.4.2 (θ, Φ) is dissipative in $\mathcal{D}$. Since $B_0 \in \mathcal{D}$, Proposition 1.9.1 implies that $\Phi(t, \theta_{-t}\omega)b_1(t) \to 0$ as $t \to \infty$. Therefore the limit in (1.51) exists. It is easy to see that $u(\omega)$ is an equilibrium and $u(\omega) \in B_0(\omega)$. Thus $\{u(\omega)\} \in \mathcal{D}$. Using the relation

$$\varphi(t, \theta_{-t}\omega)v - u(\omega) = \Phi(t, \theta_{-t}\omega)v - \Phi(t, \theta_{-t}\omega)u(\theta_{-t}\omega) \quad (1.56)$$

and Proposition 1.9.1 we obtain (1.52). Finally, if there exists another equilibrium $v(\omega)$ with the property $\{v(\omega)\} \in \mathcal{D}$, then we have

$$v(\omega) = \Phi(t, \theta_{-t}\omega)v(\theta_{-t}\omega) + \psi(t, \theta_{-t}\omega).$$

In the limit $t \to \infty$ we obtain $v(\omega) = u(\omega)$. $\qquad\square$

Remark 1.9.2. If in Proposition 1.9.2 the universe $\mathcal{D}$ contains all bounded deterministic sets, then *any* equilibrium $v(\omega)$ coincides with $u(\omega)$ almost surely. Indeed, since (θ, Φ) is dissipative, from Proposition 1.9.1 we have that

$$\lim_{t \to \infty} \mathbb{P}(U_N^\delta) = 0$$

for every $\delta > 0$ and $N \in \mathbb{N}$, where

$$U_N^\delta := \left\{ \omega : \sup_{\|v\| \leq N} \|\Phi(t, \omega)v\| > \delta \right\} .$$

It is clear that

$$\{\omega \ : \ \|\Phi(t,\omega)v(\omega)\| > \delta\} \subset \{\omega \ : \ \|v(\omega)\| > N\} \cup U_N^\delta \ .$$

Hence

$$\limsup_{t\to\infty} \mathbb{P}\{\omega \ : \ \|\Phi(t,\omega)v(\omega)\| > \delta\} \le \mathbb{P}\{\omega \ : \ \|v(\omega)\| > N\}$$

for every $\delta > 0$ and $N \in \mathbb{N}$. Thus

$$\lim_{t\to\infty} \mathbb{P}\{\omega \ : \ \|\Phi(t,\omega)v(\omega)\| > \delta\} = 0 \ .$$

Since $v(\omega)$ is an equilibrium, this implies that

$$\lim_{t\to\infty} \mathbb{P}\{\omega \ : \ \|v(\omega) - \psi(t,\theta_{-t}\omega)\| > \delta\} = 0 \ .$$

Therefore it follows from (1.51) that $v(\omega) = u(\omega)$ almost surely.

To obtain a result on the exponential stability of an equilibrium we need the following concept.

Definition 1.9.1 (Top Lyapunov Exponent). *The* top Lyapunov exponent *for a linear RDS (θ, Φ) in a separable Banach space X is the minimal real
number λ with the following property: there exists a θ-invariant set $\Omega^* \subset \Omega$
of full measure such that*

$$\|\Phi(t,\omega)x\| \le R_\varepsilon(\omega)e^{(\lambda+\varepsilon)t}\|x\|, \quad \omega \in \Omega^*, \quad t \ge 0 , \tag{1.57}$$

*for every $\varepsilon > 0$ and all $x \in X$, where $R_\varepsilon(\omega) > 0$ is a tempered random
variable.*

We refer to ARNOLD [3, Part II] for conditions which guarantee the existence
of the top Lyapunov exponent and for a comprehensive presentation of the
theory of Lyapunov exponents for finite-dimensional RDS.

Following the line of argument given in the proof of Proposition 1.9.2 we
can easily prove the next assertion.

Proposition 1.9.3. *Let (θ, φ) be an affine RDS with the cocycle given by
(1.48). Assume that the linear RDS (θ, Φ) has top Lyapunov exponent $\lambda < 0$
and for every $\omega \in \Omega$ there exists a tempered random compact set $B_0(\omega)$ such
that*

$$\lim_{t\to\infty} \mathrm{dist}_X(\psi(t,\theta_{-t}\omega), B_0(\omega)) = 0 \ .$$

*Then the limit in (1.51) exists and belongs to $B_0(\omega)$ for all $\omega \in \Omega^*$. It is
an equilibrium on Ω^*, i.e. the property $\varphi(t,\omega)u(\omega) = u(\theta_t\omega)$ holds for all*

$\omega \in \Omega^*$. Moreover this equilibrium is unique almost surely and

$$\lim_{t \to +\infty} \left\{ e^{\gamma t} \sup_{v \in D(\theta_{-t}\omega)} \|\varphi(t, \theta_{-t}\omega)v - u(\omega)\| \right\} = 0, \quad \omega \in \Omega^*, \qquad (1.58)$$

for any tempered random closed set $D \subset X$ and $\gamma < -\lambda$ (Ω^* is described in Definition 1.9.1).

Proof. As in the proof of Proposition 1.9.2 using (1.53) we find that for any $t > 0$ there exist $b \in B_0(\omega)$ and $b_1(t) \in B_0(\theta_{-t}\omega)$ such that (1.55) holds. Since $B_0(\omega)$ is a tempered, there exists a tempered random variable $r(\omega) > 0$ such that $\|b_1(t)\| \leq r(\theta_{-t}\omega)$. Therefore it follows from (1.57) that

$$\|\Phi(t, \theta_{-t}\omega)b_1(t)\| \to 0, \quad t \to \infty, \quad \omega \in \Omega^*,$$

provided $\lambda + \varepsilon < 0$. Thus (1.55) implies that the limit in (1.51) exists for $\omega \in \Omega^*$. It is clear that $u(\omega) \in B_0(\omega)$ for all $\omega \in \Omega^*$ and it is an equilibrium on Ω^*. Using relation (1.56) with an arbitrary $v \in X$, we find that

$$\|\varphi(t, \theta_{-t}\omega)v - u(\omega)\| \leq R_\varepsilon(\theta_{-t}\omega)e^{(\lambda+\varepsilon)t} (\|v\| + r(\theta_{-t}\omega)) .$$

Since $R_\varepsilon(\omega)$, $\{D(\omega)\}$ and $r(\omega)$ are tempered, we obtain (1.58).

To prove the uniqueness of $u(\omega)$ we assume that for some random variable $w(\omega)$ we have $\varphi(t, \omega)w(\omega) = w(\theta_t\omega)$ almost surely. Therefore

$$w(\omega) - \psi(t, \theta_{-t}\omega) = \Phi(t, \theta_{-t}\omega)w(\theta_{-t}\omega)$$

almost surely. Since

$$\mathbb{P}\{\omega : \|\Phi(t, \theta_{-t}\omega)w(\theta_{-t}\omega)\| \geq \delta\} = \mathbb{P}\{\omega : \|\Phi(t, \omega)w(\omega)\| \geq \delta\} \to 0 ,$$

as $t \to \infty$, we obtain $\mathbb{P}\{\omega : \|w(\omega) - u(\omega)\| \geq \delta\} = 0$ for any $\delta > 0$. Thus $w(\omega) = u(\omega)$ almost surely. $\square$

To conclude this section we refer to ARNOLD [3, Sect.5.6] for a more detailed study of the asymptotic properties of affine systems with general hyperbolic linear parts in finite-dimensional spaces.

1.10 Connection Between Attractors and Invariant Measures

A number of interesting properties follow from the fact that the RDS (θ, φ) has a random attractor. One of them is the existence of an *invariant measure* of (θ, φ) in the sense of the theory of RDS. In this section we introduce the corresponding notions and briefly discuss the properties of these measures.

For details we refer to CRAUEL [31, 32], CRAUEL/FLANDOLI [36], ARNOLD [3], SCHMALFUSS [95] and the references therein.

As above we consider an RDS on a Polish space X and denote by $\mathcal{B}$ the Borel σ-algebra on X.

To explain the main idea of introducing of invariant measures we start with a discrete time RDS which generates a Markov chain (cf. Example 1.2.1). Let $\theta = (\Omega, \mathcal{F}, \mathbb{P}, \{\theta_t, t \in \mathbb{Z}\})$ be a discrete metric dynamical system and $\psi_n(\omega) := \psi(\theta_n \omega, \cdot)$ be independent identically distributed (i.i.d.) random continuous mappings from X into itself. In this case we can construct an RDS by defining the cocycle φ by the formula

$$\varphi(n, \omega)x = \psi_{n-1}(\omega) \circ \psi_{n-2}(\omega) \circ \ldots \circ \psi_0(\omega)x, \quad x \in X.$$

One can prove (see ARNOLD [3, p. 53]) that the family of sequences

$$\{\Phi_n^x := \varphi(n, \omega)x \; : \; n \in \mathbb{Z}_+, \, x \in X\}$$

is a homogeneous Markov chain with state space X and transition probability

$$P(x, B) := \mathbb{P}\{\Phi_{n+1} \in B \mid \Phi_n = x\} = \mathbb{P}\{\omega \; : \; \varphi(n, \omega)x \in B\}, \quad B \in \mathcal{B}(X) \, .$$

For detailed presentation of the theory of Markov chains we refer to GIH-MAN/SKOROHOD [48] and MEYN/TWEEDIE [83], for example. The central topic of Markov chain theory is the existence of a stationary (invariant) probability measure (we denote it by ν) which is defined as a measure on $(X, \mathcal{B})$ satisfying the relation

$$\nu(B) = (P^*\nu)(B) := \int_X P(x, B)\nu(dx), \quad B \in \mathcal{B} \, .$$

The main consequence of the existence of a stationary probability measure is the possibility of producing a stationary process from the Markov chain. If Φ_0 is a random variable with distribution ν, then $\{\Phi_n = \varphi(n, \omega)\Phi_0 \; : \; n \in \mathbb{Z}_+\}$ is a stationary process, i.e. all variables Φ_n have the same distribution. Stationary measures are also important because of they define the long term and ergodic behaviour of the chain (MEYN/TWEEDIE [83]).

One can prove (see ARNOLD [3, Chap. 2]) that in the above case a probability measure ν on $(X, \mathcal{B})$ is stationary for the Markov chain $\{\Phi_n\}$ if and only if the measure $\mathbb{P} \times \nu$ is invariant with respect to the skew-product semiflow π_n defined by (1.6), i.e.

$$\int_{\Omega \times X} f(\omega, x)\mathbb{P}(d\omega)\nu(dx) = \int_{\Omega \times X} f(\theta_n\omega, \varphi(n, \omega)x)\mathbb{P}(d\omega)\nu(dx)$$

for any bounded measurable function f on $\Omega \times X$. This observation is the basis for the following general definition.

Definition 1.10.1 (Invariant Measure for RDS). *Let (θ, φ) be an RDS with phase space X. A probability measure μ on $(\Omega \times X, \mathcal{F} \times \mathcal{B})$ is said to be an* invariant measure *for RDS (θ, φ) (or φ-invariant, for short) if*

(i) it is invariant with respect to the skew-product semiflow π_t (see (1.6)), i.e. $\pi_t \mu = \mu$ which means that

$$\int_{\Omega \times X} f(\omega, x) \mu(d\omega, dx) = \int_{\Omega \times X} f(\theta_t \omega, \varphi(t, \omega)x) \mu(d\omega, dx)$$

for all $t \in \mathbb{T}_+$ and $f \in L^1(\Omega \times X, \mu)$;

(ii) the basic probability measure $\mathbb{P}$ is the Ω-marginal of μ on $(\Omega, \mathcal{F})$, i.e. $\mu(A \times X) = \mathbb{P}(A)$ for any $A \in \mathcal{F}$.

The measure μ is said to be φ-ergodic if for any $C \in \mathcal{F} \times \mathcal{B}$ with the property that $\pi_t^{-1} C = C$ for all $t \geq 0$, we have either $\mu(C) = 0$ or $\mu(C) = 1$.

It is known (see, e.g., ARNOLD [3] and the references therein) that any probability measure μ on $(\Omega \times X, \mathcal{F} \times \mathcal{B})$ possesses a *disintegration* (or factorization), i.e there exists a function $(\omega, B) \mapsto \mu_\omega(B)$ from $\Omega \times \mathcal{B}$ into the interval $[0, 1]$ such that

(i) $\omega \mapsto \mu_\omega(B)$ is $\mathcal{F}$-measurable for any $B \in \mathcal{B}$;
(ii) there exists a measurable set Q_μ in Ω such that $\mathbb{P}(Q_\mu) = 1$ and $B \mapsto \mu_\omega(B)$ is a probability measure on $(X, \mathcal{B})$ for all $\omega \in Q_\mu$;
(iii) for all $f \in L^1(\Omega \times X, \mu)$ we have

$$\int_{\Omega \times X} f(\omega, x) \mu(d\omega, dx) = \int_\Omega \left(\int_X f(\omega, x) \mu_\omega(dx) \right) \mathbb{P}(d\omega) .$$

The disintegration μ_ω is unique $\mathbb{P}$-almost surely.

Example 1.10.1. It follows directly from Definition 1.7.1 that any equilibrium $u(\omega)$ for the RDS (θ, φ) generates an invariant measure by the formula

$$\int_{\Omega \times X} f(\omega, x) \mu(d\omega, dx) = \int_\Omega f(\omega, u(\omega)) \mathbb{P}(d\omega) . \tag{1.59}$$

The factorization μ_ω of this invariant measure is a *random Dirac measure*, i.e. $\mu_\omega = \delta_{u(\omega)}$, where $\delta_{u(\omega)}$ is defined by the formula

$$\int_X f(x) \delta_{u(\omega)}(dx) = f(u(\omega)), \quad f \in C_b(X) ,$$

with $C_b(X)$ the space of bounded continuous functions on X. We also note that if θ is an ergodic metric dynamical system, then every equilibrium $u(\omega)$ generates a φ-ergodic invariant measure by the formula (1.59). Indeed, let

$C \in \mathcal{F} \times \mathcal{B}$ be an invariant set, i.e. $\pi_t^{-1} C = C$ for all $t \geq 0$. Then

$$\mathcal{A} := \{\omega \,:\, (\omega, u(\omega)) \in C\} = \{\omega \,:\, (\omega, u(\omega)) \in \pi_t^{-1} C\}$$

$$= \{\omega \,:\, (\theta_t \omega, u(\theta_t \omega)) \in C\} = \theta_{-t} \mathcal{A}$$

for all $t \geq 0$. Since $\theta_{-t} = \theta_t^{-1}$, we have $\theta_t \mathcal{A} = \mathcal{A}$ for all $t \in \mathbb{R}$. The ergodicity of θ implies that we have either $\mathbb{P}(\mathcal{A}) = 0$ or $\mu(\mathcal{A}) = 1$. It is clear from (1.59) that $\mu(C) = \mathbb{P}(\mathcal{A})$. Thus μ is φ-ergodic.

The following assertion (see, e.g., CRAUEL/FLANDOLI [36], CRAUEL [32, 33] and ARNOLD [3]) describes the relation between invariant measures and forward invariant random sets.

Proposition 1.10.1. *A probability measure μ on $(\Omega \times X, \mathcal{F} \times \mathcal{B})$ is invariant for (θ, φ) if and only if its disintegration μ_ω possesses property $\varphi(t, \omega) \mu_\omega = \mu_{\theta_t \omega}$ $\mathbb{P}$-almost surely, i.e. for any $f \in C_b(X)$ we have*

$$\int_X f(\varphi(t, \omega) x) \mu_\omega(dx) = \int_X f(x) \mu_{\theta_t \omega}(dx) \quad \mathbb{P} - \text{almost surely} \,.$$

Moreover there exists a forward invariant random closed set $\{C(\omega)\}$ such that $\mu_\omega(C(\omega)) = 1$ for almost all $\omega \in \Omega$.

On the other hand for any forward invariant random compact set $\{C(\omega)\}$ there exists an invariant measure μ concentrated on $\{C(\omega)\}$, i.e. $\mu\{(\omega, x) : x \in C(\omega)\} = 1$. In particular if the RDS (θ, φ) possesses a random compact attractor $\{A(\omega)\}$ in the universe $\mathcal{D}$ which contains all bounded deterministic sets, then there exists an invariant measure μ concentrated on $\{A(\omega)\}$. Moreover in the last case every invariant probability measure is concentrated on $\{A(\omega)\}$.

Remark 1.10.1. We note that if the cocycle φ can be extended to a cocycle $\tilde{\varphi}$ with two-sided time $\mathbb{T}$, then in Proposition 1.10.1 we can choose a perfect version of disintegration μ_ω, i.e. the invariant measure μ possesses a disintegration $\tilde{\mu}_\omega$ such that

$$\varphi(t, \omega) \tilde{\mu}_\omega = \tilde{\mu}_{\theta_t \omega} \quad \text{for all} \quad t \geq 0, \; \omega \in \Omega \,.$$

We refer to SCHEUTZOW [90] for the proof of this result. We also refer to SCHENK-HOPPÉ [89] for additional properties of invariant measures in the case of invertible cocycles, i.e. for RDS with time $\mathbb{T}$ (not $\mathbb{T}_+$).

Let us define the future $\mathcal{F}_+$ and the past $\mathcal{F}_-$ σ-algebras for RDS (θ, φ) by the formulas

$$\mathcal{F}_+ = \sigma\{\omega \mapsto \varphi(\tau, \theta_t \omega) \,:\, t, \tau \geq 0\}$$

and

$$\mathcal{F}_- = \sigma\{\omega \mapsto \varphi(\tau, \theta_{-t}\omega) \ : \ 0 \leq \tau \leq t\} \, ,$$

where $\sigma\{f_\alpha(\omega) \ : \ \alpha \in \Lambda\}$ denotes the σ-algebra generated by the mappings $\{f_\alpha\}$, where $\alpha \in \Lambda$.

Definition 1.10.2 (Markov Measure). *A probability measure μ on $(\Omega \times X, \mathcal{F} \times \mathcal{B})$ is said to be a Markov measure if its disintegration μ_ω is measurable with respect to the past σ-algebra $\mathcal{F}_-$.*

The following theorem (see, e.g. CRAUEL [31, 32], CRAUEL/FLANDOLI [36] and ARNOLD [3]) shows that invariant Markov measures supported by the random attractor for the RDS (θ, φ) generate stationary probability measures in the phase space of this RDS.

Theorem 1.10.1. *Assume that the RDS (θ, φ) possesses a random compact attractor $\{A(\omega)\}$ in the universe $\mathcal{D}$ which contains all bounded deterministic sets. Then there an exists invariant Markov measure μ supported by $\{A(\omega)\}$, i.e. $\mu\{(\omega, x) : x \in A(\omega)\} = 1$. Assume additionally that the processes $\{\varphi(t, \omega)x : x \in X\}$ form a Markov family, i.e. the stochastic kernels $P_t(x, B) := \mathbb{P}\{\omega : \varphi(t, \omega)x \in B\}$ satisfy the Chapman-Kolmogorov equation*

$$P_{t+s}(x, B) = \int_X P_t(y, B)P_s(x, dy), \quad t, s \geq 0, \quad B \in \mathcal{B} \, .$$

If the σ-algebras $\mathcal{F}_-$ and $\mathcal{F}_+$ are independent, then for any invariant Markov measure μ supported by $\{A(\omega)\}$ the measure ϱ on $(X, \mathcal{B})$ defined by the formula

$$\varrho(B) = \int_\Omega \mu_\omega(B)\mathbb{P}(d\omega), \quad B \in \mathcal{B} \, ,$$

is a stationary probability measure for the Markov semigroup associated with the family $\{\varphi(t, \omega)x : x \in X\}$, i.e.

$$\varrho(B) = \int_X P_t(x, B)\varrho(dx), \quad B \in \mathcal{B} \, ,$$

or, in equivalent form,

$$\int_X g(x)\varrho(dx) = \int_X \mathbb{E}g(\varphi(t, \cdot)x)\varrho(dx), \quad g(x) \in C_b(X) \, .$$

In particular under the conditions of this theorem every $\mathcal{F}_-$-measurable equilibrium $u(\omega)$ with the property $\{u(\omega)\} \in \mathcal{D}$ generates a stationary measure on $(X, \mathcal{B})$ by the formula $\varrho(B) = \mathbb{E}\chi_B(u)$, where $\chi_B(x) = 1$ for $x \in B$ and $\chi_B(x) = 0$ otherwise, i.e. by the formula $\varrho(B) = \mathbb{P}\{\omega : u(\omega) \in B\}$, $B \in \mathcal{B}$.

We also note that if $\mathcal{F}_-$ and $\mathcal{F}_+$ are independent, then every invariant Markov measure is supported by the attractor of (θ, φ) in the universe consisting of all finite subsets of X (see CRAUEL [34]). Random systems generated by stochastic differential equations give examples of RDS where the future and past σ-algebras are independent (see ARNOLD [3, Sect. 2.3]).

2. Generation of Random Dynamical Systems

In this chapter we collect some results concerning those random dynamical systems generated by random and stochastic ordinary differential equations. Most of them are well-known (see ARNOLD [3, Chap. 2] and the references therein) and we include them here mainly for the sake of completeness. We also prove several assertions on the existence of deterministic invariant domains for these systems and consider relations between random and stochastic ordinary differential equations. These results are important in the study of monotone dynamical systems connected with random and stochastic differential equations.

2.1 RDS Generated by Random Differential Equations

In this section we consider a class of ordinary differential equations (ODE) whose right-hand sides contain ω as a parameter. For every fixed ω these equations can be solved as a deterministic nonautonomous ODEs. They model the so-called "real noise case" and also include periodic and almost-periodic equations as particular cases. We also refer to LADDE/LAKSHMIKANTHAM [75] for another approach to random differential equations.

Let $\theta = (\Omega, \mathcal{F}, \mathbb{P}, \{\theta_t, t \in \mathbb{R}\})$ be a metric dynamical system. We assume that $f = (f_1, \dots, f_d) : \Omega \times \mathbb{R}^d \to \mathbb{R}^d$ is a measurable function which is locally bounded and locally Lipschitz continuous with respect to x for every $\omega \in \Omega$. More precisely, we assume that for any compact set $K \subset \mathbb{R}^d$ there exists a random variable $C_K(\omega) \geq 0$ such that

$$\int_a^{a+1} C_K(\theta_t \omega)\, dt < \infty \quad \text{for all} \quad a \in \mathbb{R},\ \omega \in \Omega\,, \tag{2.1}$$

and

$$|f(\omega, x)| \leq C_K(\omega), \quad |f(\omega, x) - f(\omega, y)| \leq C_K(\omega) \cdot |x - y| \tag{2.2}$$

for any $x, y \in K$ and $\omega \in \Omega$. Here and below $|\cdot|$ is the Euclidean distance in $\mathbb{R}^d$.

We emphasize that assumptions (2.1) and (2.2) are stated here for *all* $\omega \in \Omega$. This does not spoil generality because we can apply the following simple perfection procedure. Assume that (2.1) and (2.2) hold almost surely and consider the sets

$$\Omega_N = \left\{ \omega : \int_a^b C_{K_N}(\theta_t\omega)\, dt < \infty \text{ for all } a < b \right\},$$

where $K_N = \{x \in \mathbb{R}^d : |x| \leq N\}$. It is clear that Ω_N is a θ-invariant subset of Ω and $\mathbb{P}(\Omega_N) = 1$. The set $\Omega^* = \cap_{N \in \mathbb{N}} \Omega_N$ possesses the same properties. Relations (2.1) and (2.2) are valid for all $\omega \in \Omega^*$. Therefore instead of θ we can consider the metric dynamical system $\theta^* = (\Omega^*, \mathcal{F}^*, \mathbb{P}, \{\theta_t, t \in \mathbb{R}\})$, where $\mathcal{F}^*$ is the σ-algebra induced by $\mathcal{F}$ on Ω^*. Another way to obtain a perfect version of relations (2.1) and (2.2) would be to redefine $f(\omega, x)$ on $\Omega \setminus \Omega^*$ in an appropriate way.

We consider the random differential equation (RDE) in $\mathbb{R}^d$

$$\dot{x}(t) = f(\theta_t\omega, x(t)), \quad x(0) = x_0 \in \mathbb{R}^d, \tag{2.3}$$

driven by the metric dynamical system θ. In applications random differential equations usually arise in the following way. Assume that $g(\cdot, \cdot) : \mathbb{R}^m \times \mathbb{R}^d \to \mathbb{R}^d$ is a continuous function such that for any compact set $K \subset \mathbb{R}^d$ there exists a constant $C_K \geq 0$ such that

$$|g(\lambda, x)| \leq C_K \cdot (1 + |\lambda|^p), \quad \lambda \in \mathbb{R}^m, x \in K,$$

and

$$|g(\lambda, x) - g(\lambda, y)| \leq C_K \cdot (1 + |\lambda|^p) \cdot |x - y|, \quad \lambda \in \mathbb{R}^m, x, y \in K,$$

for some $p \geq 1$. Let $\{\xi_t(\omega) : t \in \mathbb{R}\}$ be a stationary random process in $\mathbb{R}^d$ with continuous trajectories on a probability space $(\Omega, \mathcal{F}, \mathbb{P})$. For this process there is the standard realization such that the functions $t \mapsto \xi_t(\omega)$ are continuous for *all* $\omega \in \Omega$ (see ARNOLD [3, Appendix]). Let θ be the metric dynamical system generated by $\xi_t(\omega)$. In this case $\xi_t(\omega) = \xi_0(\theta_t\omega)$. If $\mathbb{E}|\xi|^p < \infty$, then the random function $f(\omega, x) = g(\xi_0(\omega), x)$ satisfies (2.1) and (2.2) and equation (2.3) turns into RDE

$$\dot{x}(t) = g(\xi_t(\omega), x(t)), \quad x(0) = x_0 \in \mathbb{R}^d. \tag{2.4}$$

This equation can be interpreted as a model for the description of dynamics of a system governed by the equation $\dot{x} = g(\lambda, x)$ which takes into account random fluctuations of the external parameter λ around $\lambda_0 = \mathbb{E}\xi$ (cf. the discussion in the Introduction). We also note that in the case of RDE of the form (2.4) the function $t \mapsto f(\theta_t\omega, x) \equiv g(\xi_t(\omega), x)$ is continuous for all $\omega \in \Omega$ and $x \in \mathbb{R}^d$. Several results of Chap. 5 rely on this (or a weaker) continuity property.

Definition 2.1.1. *A function* $x(t, \omega) = (x_1(t, \omega), \dots, x_d(t, \omega))$ *is said to be a* local solution *to problem (2.3) if for every* $\omega \in \Omega$ *there exists* $t_0 = t_0(\omega, x_0) > 0$ *such that* $x(t, \omega)$ *is continuous with respect to t from the interval* $(0, t_0(\omega, x_0))$ *into* $\mathbb{R}^d$ *for each* $\omega \in \Omega$ *and satisfies the equation*

$$x(t, \omega) = x_0 + \int_0^t f(\theta_\tau \omega, x(\tau, \omega)) \, d\tau, \quad 0 < t < t_0(\omega, x_0), \quad \omega \in \Omega . \quad (2.5)$$

If $t_0(\omega, x_0) = \infty$ *for all* $\omega \in \Omega$, *then* $x(t, \omega)$ *is said to be a* global solution *to (2.3).*

Remark 2.1.1. It is clear from (2.5) that $x(t, \omega)$ is absolutely continuous on the segment $[0, (1 - \delta) \cdot t_0(\omega, x_0)]$ for any $\omega \in \Omega$ and $0 < \delta < 1$ and for each $\omega \in \Omega$ it satisfies differential equation (2.3) for almost all $t \in (0, t_0(\omega, x_0))$.

Proposition 2.1.1. *Under conditions (2.1) and (2.2) problem (2.3) has a unique local solution* $x(t, \omega) \equiv x(t, \omega; x_0)$ *for any initial data* $x_0 \in \mathbb{R}^d$. *This solution depends continuously on* x_0 *for every* $\omega \in \Omega$. *If we assume additionally that* $f(\omega, \cdot) \in C^1(\mathbb{R}^d)$ *for each* $\omega \in \Omega$, *then* $x(t, \omega; x_0)$ *is continuously differentiable with respect to the initial data* x_0 *and the Jacobian*

$$D_{x_0} x(t, \omega) \equiv D_{x_0} x(t, \omega; x_0) = \left\{ \frac{\partial x_i(t, \omega; x_0)}{\partial x_{0j}} \right\}_{i,j=1}^d$$

solves the variational equation

$$D_{x_0} x(t, \omega) = I + \int_0^t D_x f(\theta_\tau \omega, x(\tau, \omega)) D_{x_0} x(\tau, \omega) \, d\tau \quad (2.6)$$

for all t from the interval $(0, t_0(\omega, x_0))$ *of the existence of the solution* $x(t, \omega)$. *Moreover the determinant* $\det D_{x_0} x(t, \omega)$ *satisfies Liouville's equation*

$$\det D_{x_0} x(t, \omega) = \exp \left\{ \int_0^t \mathrm{tr}\{D_x f(\theta_\tau \omega, x(\tau, \omega))\} \, d\tau \right\} \quad (2.7)$$

for all $t \in (0, t_0(\omega, x_0))$.

Proof. This is a direct ω-wise adaptation of the corresponding deterministic proof (see, e.g., CODDINGTON/LEVINSON [28] and AMANN [2]). □

Theorem 2.1.1. *Let (2.1) and (2.2) be valid. Assume that the solution* $x(t, \omega; x_0)$ *to problem (2.3) given by Proposition 2.1.1 is global for all* $x \in \mathbb{R}^d$ *and* $\omega \in \Omega$ *(see Definition 2.1.1). Then the RDE (2.3) generates a RDS* (θ, φ) *with the cocycle* φ *defined by the formula*

$$\varphi(t, \omega, x_0) = x(t, \omega; x_0), \quad t > 0, \ \omega \in \Omega, \ x_0 \in \mathbb{R}^d ,$$

where $x(t, \omega; x_0)$ *is the global solution to problem (2.3) for the initial data* $x_0 \in \mathbb{R}^d$. *Moreover the mapping* $(t, x) \mapsto \varphi(t, \omega, x)$ *is continuous for all*

$\omega \in \Omega$. If we assume additionally that $f(\omega, \cdot) \in C^1(\mathbb{R}^d)$ for each $\omega \in \Omega$, then (θ, φ) is a C^1 RDS and the Jacobian

$$D_x\varphi(t, \omega, x) = \left\{ \frac{\partial[\varphi(t, \omega, x)]_i}{\partial x_j} \right\}_{i,j=1}^d$$

uniquely solves the variational equation

$$D_x\varphi(t, \omega, x) = I + \int_0^t D_x f(\theta_\tau\omega, \varphi(\tau, \omega, x)) D_x\varphi(\tau, \omega, x)\, d\tau, \quad t > 0,\ \omega \in \Omega\,.$$

$$(2.8)$$

Moreover the determinant $\det D_x\varphi(t, \omega, x)$ satisfies Liouville's equation

$$\det D_x\varphi(t, \omega, x) = \exp\left\{ \int_0^t \text{tr}\{D_x f(\theta_\tau\omega, \varphi(\tau, \omega, x))\}\, d\tau \right\}, \quad t > 0\,. \quad (2.9)$$

Proof. This follows from Proposition 2.1.1. We refer to ARNOLD [3] for details. □

Corollary 2.1.1. *Assume that* $f(\omega, x)$ *satisfies (2.1) and (2.2) and there exist random variables* $c_1(\omega)$ *and* $c_2(\omega)$ *such that* $t \mapsto c_j(\theta_t\omega)$ *is locally integrable and*

$$\langle x, f(\omega, x) \rangle \le c_1(\omega)|x|^2 + c_2(\omega)\,, \quad (2.10)$$

where $\langle \cdot, \cdot \rangle$ *is the inner product in* $\mathbb{R}^d$ *generated by the Euclidean norm* $|\cdot|$. *Then the conclusions of Theorem 2.1.1 are true.*

Proof. Under condition (2.10) we obviously have (cf. Remark 2.1.1) that

$$\frac{1}{2} \cdot \frac{d}{dt}|x(t, \omega)|^2 \le c_1(\theta_t\omega)|x(t, \omega)|^2 + c_2(\theta_t\omega)$$

on the existence semi-interval $[0, t_0(\omega, x_0))$. Consequently the Gronwall lemma gives

$$|x(t, \omega)|^2 \le \left[|x_0|^2 + 2 \int_0^t |c_2(\theta_\tau\omega)|\, d\tau \right] \cdot \exp\left\{ 2 \int_0^t |c_1(\theta_\tau\omega)|\, d\tau \right\}\,. \quad (2.11)$$

Therefore the standard result on continuation of solutions (see, e.g., HARTMAN [51] or HALE [49]) gives that the solution $x(t, \omega)$ can be continued to the whole semi-axis $\mathbb{R}_+$. □

Example 2.1.1 (Binary Biochemical Model). The equations

$$\begin{aligned}
\dot{x}_1 &= g(x_2) - \alpha_1(\theta_t\omega)x_1\,, \\
\dot{x}_2 &= x_1 - \alpha_2(\theta_t\omega)x_2\,,
\end{aligned} \quad (2.12)$$

generate an RDS in $\mathbb{R}^2$ provided that $g(x) \in C^1(\mathbb{R})$, $g'(x)$ is bounded and the function $t \mapsto \alpha_i(\theta_t)$ is locally integrable for each $\omega \in \Omega$ and $i = 1, 2$. The cocycle of this RDS has the form $\varphi(t, \omega)x = x(t)$, where $x(t) = (x_1(t), x_2(t))$ is the solution to (2.12) with $x(0) = x$.

Remark 2.1.2. (i) Under the hypotheses of Corollary 2.1.1 the cocycle $\varphi(t, \omega)$ possesses the following property which is important in the study of pull back trajectories (cf. Proposition 1.5.2): for every $x \in \mathbb{R}^d$ and $\omega \in \Omega$ the function $t \mapsto \varphi(t, \theta_{-t}\omega)x$ is right continuous on $\mathbb{R}_+$. To prove this we note that (2.5) implies the relation

$$\varphi(t, \theta_{-t}\omega)x - x = \int_{-t}^{0} f(\theta_s\omega, \varphi(t + s, \theta_{-t}\omega)x) \; ds \; .$$

It follows from (2.11) that for any $T > 0$ there exists $C_T(\omega) > 0$ such that

$$|\varphi(t + s, \theta_{-t}\omega)x| \leq C_T(\omega) \quad \text{for all} \quad -t \leq s \leq 0, \; 0 \leq t \leq T \; .$$

Therefore from (2.2) we have

$$|\varphi(t, \theta_{-t}\omega)x - x| \leq \int_{-t}^{0} C_{K(\omega)}(\theta_s\omega) \; ds \; ,$$

where $K(\omega) = \{x \in \mathbb{R}^d : |x| \leq C_T(\omega)\}$. Therefore (2.1) implies that

$$\lim_{t \to +0} |\varphi(t, \theta_{-t}\omega)x - x| = 0 \quad \text{for any} \quad x \in \mathbb{R}^d, \; \omega \in \Omega \; . \tag{2.13}$$

By the cocycle property we have $\varphi(s + t, \theta_{-s-t}\omega) = \varphi(s, \theta_{-s}\omega)\varphi(t, \theta_{-s-t}\omega)$ for any $t, s \geq 0$. Therefore (2.13) implies that

$$\lim_{t \to s+0} |\varphi(t, \theta_{-t}\omega)x - \varphi(s, \theta_{-s}\omega)x| = 0 \quad \text{for any} \quad s > 0, \; x \in \mathbb{R}^d, \; \omega \in \Omega \; .$$

We note it is also possible to prove the continuity of the mapping $(t, x) \mapsto \varphi(t, \theta_{-t}\omega)x$ for every $\omega \in \Omega$ (see ARNOLD [3, Part I]). However we do not use this in what follows.

(ii) Assume that in (2.10) the random variable $c_1(\omega)$ satisfies the condition

$$\lim_{t \to +\infty} \frac{1}{t} \int_0^t c_1(\theta_\tau\omega) \; d\tau = \lim_{t \to +\infty} \frac{1}{t} \int_{-t}^0 c_1(\theta_\tau\omega) \; d\tau = \alpha$$

for all $\omega \in \Omega$ with a negative constant α and that the variable $\max\{0, c_2(\omega)\}$ is tempered. Then under the hypotheses of Corollary 2.1.1 we can apply Proposition 1.4.1 (see also Remark 1.4.1) with the function $V(x) = |x|^2$ to prove that the RDS generated by (2.3) is dissipative in the universe of all tempered subsets of $\mathbb{R}^d$ and possesses a random attractor in this universe (see Corollary 1.8.2). In particular (2.12) generates a dissipative RDS provided that the assumptions of Example 1.4.4 hold.

Now we consider the affine (linear nonhomogenious) RDE in $\mathbb{R}^d$

$$\dot{x}(t) = A(\theta_t\omega)x(t) + b(\theta_t\omega), \quad x(0) = x_0 \in \mathbb{R}^d \; , \tag{2.14}$$

driven by the metric dynamical system θ. Here $A(\omega) = \{a_{ij}(\omega)\}_{i,j=1}^d$ is a random matrix and $b(\omega) = (b_1(\omega), \ldots, b_d(\omega))$ is a random vector in $\mathbb{R}^d$.

If $\|A(\theta_t\omega)\|$ and $|b(\theta_t\omega)|$ belong to the space $L_{loc}^1(\mathbb{R})$ for all $\omega \in \Omega$, then we can apply Corollary 2.1.1 to construct an affine RDS (θ, φ) in $\mathbb{R}^d$. It is easy to see that the cocycle φ can be represented in the form

$$\varphi(t,\omega)x = \Phi(t,\omega)x + \int_0^t \Phi(t-s, \theta_s\omega)b(\theta_s\omega)\, ds\,, \tag{2.15}$$

where $\Phi(t,\omega)$ is the linear cocycle in $\mathbb{R}^d$ generated by the linear RDE

$$\dot{x}(t) = A(\theta_t\omega)x(t), \quad x(0) = x_0 \in \mathbb{R}^d\,. \tag{2.16}$$

The following result contains useful information on the top Lyapunov exponent of the linear RDS (θ, Φ). It is an easy consequence of the multiplicative ergodic theorem (see, e.g., ARNOLD [3, Chaps. 3,4]).

Theorem 2.1.2. *Assume that the matrix $A(\omega)$ satisfies $\|A(\cdot)\| \in L^1(\Omega, \mathcal{F}, \mathbb{P})$ and $\|A(\theta_t\omega)\| \in L_{loc}^1(\mathbb{R})$ for all $\omega \in \Omega$. Let $\Phi(t,\omega)$ be the linear cocycle in $\mathbb{R}^d$ generated by (2.16). Then there exists a θ-invariant set $\Omega^* \subset \Omega$ of full measure such that for each $x \in \mathbb{R}^d \setminus \{0\}$ the Lyapunov exponent*

$$\lambda(\omega, x) := \lim_{t\to+\infty} \frac{1}{t} \log |\Phi(t,\omega)x| \tag{2.17}$$

exists for all $\omega \in \Omega^*$. For every $\omega \in \Omega^*$ the image of the function $x \mapsto \lambda(\omega, x)$ is a finite set. If θ is an ergodic metric dynamical system, then $\lambda := \max_{x\in\mathbb{R}^d\setminus\{0\}} \lambda(\omega, x)$ is a constant on Ω^* and it is the top Lyapunov exponent in the sense of Definition 1.9.1. Moreover in this case we have

$$\mathbb{E}\mu_{\min} \leq \lambda(\omega, x) \leq \mathbb{E}\mu_{\max}, \quad x \in \mathbb{R}^d \setminus \{0\}, \quad \omega \in \Omega^*\,, \tag{2.18}$$

where $\mu_{\min}(\omega)$ and $\mu_{\max}(\omega)$ are the least and the greatest eigenvalues of the Hermitian part of the matrix $A(\omega)$. In particular the top Lyapunov exponent λ belongs to the interval $[\mathbb{E}\mu_{\min}, \mathbb{E}\mu_{\max}]$.

Proof. This follows directly from ARNOLD [3, Theorem 3.4.1] (see also ARNOLD [3, Example 3.4.15]). Relation (2.18) follows from the Birkhoff–Khinchin ergodic theorem (see, e.g., SINAI YA. G. [100]) and the argument given in HARTMAN [51, p. 56], see also ARNOLD [3, Theorem 6.2.8]. □

We also note that if $\mathbb{E}\mu_{\max} < 0$ and $b(\omega)$ is tempered, then the affine RDS (θ, φ) generated by (2.14) over an ergodic θ is dissipative in the universe of all tempered subsets of $\mathbb{R}^d$ (see Remark 2.1.2(ii)) and both Propositions 1.9.2 and 1.9.3 can be applied here.

Example 2.1.2 (1D Affine RDE). Consider the one-dimensional RDE

$$\dot{x} = \alpha(\theta_t\omega)x + \beta(\theta_t\omega)$$

over an ergodic metric dynamical system θ, where $\alpha(\omega)$ and $\beta(\omega)$ are random variables such that $t \mapsto \alpha(\theta_t\omega)$ and $t \mapsto \beta(\theta_t\omega)$ are locally integrable. This equation generates an affine RDS in $\mathbb{R}$. The cocycle φ has the form (2.15) with $\Phi(t, \omega)x = x \exp\left\{ \int_0^t \alpha(\theta_\tau\omega)d\tau \right\}$. If $\alpha \in L^1(\Omega, \mathcal{F}, \mathbb{P})$, then the Birkhoff–Khinchin ergodic theorem implies that the (top) Lyapunov exponent for (θ, Φ) is $\lambda = \mathbb{E}\alpha$ (see Remark 1.4.1). The RDS (θ, φ) is dissipative in the universe $\mathcal{D}$ of all tempered subsets of $\mathbb{R}$ provided that $\mathbb{E}\alpha < 0$ and $\beta(\omega)$ is a tempered random variable. In this case

$$\psi(t, \theta_{-t}\omega) = \int_{-t}^0 \beta(\theta_s\omega) \exp\left\{ \int_s^0 \alpha(\theta_\tau\omega)d\tau \right\} ds$$

in representation (1.48) and therefore (see Propositions 1.9.2 and 1.9.3 and also Remark 1.9.2) the RDS (θ, φ) possesses a unique exponentially stable equilibrium

$$u(\omega) = \int_{-\infty}^0 \beta(\theta_s\omega) \exp\left\{ \int_s^0 \alpha(\theta_\tau\omega)d\tau \right\} ds \ .$$

In the case $\mathbb{E}\alpha > 0$ the RDS (θ, φ) is not dissipative in $\mathcal{D}$ (see Remark 1.9.1). Nevertheless a simple calculation shows that

$$v(\omega) = -\int_0^\infty \beta(\theta_s\omega) \exp\left\{ -\int_0^s \alpha(\theta_\tau\omega)d\tau \right\} ds$$

is an equilibrium for (θ, φ) provided that $\mathbb{E}\alpha > 0$.

2.2 Deterministic Invariant Sets

In this section we give a result concerning deterministic invariant sets for RDS generated by random differential equations. We will use it in Chap. 5 to prove positivity of solutions to problem (2.3) under some conditions concerning $f(\omega, x)$. We note that there are many results concerning these invariance properties for nonautonomous ODE (see, e.g., MARTIN [81] and DEIMLING [41] and the references therein). However all of them assume continuous dependence of the right hand sides on t and x. This assumption looks rather restrictive for random ODE. Below we show that it can be avoided.

We do not assume the smoothness of invariant sets and we use the following definition of an outer normal vector.

Definition 2.2.1. *Let $\mathbb{D}$ be a closed set in $\mathbb{R}^d$. Assume that x_0 belongs to the boundary $\partial\mathbb{D}$ of the set $\mathbb{D}$. A unit vector ν is said to be an* outer normal *to $\mathbb{D}$ at the point x_0, if there exists a ball $B(x_1)$ with center at x_1 such that $B(x_1) \cap \mathbb{D} = \{x_0\}$ and $\nu = \lambda \cdot (x_1 - x_0)$ for some positive λ.*

We use the following concept of an invariant set for an RDE (2.3).

Definition 2.2.2. *The set $\mathbb{F}$ is said to be a deterministic forward invariant set for the RDE (2.3) if its local solution $x(t, \omega; x_0)$ lies in $\mathbb{F}$ for every $x_0 \in \mathbb{F}$, $t \in (0, t(\omega, x_0))$ and $\omega \in \Omega$. Here $(0, t(\omega, x_0))$ is the maximal interval of the existence of the solution $x(t, \omega; x_0)$.*

We have the following result on the existence of invariant sets.

Theorem 2.2.1. *Assume that (2.1) and (2.2) are valid. Let $\mathbb{D}$ be a closed set in $\mathbb{R}^d$ possessing the properties: (i) the set $\mathbb{D}$ has an outer normal at every point of the boundary $\partial \mathbb{D}$ and (ii) for any $x \in \partial \mathbb{D}$ we have the relation*

$$\langle f(\omega, x), \nu_x \rangle \leq 0, \quad \omega \in \Omega, \tag{2.19}$$

for every outer normal ν_x at x. Then the set $\mathbb{D}$ is a deterministic forward invariant set for the RDE (2.3).

This theorem immediately implies the following assertion.

Corollary 2.2.1. *Under the conditions of Theorems 2.1.1 and 2.2.1 the set $\mathbb{D}$ is a deterministic forward invariant set for the RDS (θ, φ) generated by problem (2.3).*

The argument given in the proof of Corollary 2.1.1 makes it possible to obtain the following result.

Corollary 2.2.2. *Let the hypotheses of Theorem 2.2.1 hold. Assume that there exist random variables $c_1(\omega)$ and $c_2(\omega)$ such that $t \mapsto c_j(\theta_t \omega)$ is locally integrable and inequality (2.10) holds for any $x \in \mathbb{D}$ and for all $\omega \in \Omega$. Then for any $x_0 \in \mathbb{D}$ problem (2.3) possesses a unique global solution $x(t, \omega; x)$ such that $x(t, \omega; x) \in \mathbb{D}$ for all $t \geq 0$ and $\omega \in \Omega$. This solution generates an RDS with phase space $\mathbb{D}$.*

Example 2.2.1. The one-dimensional RDE

$$\dot{x} = \alpha + \beta(\theta_t \omega) \cdot x - x^3$$

satisfies the hypotheses of Corollary 2.2.2 with $\mathbb{D} = \mathbb{R}_+$ if $\alpha \geq 0$ and the function $t \mapsto \beta(\theta_t \omega)$ is locally integrable for every $\omega \in \Omega$. We can also apply Corollary 2.2.2 with $\mathbb{D} = [0, 1]$ to the equation

$$\dot{x} = \beta(\theta_t \omega) \cdot x(1 - x) \, .$$

Example 2.2.2 (Binary Biochemical Model). Let (θ, φ) be the RDS considered in Example 2.1.1. If $g(0) \geq 0$, then $\mathbb{R}_+^2 = \{x = (x_1, x_2) : x_i \geq 0\}$ is a forward invariant set for (θ, φ). This property is important because x_1 and x_2 represent concentrations of macro-molecules.

Theorem 2.2.1 is a particular case of the following assertion which is also important in what follows.

Theorem 2.2.2. *Assume that (2.1) and (2.2) hold. Let $\mathbb{O} \subseteq \mathbb{R}^d$ be a deterministic forward invariant open set for the RDE (2.3) and $\mathbb{D}$ be a closed set in $\mathbb{R}^d$ such that (i) $\mathbb{D} \cap \mathbb{O} \neq \emptyset$, (ii) $\mathbb{D}$ has an outer normal at every point of the set $\partial\mathbb{D} \cap \mathbb{O}$ and (iii) relation (2.19) holds for any $x \in \partial\mathbb{D} \cap \mathbb{O}$. Then the set $\mathbb{D} \cap \mathbb{O}$ is a deterministic forward invariant set for the RDE (2.3).*

In the proof of Theorem 2.2.2 we rely on some ideas presented in BONY [17]. We start with the following deterministic lemma.

Lemma 2.2.1. *Assume that $f(t)$ is a continuous function on the segment $[a, b]$ such that*

$$\liminf_{h \to 0,\ h<0} \frac{1}{|h|} \left(f(t+h) - f(t) \right) \geq -m(t)$$

for almost all $t \in (a, b)$, where $m(t) \in L^1(a, b)$. Then

$$f(t_2) - f(t_1) \leq \int_{t_1}^{t_2} m(\tau)\, d\tau \quad \text{for all} \quad a \leq t_1 < t_2 \leq b. \tag{2.20}$$

Proof. It is clear that

$$g(t) \equiv f(t) - \int_a^t m(\tau)\, d\tau \in C[a, b]$$

and satisfies the relation

$$\liminf_{h \to 0,\ h<0} \frac{1}{|h|} \left(g(t+h) - g(t) \right) \geq 0 \tag{2.21}$$

for all $t \in \mathcal{B}$, where $\mathcal{B}$ is a measurable set of full measure in (a, b). To obtain (2.20) we should prove that $g(t)$ is a nonincreasing function on $[a, b]$. It is sufficient to prove that the function $\Phi(t) = g(t) - \gamma t$ is nonincreasing for any $\gamma > 0$. From (2.21) we have

$$\liminf_{h \to 0,\ h<0} \frac{1}{|h|} \left(\Phi(t+h) - \Phi(t) \right) \geq \gamma > 0, \quad t \in \mathcal{B}. \tag{2.22}$$

This implies that for every $t \in \mathcal{B}$ there exists $h(t) > 0$ such that

$$\Phi(t - \tau) \geq \Phi(t), \quad 0 \leq \tau < h(t),\ t \in \mathcal{B}. \tag{2.23}$$

Let $t_1 < t_2$ be points from $\mathcal{B}$. Consider the covering of the segment $[t_1, t_2]$ by intervals $(t - \min\{h(t_2), h(t)\}, t)$, where $t \in \mathcal{B}$. It is clear that there exists a finite subcovering. Moreover we can choose the points $\tau_1 < \tau_2 < \ldots < \tau_N$ from $\mathcal{B} \cap (t_1, t_2)$ such that

$$t_1 \in (\tau_1 - h(\tau_1), \tau_1), \quad \tau_N \in (t_2 - h(t_2), t_2)$$

and

$$\tau_k \in (\tau_{k+1} - h(\tau_{k+1}), \tau_{k+1}), \quad k = 1, \ldots N - 1 .$$

Therefore from (2.23) we have

$$\Phi(t_1) \geq \Phi(\tau_1); \quad \Phi(\tau_k) \geq \Phi(\tau_{k+1}), \ k = 1, \ldots N - 1; \quad \Phi(\tau_N) \geq \Phi(t_2) .$$

This implies that $\Phi(t_1) \geq \Phi(t_2)$.   □

Proof of Theorem 2.2.2. Let $x(t)$ be a local solution to (2.3) for some fixed ω with initial data from $\mathbb{D} \cap \mathbb{O}$. Assume that this solution may leave the set $\mathbb{D} \cap \mathbb{O}$. Since $\mathbb{O}$ is forward invariant, there exist a point $x^* \in \partial \mathbb{D} \cap \mathbb{O}$ and a semiinterval $(t_0, t_1]$ such that $x(t_0) = x^*$, $x(t) \in B_r(x^*)$ and $x(t) \notin \mathbb{D} \cap \mathbb{O}$ for $t \in (t_0, t_1]$. Here $B_r(x^*)$ is an open ball with center x^* and with radius r chosen such that $\overline{B_{2r}(x^*)} \subset \mathbb{O}$.

Let $h_n < 0$ and $h_n \to 0$. Assume that $t, t + h_n \in (t_0, t_1]$ and denote $x = x(t)$ and $x_n = x(t + h_n)$ for short. Let $\delta(t) = \text{dist}(x(t), \mathbb{D} \cap \overline{B_{2r}(x^*)})$. Since $x_n \to x$, it is clear that we can suppose that there exists a sequence $\{y_n\} \subset \partial \mathbb{D} \cap B_{2r}(x^*)$ which converges to some element $y \in \partial \mathbb{D} \cap B_{2r}(x^*)$ such that $\delta(t + h_n) = |x_n - y_n|$ and $\delta(t) = |x - y|$. Therefore we obtain the relation

$$\delta(t + h_n) - \delta(t) \geq |x_n - y_n| - |x - y_n| = \frac{|x_n - y_n|^2 - |x - y_n|^2}{|x_n - y_n| + |x - y_n|} .$$

Thus we have

$$\delta(t + h_n) - \delta(t) \geq \frac{|w_n|^2 - |w_n + v_n|^2}{|w_n| + |w_n + v_n|} = \frac{-2\langle w_n, v_n \rangle - |v_n|^2}{|w_n| + |w_n + v_n|} ,$$

where $w_n = x_n - y_n$ and $v_n = x - x_n$. We have that $w_n \to x - y \neq 0$ and $v_n \cdot h_n^{-1} \to -f(\theta_t \omega, x)$ for almost all $t \in (t_0, t_1)$. Therefore

$$\liminf_{n \to \infty} \frac{1}{|h_n|} \{\delta(t + h_n) - \delta(t)\} \geq - \lim_{n \to \infty} \left\langle \frac{w_n}{|w_n|}, \frac{v_n}{|h_n|} \right\rangle$$

$$= -\left\langle \frac{x - y}{|x - y|}, f(\theta_t \omega, x) \right\rangle$$

for almost all $t \in (t_0, t_1)$. It is clear that the vector $\frac{x-y}{|x-y|}$ is an outer normal to $\mathbb{D}$ at $y \in \partial \mathbb{D} \cap \mathbb{O}$. Consequently from (2.19) we obtain that

$$\liminf_{h \to 0, \ h < 0} \frac{1}{|h|} \{\delta(t + h) - \delta(t)\} \geq -\left\langle \frac{x - y}{|x - y|}, f(\theta_t \omega, x) - f(\theta_t \omega, y) \right\rangle .$$

It follows from (2.2) that there exists a constant $C(\omega, t) > 0$ such that

$$\int_{t_0}^{t_1} C(\omega, t) dt < \infty, \quad \text{and} \quad |f(\theta_t \omega, x) - f(\theta_t \omega, y)| \leq C(\omega, t)|x - y| .$$

Therefore

$$\liminf_{h \to 0,\ h < 0} \frac{1}{|h|} \{\delta(t+h) - \delta(t)\} \geq -C(\omega, t) \cdot \delta(t).$$

Using Lemma 2.2.1 we obtain

$$\delta(t) - \delta(s) \leq \int_s^t C(\omega, \tau) \delta(\tau) \, d\tau \quad \text{for all} \quad t_0 \leq s < t \leq t_1 \,.$$

Since $\delta(t_0) = 0$, Gronwall's lemma implies that $\delta(t) = 0$ for all $t_0 \leq t \leq t_1$. This contradicts to the assumption $x(t) \notin \mathbb{D} \cap \mathbb{O}$ for $t \in (t_0, t_1]$. Thus Theorem 2.2.2 is proved. □

It is clear from the proof of Theorem 2.2.2 that the following assertion concerning deterministic nonautonomous equations holds.

Proposition 2.2.1. *Suppose that a measurable function* $f(t, x) : \mathbb{R}_+ \times \mathbb{R}^d \to \mathbb{R}^d$ *possesses the following property: for every compact set* $K \subset \mathbb{R}^d$ *there exists a nonnegative function* $C_K(t) \in L^1_{loc}(\mathbb{R}_+)$ *such that*

$$|f(t, x)| \leq C_K(t) \quad and \quad |f(t, x) - f(t, y)| \leq C_K(t) \cdot |x - y|$$

for any $x, y \in K$. *Let* $\mathbb{D}$ *be a closed set in* $\mathbb{R}^d$ *which possesses an outer normal at every point of the boundary* $\partial \mathbb{D}$ *such that for any* $x \in \partial \mathbb{D}$ *and* $t \in \mathbb{R}_+$ *we have the relation*

$$\langle f(t, x), \nu_x \rangle \leq 0$$

for every outer normal ν_x *at* x. *Then the set* $\mathbb{D}$ *is a forward invariant set for the ordinary differential equation*

$$\dot{x}(t) = f(t, x(t)), \quad x(0) = x_0 \,, \tag{2.24}$$

i.e. for any local solution $x(t; x_0)$ *to problem (2.24) the property* $x_0 \in \mathbb{D}$ *implies that* $x(t; x_0) \in \mathbb{D}$ *for all* $t \in (0, t(x_0))$, *where* $(0, t(\omega, x_0))$ *is the maximal interval of the existence* $x(t; x_0)$.

We use Proposition 2.2.1 in Chap. 5 to study monotonicity properties of RDS generated by cooperative equations.

2.3 The Itô and Stratonovich Stochastic Integrals

In this section we recall several standard definitions and facts from stochastic analysis and give a short description of stochastic integration. We need this to construct RDS generated by Itô and Stratonovich stochastic differential equations in the next section. For details concerning stochastic integration we refer to CHUNG/WILLIAMS [27], IKEDA/WATANABE [57], KUNITA [74], McKEAN [82], for instance.

Let $W_t(\omega) = (W_t^1(\omega), \ldots, W_t^m(\omega))$ be a Wiener process with values in $\mathbb{R}^m$ for which we take two-sided time $t \in \mathbb{R}$, $m \geq 1$. The realization of this process in the space $C(\mathbb{R}; \mathbb{R}^m)$ of continuous functions was discussed in Example 1.1.7. Let $(\Omega, \mathcal{F}, \mathbb{P})$ be the corresponding canonical Wiener space and let $\{\mathcal{F}_s^t, -\infty \leq s < t \leq \infty\}$ be the filtration of the σ-algebras defined by the formula

$$\mathcal{F}_s^t = \sigma\{W_{\tau_1} - W_{\tau_2} : s \leq \tau_1, \tau_2 \leq t\} \vee \mathcal{N},$$

where $\sigma\{\xi\}$ denotes σ-algebra generated by the random variable ξ and $\mathcal{N}$ is the collection of null sets of $\mathcal{F}$. Below we denote $\mathcal{F}_t = \mathcal{F}_{-\infty}^t$. The process W_t satisfies

(i) W_t^i is an $\mathcal{F}_t$-measurable Gaussian variable, $i = 1, \ldots, m$;
(ii) $W_{t+s} - W_t$ is independent of $\mathcal{F}_t$ for $s > 0$;
(iii) $\mathbb{E}W_t = 0$ and $\mathbb{E}(W_t^i - W_s^i)(W_t^j - W_s^j) = \delta_{ij} \cdot |t - s|$.

Below we consider random dynamical systems over the ergodic metric dynamical system $\theta = (\Omega, \mathcal{F}, \mathbb{P}, \{\theta_t, t \in \mathbb{R}\})$ connected with this Wiener process (see Example 1.1.7). The transformations θ_t are defined such that

$$W_t(\theta_\tau \omega) = W_{t+\tau}(\omega) - W_\tau(\omega), \quad t, \tau \in \mathbb{R}, \ \omega \in \Omega. \tag{2.25}$$

Definition 2.3.1 (Continuous, Adapted and Predictable Processes).
A mapping f from $[a, b] \times \Omega$ into $\mathbb{R}^m$, where $[a, b] \subseteq \mathbb{R}$, is said to be a random process on $[a, b]$ (with values in $\mathbb{R}^m$) if $\omega \mapsto f(t, \omega)$ is measurable for every $t \in [a, b]$. This process is said to be

(i)   a continuous random process on $[a, b]$ if $f(t, \omega)$ is a continuous function with respect to $t \in [a, b]$ for almost all ω, i.e.

$$\mathbb{P}\left\{ \bigcup_{t \in [a,b]} \left\{ \omega : \lim_{h \to 0} |f(t + h, \omega) - f)t, \omega)| \neq 0 \right\} \right\} = 0 ;$$

(ii)   an $\mathcal{F}_t$-adapted random process on $[a, b]$ if $f(t, \omega)$ is $\mathcal{F}_t$-measurable for every fixed $t \in [a, b]$;
(iii) a predictable random process on $[a, b]$ if $f(t, \omega)$ is measurable with respect to σ-algebra generated in $[a, b] \times \Omega$ by all $\mathcal{F}_t$-adapted continuous random processes on $[a, b]$.

Below we denote by $\mathcal{L}^2[a, b]$ the set of all predictable processes $f(t, \omega)$ with the property

$$\int_a^b |f(t, \omega)|^2 \, dt < \infty \quad \text{almost surely} .$$

We note that the set of all $\mathcal{F}_t$-adapted continuous random processes on $[a, b]$ are dense in $\mathcal{L}^2[a, b]$, i.e. for any $f \in \mathcal{L}^2[a, b]$ there exists a sequence $\{f_n\}$ $\mathcal{F}_t$-adapted continuous processes such that

$$\lim_{n \to \infty} \int_a^b |f(t, \omega) - f_n(t, \omega)|^2 \, dt = 0 \quad \text{almost surely}.$$

For any $f \in \mathcal{L}^2[a, b]$ we can uniquely define the *Itô stochastic integral*

$$I_a^t(f) = \int_a^t \langle f(\tau, \omega), dW_\tau(\omega) \rangle = \sum_{i=1}^m \int_a^t f_i(\tau, \omega) dW_\tau^i(\omega), \quad t \in [a, b],$$

as an $\mathcal{F}_t$-adapted continuous process on $[a, b]$ with the properties:

(i) If $f(t, \omega)$ is a continuous $\mathcal{F}_t$-adapted process, then

$$I_a^t(f) = (\mathbb{P})\text{-} \lim_{|\Delta| \to 0} \sum_{k=1}^n \langle f(t \wedge t_k), W_{t \wedge t_{k+1}} - W_{t \wedge t_k} \rangle \tag{2.26}$$

uniformly with respect to $t \in [a, b]$ for any partition

$$\Delta = \{a = t_1 < t_2 \ldots < t_{n+1} = b\}$$

with diameter $|\Delta| \to 0$. As usual $u \wedge v = \min\{u, v\}$ and the symbol $(\mathbb{P})\text{-}\lim$ denotes the limit in probability.

(ii) The relations

$$\mathbb{E} I_a^t(f) = 0 \quad \text{and} \quad \mathbb{E}|I_a^t(f)|^2 = \int_a^t \mathbb{E}|f(\tau, \cdot)|^2 \, d\tau$$

are valid, if the integral on the right-hand side exists.

In many situations it is convenient to use the Stratonovitch stochastic integral. To define this we have to assume that the integrand f is a continuous semimartingale. Let us introduce the following concepts.

Definition 2.3.2 (Stopping Time). *A random variable $\tau(\omega)$ with values in $\mathbb{R}_+$ is said to be a stopping time on $[a, b]$ if $\{\omega : \tau(\omega) \leq t\} \in \mathcal{F}_t$ for all $t \in [a, b]$.*

Definition 2.3.3 (Martingale). *An $\mathcal{F}_t$-adapted random process $m(t, \omega)$ with values in $\mathbb{R}$ is said to be a martingale on $[a, b]$ if $\mathbb{E}|m(t)|^2 < \infty$ for all $t \in [a, b]$ and $\mathbb{E}\{m(t) \,|\, \mathcal{F}_s\} = m(s)$ almost surely for any $a \leq s < t \leq b$. Here $\mathbb{E}\{m \,|\, \mathcal{F}_s\}$ denotes the conditional expectation of the random variable m with respect to σ-algebra $\mathcal{F}_s$.*

Definition 2.3.4 (Local Martingale). *A random process $m(t, \omega)$ in $\mathbb{R}$ is said to be a* local martingale *on $[a, b]$ if there exists an increasing sequence of stopping times $\{\tau_n(\omega)\}$ such that $\mathbb{P}\{\tau_n < b\} \to 1$, $n \to \infty$, and $m_n(t, \omega) = m(t \wedge \tau_n(\omega), \omega)$ are martingales for each n, where $u \wedge v = \min\{u, v\}$.*

Definition 2.3.5 (Continuous Semimartingale). *An $\mathcal{F}_t$-adapted continuous random process $f(t, \omega)$ on $[a, b]$ is said to be a* continuous semimartingale *on $[a, b]$ if it has a decomposition $f(t, \omega) = m(t, \omega) + a(t, \omega)$, where $m(t, \omega)$ is a local martingale on $[a, b]$ and $a(t, \omega)$ is a process of bounded variation, i.e.*

$$\sup\left\{\sum_{k=1}^{n} |a(t_{k+1}, \omega) - a(t_k, \omega)| \; : \; a = t_1 < t_2 \ldots < t_{n+1} = b, \; n \geq 1\right\} < \infty$$

almost surely. A process $f(t, \omega) = (f_1, \ldots, f_d)$ with values in $\mathbb{R}^d$ is called continuous semimartingale, *if its components f_i possess this property.*

As an example of a continuous semimartingale we can consider the process

$$X(t, \omega) = c_0(\omega) + \int_a^t h_0(\tau, \omega) d\tau + \sum_{i=1}^{m} \int_a^t h_i(\tau, \omega) dW_\tau^i(\omega) \,,$$

where $h_k \in \mathcal{L}^2[a, b]$, $k = 0, 1, \ldots, m$, and $c_0(\omega)$ is a random $\mathcal{F}_a$-measurable variable.

Suppose that $X(t, \omega)$ is a continuous semimartingale on $[a, b]$ with values in $\mathbb{R}^d$ and $G(x)$ is C^2-mapping from $\mathbb{R}^d$ into $\mathbb{R}^m$. Let $g(t, \omega) = G(X(t, \omega))$. Then (see, e.g., IKEDA/WATANABE [57] or KUNITA [74]) the limit (cf. (2.26))

$$S_a^t(g) = (\mathbb{P})\text{-} \lim_{|\Delta| \to 0} \sum_{k=1}^{n} \left\langle \frac{g(t \wedge t_{k+1}) + g(t \wedge t_k)}{2}, W_{t \wedge t_{k+1}} - W_{t \wedge t_k} \right\rangle$$

exists, uniformly with respect to $t \in [a, b]$ for any partition $\Delta = \{a = t_1 < t_2 \ldots < t_{n+1} = b\}$ with diameter $|\Delta| \to 0$. This random process $S_a^t(g)$ is called the *Stratonovich stochastic integral* of g by $\circ dW_t$ and it is denoted by

$$S_a^t(g) = \int_a^t \langle g(\tau, \omega), \circ dW_\tau(\omega) \rangle = \sum_{i=1}^{m} \int_a^t G_i(X(\tau, \omega)) \circ dW_\tau^i(\omega), \quad t \in [a, b] \,.$$

The Stratonovich integral can be defined for more general mappings G (see, e.g., KUNITA [74]). However this will not be necessary in our subsequent considerations.

We have the following relation between the Stratonovich and Itô integrals (see, e.g., KUNITA [74]). Let $G(x)$ be a C^2-mapping from $\mathbb{R}^d$ into $\mathbb{R}^m$ and let $X(t, \omega) = (X_1(t, \omega), \ldots, X_d(t, \omega))$ be continuous semimartingale with values in $\mathbb{R}^d$. Then

$$\int_a^t \langle G(X(\tau,\omega)), \circ dW_\tau(\omega) \rangle = \int_a^t \langle G(X(\tau,\omega)), dW_\tau(\omega) \rangle$$

$$+ \frac{1}{2} \sum_{i=1}^m \sum_{j=1}^d \int_a^t \frac{\partial G_i}{\partial x_j}(X(\tau,\omega)) d\{W^i, X_j\}_\tau,$$

where $\{M, N\}_t$ is the joint quadratic variation of continuous semimartingales M_t and N_t which is defined by the formula

$$\{M, N\}_t = (\mathbb{P})\text{-}\lim_{|\Delta| \to 0} \sum_{k=1}^n \left(M_{t \wedge t_{k+1}} - M_{t \wedge t_k}\right) \cdot \left(N_{t \wedge t_{k+1}} - N_{t \wedge t_k}\right),$$

where $\Delta = \{a = t_1 < t_2 \ldots < t_{n+1} = b\}$ is a partition with diameter $|\Delta| \to 0$. In many cases the value $\{M, N\}_t$ can be calculated. For instance, if

$$X_i(t,\omega) = c_i(\omega) + \int_a^t h_{i0}(\tau,\omega) d\tau + \sum_{j=1}^m \int_a^t h_{ij}(\tau,\omega) dW_\tau^j(\omega), \qquad (2.27)$$

where $h_{ij} \in \mathcal{L}^2[a, b]$ for all $i = 1, \ldots, d$ and $j = 0, 1, \ldots, m$ and $c_i(\omega)$ are random $\mathcal{F}_a$-measurable variables, then

$$\{X_i, X_j\}_t = \sum_{k=1}^m \int_a^t h_{ik}(\tau,\omega) h_{jk}(\tau,\omega) d\tau.$$

Thus in this case we have

$$\int_a^t \langle G(X(\tau,\omega)), \circ dW_\tau(\omega) \rangle = \int_a^t \langle G(X(\tau,\omega)), dW_\tau(\omega) \rangle$$

$$+ \frac{1}{2} \sum_{i=1}^m \sum_{j=1}^d \int_a^t \frac{\partial G_i}{\partial x_j}(X(\tau,\omega)) h_{ji}(\tau,\omega) d\tau.$$

In particular both integrals coincide if $X_i(t,\omega)$ are absolutely continuous functions for almost all ω and the derivatives $\dot{X}_i(t,\omega)$ belong to $\mathcal{L}^2[a, b]$ (the case $h_{ij} \equiv 0$ for $j = 1, \ldots, m$).

The main advantage of the Stratonovich integral in comparison with the Itô integral is connected with the differentiation rule which is stated in the following well-known assertion (see, e.g., IKEDA/WATANABE [57] or KUNITA [74]).

Theorem 2.3.1 (Itô's Formula). *Let $G(x) \in C^2(\mathbb{R}^d)$ and $X(t,\omega) = (X_1(t,\omega), \ldots, X_d(t,\omega))$ be given by (2.27). Then $G(X(t,\omega))$ is a continuous semimartingale and it satisfies the formula*

$$G(X(t,\omega)) - G(X(a,\omega)) = \sum_{i=1}^{d} \int_a^t \frac{\partial G}{\partial x_i}(X(\tau))dX_i(\tau)$$

$$+ \frac{1}{2} \sum_{l=1}^{m} \sum_{i,j=1}^{d} \int_a^t \frac{\partial^2 G}{\partial x_i x_j}(X(\tau,\omega))h_{il}(\tau,\omega)h_{jl}(\tau,\omega)d\tau.$$

If $G(x) \in C^3(\mathbb{R}^d)$ and

$$X_i(t,\omega) = c_i(\omega) + \int_a^t h_{i0}(\tau,\omega)d\tau + \sum_{j=1}^{m} \int_a^t h_{ij}(\tau,\omega) \circ dW_\tau^j(\omega),$$

where h_{ij} and $c_i(\omega)$ are the same as above, then we have

$$G(X(t,\omega)) - G(X(a,\omega)) = \sum_{i=1}^{d} \int_a^t \frac{\partial G}{\partial x_i}(X(\tau)) \circ dX_i(\tau).$$

Here we have used the notation

$$\int_a^t g(\tau)dX_i(\tau) = \int_a^t g(\tau)h_{i0}(\tau)d\tau + \sum_{j=1}^{m} \int_a^t g(\tau)h_{ij}(\tau)dW_\tau^j,$$

similarly for the Stratonovich integral $\int_a^t g(\tau) \circ dX_i(\tau)$.

2.4 RDS Generated by Stochastic Differential Equations

In this section we consider stochastic differential equations (SDE) in the sense of Itô and Stratonovich. The main attention is paid to the case of Stratonovich SDEs because in Chap. 6 we deal mainly with applications of the general theory to Stratonovich equations. Since there is a simple relation between the Itô and Stratonovich cases (see Theorem 2.4.2 below), this means that we assume some additional smoothness properties concerning the coefficients in comparison with the standard theory of Itô differential equations. We prefer to deal with the Stratonovich case because it leads to some simplifications in formulas. We also refer to HORSTHEMKE/LEFEVER [55] for a discussion of the relation between Itô and Stratonovich SDEs from an applied point of view.

We need the following functional spaces (cf., e.g., ARNOLD [3] and KUNITA [74]).

Definition 2.4.1 (Spaces $C_b^{k,\delta}$). *For any $k \in \mathbb{Z}_+$ and $0 < \delta \leq 1$ we introduce the space $C_b^{k,\delta} \equiv C_b^{k,\delta}(\mathbb{R}^d)$ as the set of continuous functions $f(x)$ from*

$\mathbb{R}^d$ *into* $\mathbb{R}$ *such that* $\|f\|_{k,\delta} < \infty$, *where*

$$\|f\|_{k,0} = \sup_{x \in \mathbb{R}^d} \frac{|f(x)|}{1 + |x|} + \sum_{1 \leq |\alpha| \leq k} \sup_{x \in \mathbb{R}^d} |D^\alpha f(x)|,$$

$$\|f\|_{k,\delta} = \|f\|_{k,0} + \sum_{|\alpha|=k} \sup_{x \neq y} \frac{|D^\alpha f(x) - D^\alpha f(x)|}{|x - y|^\delta}, \quad 0 < \delta \leq 1.$$

Here $\alpha = (\alpha_1, \ldots, \alpha_d)$, $|\alpha| = \sum \alpha_i$ *and* $D^\alpha = \frac{\partial^{|\alpha|}}{\partial x_1^{\alpha_1} \ldots \partial x_d^{\alpha_d}}$. *For a closed set* $\mathbb{D} \subset \mathbb{R}^d$ *we denote by* $C_b^{k,\delta}(\mathbb{D})$ *the space of restrictions to* $\mathbb{D}$ *of functions from* $C_b^{k,\delta}(\mathbb{R}^d)$.

Now we consider the following system of Itô SDEs

$$dx_i = h_i(x_1, \ldots, x_d)dt + \sum_{j=1}^{m} \sigma_{ij}(x_1, \ldots, x_d)dW_t^j, \quad i = 1, \ldots, d. \quad (2.28)$$

We assume that $h_i(x)$ and $\sigma_{ij}(x)$ are functions from $C_b^{0,1}(\mathbb{R}^d)$.

Definition 2.4.2 (Solutions to Itô SDE). *A random process* $x(t, \omega)$ *on* $\mathbb{R}_+$ *with values in* $\mathbb{R}^d$ *is said to be a solution to the system of Itô SDEs (2.28) with initial data* $x^*(\omega) = (x_1^*(\omega), \ldots, x_d^*(\omega))$ *if it is an* $\mathcal{F}_t$-*adapted continuous process on* $\mathbb{R}_+$ *satisfying the integral equation*

$$\begin{aligned} x_i(t, \omega) = x_i^*(\omega) + \int_0^t h_i(x_1(\tau, \omega), \ldots, x_d(\tau, \omega))d\tau \\ + \sum_{j=1}^{m} \int_0^t \sigma_{ij}((x_1(\tau, \omega), \ldots, x_d(\tau, \omega))dW_\tau^j(\omega) \end{aligned} \quad (2.29)$$

almost surely for all $i = 1, \ldots, d$ *and* $t > 0$.

The following theorem shows that solutions to (2.28) generate an RDS in $\mathbb{R}^d$ (for the proof and discussions we refer to ARNOLD [3, Chap. 2]).

Theorem 2.4.1 (Generation by Itô SDE). *Let* $h_i(x)$ *and* $\sigma_{ij}(x)$ *be functions from* $C_b^{0,1}(\mathbb{R}^d)$. *Then there exists a unique (up to indistinguishability) continuous RDS* (θ, φ) *over the metric dynamical system* θ *connected with the Wiener process* W_t *such that* $x(t, \omega) = \varphi(t, \omega, x^*(\omega))$ *is a solution to the system of Itô SDEs (2.28) for every* $\mathcal{F}_0$-*measurable initial data* $x^*(\omega) \in L^2(\Omega, \mathcal{F}, \mathbb{P})$. *Moreover* $\sup_{[0,T]} \mathbb{E}|x(t, \cdot)|^2 < \infty$ *for every* $T > 0$, *the function* $(t, x) \mapsto \varphi(t, \omega, x)$ *is a continuous mapping from* $\mathbb{R}_+ \times \mathbb{R}^d$ *into* $\mathbb{R}^d$ *for every* $\omega \in \Omega$, *and the processes* $\{\varphi(t, \omega)x : x \in \mathbb{R}^d\}$ *form a Markov family. If* $h_i(x)$ *and* $\sigma_{ij}(x)$ *are from* $C_b^{k,\delta}(\mathbb{R}^d)$ *for some* $k \geq 1$ *and* $\delta > 0$, *then* (θ, φ) *is* C^k *RDS.*

Example 2.4.1 (Binary Biochemical Model). Let (W_t^1, W_t^2) be a Wiener process in $\mathbb{R}^2$. Consider the system of Itô ordinary differential equations

$$
\begin{aligned}
dx_1 &= (g(x_2) - \alpha_1 x_1)\, dt + \sigma_1 x_1 dW_t^1\ , \\
dx_2 &= (x_1 - \alpha_2 x_2)\, dt + \sigma_2 x_2 dW_t^2\ ,
\end{aligned}
\tag{2.30}
$$

where α_i and σ_i are constants and $g(x)$ is a Lipschitz continuous function. Equations (2.30) generate an RDS in $\mathbb{R}^2$. This is a C^k RDS provided that $g(x) \in C_b^{k,1}(\mathbb{R})$.

Now we consider the following system of Stratonovich SDEs

$$
dx_i = f_i(x_1, \dots, x_d)dt + \sum_{j=1}^m \sigma_{ij}(x_1, \dots, x_d) \circ dW_t^j, \quad i = 1, \dots, d\ . \tag{2.31}
$$

We assume that $f_i(x) \in C_b^{1,\delta}(\mathbb{R}^d)$ and $\sigma_{ij}(x) \in C_b^{2,\delta}(\mathbb{R}^d)$.

Definition 2.4.3 (Solutions to Stratonovich SDE). *A continuous semimartingale $x(t, \omega)$ on $\mathbb{R}_+$ with values in $\mathbb{R}^d$ is said to be a solution to the system (2.31) with initial data $x^*(\omega) = (x_1^*(\omega), \dots, x_d^*(\omega))$ if it satisfies the integral equation*

$$
\begin{aligned}
x_i(t, \omega) &= x_i^*(\omega) + \int_0^t f_i(x_1(\tau, \omega), \dots, x_d(\tau, \omega))d\tau \\
&\quad + \sum_{j=1}^m \int_0^t \sigma_{ij}((x_1(\tau, \omega), \dots, x_d(\tau, \omega)) \circ dW_\tau^j(\omega)
\end{aligned}
\tag{2.32}
$$

almost surely for all $i = 1, \dots, d$ and $t > 0$.

The proof of the following theorem can be found in KUNITA [74, Chap. 3].

Theorem 2.4.2 (Existence for Stratonovich SDE). *Let $f_i(x) \in C_b^{1,\delta}(\mathbb{R}^d)$ and $\sigma_{ij}(x) \in C_b^{2,\delta}(\mathbb{R}^d)$ for some $0 < \delta \le 1$ and for all $i = 1, \dots, d$, $j = 1, \dots, m$. Assume that*

$$
c_k(x) \equiv \frac{1}{2} \sum_{i=1}^d \sum_{j=1}^m \sigma_{ij}(x) \cdot \frac{\partial \sigma_{kj}(x)}{\partial x_i} \in C_b^{1,\delta}(\mathbb{R}^d), \quad k = 1, \dots, d\ . \tag{2.33}
$$

Then the system of Stratonovich SDEs (2.31) has a unique solution for every $\mathcal{F}_0$-measurable initial data $x^*(\omega) \in L^2(\Omega, \mathcal{F}, \mathbb{P})$. Further this solution satisfies Itô's SDEs (2.28) with $h_i(x) = f_i(x) + c_i(x)$, where $c_i(x)$ is given by (2.33), $i = 1, \dots, d$.

Conversely, if $x(t, \omega)$ is a solution to Itô SDEs (2.28) with $h_i(x) \in C_b^{1,\delta}(\mathbb{R}^d)$ and $\sigma_{ij}(x)$ as above, then $x(t, \omega)$ solves the system of Stratonovich SDEs (2.31) with $f_i(x) = h_i(x) - c_i(x)$, $i = 1, \dots, d$.

Example 2.4.2 (Binary Biochemical Model). Let (θ, φ) be the RDS generated by (2.30). If $g(x) \in C_b^{1,\delta}(\mathbb{R})$, then $x(t) = \varphi(t, \omega, x)$ solves the Stratonovich equations

$$dx_1 = \left(g(x_2) - \left(\alpha_1 + \frac{\sigma_1^2}{2} \right) x_1 \right) dt + \sigma_1 x_1 \circ dW_t^1 \,,$$

$$dx_2 = \left(x_1 - \left(\alpha_2 + \frac{\sigma_2^2}{2} \right) x_2 \right) dt + \sigma_2 x_2 \circ dW_t^2 \,.$$

The following theorem shows that solutions to (2.31) generate an RDS in $\mathbb{R}^d$ (for the proof and discussions we refer to ARNOLD [3, Chap. 2]).

Theorem 2.4.3 (Generation by Stratonovich SDE). *Assume that the functions $f_i(x)$, $\sigma_{ij}(x)$ and $c_i(x)$ satisfy the hypotheses of Theorem 2.4.2. Then there exists a unique (up to indistinguishability) continuous C^1 RDS (θ, φ) over the metric dynamical system θ connected with the Wiener process W_t such that $x(t, \omega) = \varphi(t, \omega)x^*$ is a solution to the system of Stratonovich SDEs (2.31) with initial data $x^* \in \mathbb{R}^d$ and the function $(t, x) \mapsto \varphi(\omega, x)$ is a continuous mapping from $\mathbb{R}_+ \times \mathbb{R}^d$ into $\mathbb{R}^d$ for every $\omega \in \Omega$. The Jacobian*

$$D_x \varphi(t, \omega, x) = \left\{ \frac{\partial [\varphi(t, \omega, x)]_i}{\partial x_j} \right\}_{i,j=1}^d$$

uniquely solves the variational equation

$$
\begin{aligned}
D_x \varphi(t, \omega, x) = I &+ \int_0^t D_x f(\varphi(\tau, \omega, x)) D_x \varphi(\tau, \omega, x)\, d\tau \\
&+ \sum_{j=1}^m \int_0^t D_x \sigma_j(\varphi(\tau, \omega, x)) D_x \varphi(\tau, \omega, x) \circ dW_\tau^j
\end{aligned}
\tag{2.34}
$$

for all $t > 0$. Here $f = (f_1 \ldots, f_d)$ and $\sigma_j = (\sigma_{1j} \ldots, \sigma_{dj})$ are mappings from $\mathbb{R}^d$ into itself. Moreover for every $t > 0$ the determinant $\det D_x \varphi(t, \omega, x)$ satisfies Liouville's equation

$$
\begin{aligned}
\det D_x \varphi(t, \omega, x) = \exp \Bigg(&\int_0^t \operatorname{tr}\{D_x f(\varphi(\tau, \omega, x))\} d\tau \\
&+ \sum_{j=1}^m \int_0^t \operatorname{tr}\{D_x \sigma_j(\varphi(\tau, \omega, x))\} \circ dW_\tau^j \Bigg).
\end{aligned}
\tag{2.35}
$$

Furthermore for all $t_0 \le t_1 \le \ldots \le t_n$ the random variables

$$\{ \varphi(t_k - t_{k-1}, \theta_{t_{k-1}} \omega) \ : \ k = 1, \ldots, n \}$$

are independent (in particular, the past and future σ-algebras are independent) and the processes $\{\varphi(t, \omega)x \ : \ x \in \mathbb{R}^d\}$ form a Markov family.

Example 2.4.3 (Binary Biochemical Model). Consider the following Stratonovich version of equations (2.30)

$$
\begin{aligned}
dx_1 &= (g(x_2) - \alpha_1 x_1)\, dt + \sigma_1 x_1 \circ dW_t^1 \ , \\
dx_2 &= (x_1 - \alpha_2 x_2)\, dt + \sigma_2 x_2 \circ dW_t^2 \ .
\end{aligned}
\tag{2.36}
$$

If $g(x) \in C_b^{1,\delta}(\mathbb{R})$, $0 < \delta \le 1$, then these equations generate a C^1 RDS in $\mathbb{R}^2$. By Theorem 2.4.2 equations (2.36) can be rewritten in the Itô form

$$
dx_1 = \left(g(x_2) - \left(\alpha_1 - \frac{\sigma_1^2}{2} \right) x_1 \right) dt + \sigma_1 x_1 dW_t^1 \ ,
$$

$$
dx_2 = \left(x_1 - \left(\alpha_2 - \frac{\sigma_2^2}{2} \right) x_2 \right) dt + \sigma_2 x_2 dW_t^2 \ .
$$

Remark 2.4.1. It is possible (see ARNOLD [3, Chap. 2]) to prove in the two cases described in Theorems 2.4.1 and 2.4.3 that the cocycle $\varphi(t, \omega)$ can be extended to a cocycle $\tilde{\varphi}$ with two-sided time. In particular this implies (see ARNOLD [3, Theorem 1.1.6]) that the function $(t, x) \mapsto \varphi(t, \theta_{-t}\omega)x = [\tilde{\varphi}(-t, \omega)]^{-1}x$ is continuous for every $\omega \in \Omega$. Therefore the mapping $t \mapsto \varphi(t, \theta_{-t}\omega)v(\theta_{-t}\omega)$ is continuous for a dense (with respect to convergence in probability) set of random variables $v(\omega)$ (cf. Remark 1.5.1).

Theorem 2.4.3 can be applied to the affine (linear nonhomogeneous) Stratonovich SDE

$$
dx(t) = (A_0 x(t) + b_0)\, dt + \sum_{j=1}^{m} (A_j x(t) + b_j) \circ W_t^j \ ,
\tag{2.37}
$$

where $A_j = \{a_{ik}^j\}_{i,k=1}^d$ is a $d \times d$ matrix and b_j is a vector in $\mathbb{R}^d$, $j = 0, 1, \ldots, m$. Thus (2.37) generates an affine RDS (θ, φ) in $\mathbb{R}^d$ (see ARNOLD [3, Sect.2.3]). The cocycle φ can be represented in the form

$$
\varphi(t, \omega)x = \Phi(t, \omega)x + \psi(t, \omega) \ ,
\tag{2.38}
$$

where $\psi(t, \omega) = \varphi(t, \omega)0$ and $\Phi(t, \omega)$ is the linear cocycle in $\mathbb{R}^d$ generated by the linear SDE

$$
dx(t) = A_0 x(t) dt + \sum_{j=1}^{m} A_j x(t) \circ W_t^j \ .
\tag{2.39}
$$

As in the random case (cf. Theorem 2.1.2) from the multiplicative ergodic theorem (see ARNOLD [3, Chaps. 3,4]) we can derive the following assertion on the top Lyapunov exponent of the linear RDS (θ, Φ).

Theorem 2.4.4. *Let $\Phi(t,\omega)$ be the linear cocycle in $\mathbb{R}^d$ generated by (2.39). Then there exists a θ-invariant set $\Omega^* \subset \Omega$ of full measure such that for each $x \in \mathbb{R}^d \setminus \{0\}$ the Lyapunov exponent*

$$\lambda(x) := \lim_{t \to +\infty} \frac{1}{t} \log |\Phi(t,\omega)x| \tag{2.40}$$

exists for all $\omega \in \Omega^*$ and is independent of $\omega \in \Omega^*$. The image of the function $x \mapsto \lambda(x)$ is a finite set and $\lambda := \max_{x \in \mathbb{R}^d \setminus \{0\}} \lambda(x)$ is the top Lyapunov exponent in the sense of Definition 1.9.1. Moreover there exists a probability measure ϱ on $S^{d-1} := \{x : |x| = 1\}$ such that

$$\lambda = \int_{S^{d-1}} \left\{ \langle A_0 s, s \rangle + \frac{1}{2} \sum_{j=1}^m \left(\langle (A_j + A_j^*)s, A_j s \rangle - 2\langle A_j s, s \rangle^2 \right) \right\} \varrho(ds) . \tag{2.41}$$

Proof. The existence of Lyapunov exponents follows directly from ARNOLD [3, Theorem 3.4.1] (see also ARNOLD [3, Example 3.4.19]). To obtain (2.41) we apply Proposition 6.2.11 and Remark 6.2.4 from ARNOLD [3].   □

Remark 2.4.2. Relation (2.41) is known as the Furstenberg-Khasminskii formula. Under some generic conditions the measure ϱ can be found as a solution of the Fokker-Plank equation (ARNOLD [3, Sect. 6.2]). We also refer to KHASMINSKII [64] and MAO [80] for examples of calculations of bounds for the top Lyapunov exponent.

Example 2.4.4 (1D Affine SDE). Consider the following one-dimensional Stratonovich SDE

$$dx = (\lambda x + \beta)dt + \sigma \cdot x \circ dW_t,$$

where W_t is a Wiener process in $\mathbb{R}$ and λ, β, σ are constants. This equation generates an affine RDS (θ, φ) in $\mathbb{R}$. The cocycle φ has the form (2.38), where $\Phi(t,\omega)x = x \exp\{\lambda t + \sigma W_t(\omega)\}$ and

$$\psi(t,\omega) = \beta \int_0^t \exp\{\lambda(t-\tau) + \sigma(W_t(\omega) - W_\tau(\omega))\} \, d\tau . \tag{2.42}$$

It is clear that the number λ is the (top) Lyapunov exponent for (θ, Φ). If $\lambda < 0$, then the RDS (θ, φ) is dissipative in the universe of all tempered subsets of $\mathbb{R}$. By Proposition 1.9.2 and Remark 1.9.2 the RDS (θ, φ) has a unique equilibrium $u(\omega)$. Relations (1.51) and (2.42) imply that

$$u(\omega) = \beta \int_{-\infty}^0 \exp\{-\lambda\tau - \sigma W_\tau(\omega)\} \, d\tau .$$

This equilibrium is measurable with respect to the past σ-algebra $\mathfrak{F}_-$ (see the definition in Sect. 1.10) and exponentially stable (see Proposition 1.9.3).

If $\lambda > 0$, then the RDS (θ, φ) possesses the equilibrium

$$v(\omega) = -\beta \int_0^\infty \exp\{-\lambda\tau - \sigma W_\tau(\omega)\} \, d\tau \;,$$

which is measurable with respect to the future σ-algebra $\mathcal{F}_+$.

A similar picture is observed for the Ornstein-Uhlenbeck type equation

$$dx = \lambda x \, dt + \sigma dW_t \;,$$

which generates an RDS with the cocycle

$$\varphi(t, \omega)x = e^{\lambda t}x + \sigma \int_0^t e^{\lambda(t-\tau)} dW_\tau, \quad x \in \mathbb{R} \;.$$

This RDS has an exponentially stable $\mathcal{F}_-$-measurable equilibrium

$$u(\omega) = \sigma \int_{-\infty}^0 e^{-\lambda\tau} dW_\tau$$

provided that $\lambda < 0$. In the case $\lambda > 0$ it possesses an unstable $\mathcal{F}_+$-measurable equilibrium

$$v(\omega) = -\sigma \int_0^\infty e^{-\lambda\tau} dW_\tau \;.$$

2.5 Relations Between Random and Stochastic Differential Equations

In this section we first consider approximations of solutions to Stratonovich stochastic differential equations by solutions to random differential equations. Apparently the first result in this direction was obtained by WONG/ZAKAÏ [109] (see also WONG [108]). There are now many expansions and generalizations of the Wong–Zakaï theorem (see, e.g., BELOPOLSKAYA/DALECKY [15], IKEDA/WATANABE [57], KUNITA [74] and also the survey TWARDOWSKA [105] and the references therein).

We introduce the following smooth approximation of the Wiener process $W_t = (W_t^1, \ldots, W_t^j)$. Let $\phi(t)$ be a nonnegative function on $\mathbb{R}$ with the properties

$$\phi(t) \in C^1(\mathbb{R}), \quad \operatorname{supp}\phi(t) \subset [0, 1], \quad \int_0^1 \phi(t) \, dt = 1 \;.$$

We set $\phi_\varepsilon(t) = \varepsilon^{-1}\phi(t/\varepsilon)$ for $\varepsilon > 0$ and

$$W_t^{j,\varepsilon}(\omega) = \int_{-\infty}^\infty \phi_\varepsilon(\tau - t)W_\tau^j(\omega) \, d\tau = \int_0^\varepsilon \phi_\varepsilon(\tau)W_{\tau+t}^j(\omega) \, d\tau \;. \tag{2.43}$$

Using the relation $W_\tau(\theta_t\omega) = W_{t+\tau}(\omega) - W_t(\omega)$, it is easy to see that $\frac{d}{dt}W_t^{j,\varepsilon}(\omega) = \eta_j^\varepsilon(\theta_t\omega)$, where

$$\eta_j^\varepsilon(\omega) = -\int_0^\varepsilon \dot\phi_\varepsilon(\tau)W_\tau^j(\omega)\,d\tau\;.$$

We consider the following approximation of the Stratonovich SDE (2.31):

$$\frac{dx_i}{dt} = f_i(x_1,\dots,x_d) + \sum_{j=1}^m \sigma_{ij}(x_1,\dots,x_d)\cdot\eta_j^\varepsilon(\theta_t\omega),\quad i = 1,\dots,d\;.\quad (2.44)$$

If $f_i(x) \in C_b^{1,\delta}(\mathbb{R}^d)$ and $\sigma_{ij}(x) \in C_b^{2,\delta}(\mathbb{R}^d)$, then Corollary 2.1.1 implies that the random differential equations (2.44) generates an RDS in $\mathbb{R}^d$. In particular equations (2.44) are uniquely solved on $\mathbb{R}^d$ for any initial data $x^* \in \mathbb{R}^d$.

We have the following Wong–Zakaï type approximation theorem (see, e.g., IKEDA/WATANABE [57] and KUNITA [74]).

Theorem 2.5.1. *Let the hypotheses of Theorem 2.4.2 hold. Let a function $x(t,\omega;x^*)$ be the solution to the Stratonovich SDE (2.31) with initial data $x^* \in \mathbb{R}^d$ and $x^\varepsilon(t,\omega;x^*)$ be the solution to the RDE (2.44) with the same initial data. Then we have*

$$\lim_{\varepsilon\to 0}\mathbb{E}\left\{\sup_{[0,T]}\sup_{|x^*|\le R}|x(t;x^*) - x^\varepsilon(t;x^*)|^2\right\} = 0 \quad (2.45)$$

for any positive T and R.

Remark 2.5.1. Assume that the hypotheses of Theorem 2.4.2 hold. Let $\varphi(t,\omega)$ and $\varphi^\varepsilon(t,\omega)$ be the cocycles of the RDS generated by (2.31) and (2.44) respectively. From (2.45) we have

$$\lim_{\varepsilon\to 0}\mathbb{E}\int_{-m}^m d\tau\left\{\sup_{t\in[0,l]}\sup_{|x^*|\le r}|\varphi(t,\theta_\tau\omega)x^* - \varphi^\varepsilon(t,\theta_\tau\omega)x^*|^2\right\} = 0$$

for all $m,l,r \in \mathbb{N}$. This implies that there exists a sequence $\varepsilon_n \to 0$ such that the set Ω^* of all $\omega \in \Omega$ satisfying

$$\lim_{n\to 0}\int_\alpha^\beta d\tau\left\{\sup_{[0,T]}\sup_{|x^*|\le R}|\varphi(t,\theta_\tau\omega)x^* - \varphi^{\varepsilon_n}(t,\theta_\tau\omega)x^*|^2\right\} = 0 \quad (2.46)$$

for all $\alpha < \beta$, $T > 0$ and $R > 0$, has full measure, i.e. $\mathbb{P}(\Omega^*) = 1$. Moreover the set Ω^* is θ-invariant, i.e. $\theta_t\Omega^* = \Omega^*$ for all $t \in \mathbb{R}$.

This remark along with Theorems 2.2.1 and 2.5.1 allows us to obtain the following result concerning deterministic invariant domains for RDS generated by Stratonovich SDE (2.31).

Corollary 2.5.1. *Assume that the hypotheses of Theorem 2.4.2 hold. Let $\mathbb{D}$ be a closed set in $\mathbb{R}^d$ such that (i) $\mathbb{D}$ has an outer normal at every point of the boundary $\partial\mathbb{D}$ and (ii) for any $x \in \partial\mathbb{D}$ we have*

$$\langle f(x), \nu_x \rangle \leq 0 \quad and \quad \langle \sigma_j(x), \nu_x \rangle = 0, \quad j = 1, \ldots, m, \qquad (2.47)$$

for every outer normal ν_x to $\mathbb{D}$ at x (see Definition 2.2.1), where $f = (f_1, \ldots, f_d)$ and $\sigma_j = (\sigma_{1j}, \ldots, \sigma_{dj})$. Then the set $\mathbb{D}$ is a deterministic forward invariant set for the Stratonovich SDE (2.31), i.e. the property $x(0) = x^* \in \mathbb{D}$ implies that $x(t, \omega; x^*) \in \mathbb{D}$ for all $t > 0$ and for almost all $\omega \in \Omega$. Moreover there exists a measurable set $\Omega^* \subset \Omega$ such that $\mathbb{P}(\Omega^*) = 1$, $\theta_t \Omega^* = \Omega^*$ for all $t \in \mathbb{R}$ and

$$\varphi(t, \omega)x \in \mathbb{D} \quad for\ all \quad t \in \mathbb{R}_+, \ x \in \mathbb{D}, \ \omega \in \Omega^*, \qquad (2.48)$$

where $\varphi(t, \omega)$ is the cocycle of RDS generated by (2.31).

Proof. It is sufficient to prove (2.48). Let Ω^* be the set of $\omega \in \Omega$ satisfying (2.46). Then it is clear from (2.46) that

$$\lim_{n \to 0} \mathbb{E} \int_0^\beta dt \sup_{|x^*| \leq R} |\varphi(t, \theta_{-t}\omega)x^* - \varphi^{\varepsilon_n}(t, \theta_{-t}\omega)x^*| = 0, \quad \omega \in \Omega^*,$$

for all positive β and R. Consequently for every $\omega \in \Omega^*$ and $x^* \in \mathbb{R}^d$ there exists a subsequence $\{\varepsilon_{n(k)}\}$ such that

$$\varphi^{\varepsilon_{n(k)}}(t, \theta_{-t}\omega)x^* \to \varphi(t, \theta_{-t}\omega)x^*, \quad k \to \infty, \qquad (2.49)$$

for almost all t from the interval $[0, \beta]$. If $x^* \in \mathbb{D}$, Theorem 2.2.1 implies that $\varphi^{\varepsilon_{n(k)}}(t, \theta_{-t}\omega)x^* \in \mathbb{D}$. Therefore it follows from (2.49) that $\varphi(t, \theta_{-t}\omega)x^* \in \mathbb{D}$ for almost all t from $[0, \beta]$. Since the function $t \mapsto \varphi(t, \theta_{-t}\omega)x$ is continuous for every $\omega \in \Omega$ and $x \in \mathbb{R}^d$ (see Remark 2.4.1), we have that $\varphi(t, \theta_{-t}\omega)x^* \in \mathbb{D}$ for all $t \in [0, \beta]$, $x^* \in \mathbb{D}$ and $\omega \in \Omega^*$ with arbitrary $\beta > 0$. Now the invariance of Ω^* implies (2.48). $\qquad \square$

Corollary 2.5.1 makes it possible to redefine the cocycle φ to obtain the following assertion.

Corollary 2.5.2. *Assume that the hypotheses of Corollary 2.5.1 hold. Then there exists a unique (up to indistinguishability) continuous C^1 RDS (θ, φ) over the metric dynamical system θ connected with the Wiener process W_t such that the conclusions of Theorem 2.4.3 are valid and $\varphi(t, \omega)\mathbb{D} \subset \mathbb{D}$ for all $t \in \mathbb{R}_+$ and $\omega \in \Omega$. This means that equations (2.31) generate a C^1 RDS in $\mathbb{D}$.*

Proof. Let Ω^* be the set given by Corollary 2.5.1. If we redefine the cocycle φ of the RDS generated by (2.31) (see Theorem 2.4.3) by the formula

$$\tilde{\varphi}(t,\omega) := \begin{cases} \varphi(t,\omega) & \text{if } \omega \in \Omega^*, \\ \text{id} & \text{if } \omega \notin \Omega, \end{cases} \tag{2.50}$$

then we obtain a cocycle which is indistinguishable from $\varphi(t,\omega)$. Obviously $\tilde{\varphi}(t,\omega)\mathbb{D} \subset \mathbb{D}$ for all $t \in \mathbb{R}_+$ and $\omega \in \Omega$ and the conclusions of Theorem 2.4.3 are valid for $\tilde{\varphi}(t,\omega)$. $\qquad\square$

Example 2.5.1 (1D Stochastic Equation). Let $f(x) \in C_b^{1,1}(\mathbb{R})$, $\sigma(x) \in C_b^{2,1}(\mathbb{R})$ and $\sigma(x) \cdot \sigma'(x) \in C_b^{1,1}(\mathbb{R})$. If $f(0) \geq 0$ and $\sigma(0) = 0$, then the equation

$$dx(t) = f(x(t))dt + \sigma(x(t)) \circ dW_t$$

generates an RDS in $\mathbb{R}_+$.

Example 2.5.2 (Binary Biochemical Model). If $g(x) \in C_b^{1,\delta}(\mathbb{R})$, $0 < \delta \leq 1$, and $g(0) \geq 0$, then equations (2.36) generate an RDS in $\mathbb{R}_+^2 = \{(x_1, x_2) : x_i \geq 0\}$.

Now we give a result by IMKELLER/SCHMALFUSS [59] (see also IMKELLER/LE-DERER [58]) on the conjugacy of stochastic and random differential equations.

To construct a random equation which is equivalent to the stochastic equation (2.31) we involve the stationary Ornstein-Uhlenbeck process $z(t,\omega) = (z_1(t,\omega), \ldots, z_m(t,\omega))$ in $\mathbb{R}^m$ which solves the equations

$$dz_k = -\mu z_k dt + dW_t^k, \quad k = 1, \ldots, m, \tag{2.51}$$

for some $\mu > 0$. For existence and properties of solutions to (2.51) we refer to IKEDA/WATANABE [57] or MCKEAN [82], for instance. The stationary solution $\{z_k(t,\omega)\}$ can be written in the form

$$z_k(t,\omega) = \int_{-\infty}^{t} e^{-\mu(t-\tau)}dW_t^k(\omega) \quad \text{almost surely}.$$

However to produce an RDE for *every* $\omega \in \Omega$ we need a perfect version of this process. The existence and properties of this version are stated in the following assertion which is a direct corollary of the infinite-dimensional result proved by CHUESHOV/SCHEUTZOW [23, Proposition 3.1].

Lemma 2.5.1. *On the probability space $(\Omega, \mathcal{F}, \mathbb{P})$ there exists a tempered random variable $z(\omega) = (z_1(\omega), \ldots, z_m(\omega))$ which maps Ω into $\mathbb{R}^m$ and possesses the properties:*

(i) $\{z_i(\omega)\}$ *are independent Gaussian variables with $\mathbb{E}z_i = 0$ and $\mathbb{E}z_i^2 = (2\mu)^{-1}$;*

(ii) $t \mapsto z(\theta_t\omega)$ *is continuous from $\mathbb{R}$ into $\mathbb{R}^d$ for all $\omega \in \Omega$;*

(iii) the process $z(t, \omega) \equiv z(\theta_t \omega)$ solves equations (2.51);
(iv) the relation

$$\lim_{t \to +\infty} \frac{1}{t} \int_0^t z_i(\theta_\tau \omega) \, d\tau = \lim_{t \to +\infty} \frac{1}{t} \int_{-t}^0 z_i(\theta_\tau \omega) \, d\tau = 0 \qquad (2.52)$$

holds for all $i = 1, \ldots, m$ and $\omega \in \Omega$.

Proof. The existence of the variable $z(\omega)$ with properties (i)–(iii) follows from CHUESHOV/SCHEUTZOW [23, Proposition 3.1]. To obtain (iv) we note that the ergodic theorem for stationary processes (see, e.g., GIHMAN/SKORO-HOD [48, p.140]) implies that (2.52) holds almost surely. Let Ω^* be the set of all $\omega \in \Omega$ such that (2.52) holds. It is clear that Ω^* is a θ-invariant set of full measure. Therefore we can redefine $z(\omega)$ by zero outside of Ω^*. □

Now we can state the conjugacy theorem (see IMKELLER/SCHMALFUSS [59] for the proof).

Theorem 2.5.2. *Let the hypotheses of Theorem 2.4.2 hold. Assume additionally that f_i and σ_{ij} belong to $C^\infty(\mathbb{R}^d)$ and the diffusion terms $\sigma_{ij}(x)$ in (2.31) satisfy*

$$[\sigma_k, \sigma_l]_j \equiv \sum_{i=1}^d \left(\sigma_{ik}(x) \frac{\partial \sigma_{jl}(x)}{\partial x_i} - \sigma_{il}(x) \frac{\partial \sigma_{jk}(x)}{\partial x_i} \right) = 0 \qquad (2.53)$$

for all $k, l = 1, \ldots, m$ and $j = 1, \ldots, d$. Let $u = (u_1, \ldots, u_d) : \mathbb{R}^m \times \mathbb{R}^d \mapsto \mathbb{R}^d$ be a solution to the equations

$$\frac{\partial u_j(z, x)}{\partial z_i} = \sigma_{ji}(u(z, x)), \quad u(0, x) = x \in \mathbb{R}^d, \ i = 1, \ldots, m, \ j = 1, \ldots, d \, , \tag{2.54}$$

and let $z(\omega)$ be the random variable given by Lemma 2.5.1. Then $u(z(\omega), x)$ is a tempered random variable in $\mathbb{R}^d$ for every $x \in \mathbb{R}^d$ and the mapping $x \mapsto T(\omega, x) \equiv u(z(\omega), x)$ is a diffeomorphism of $\mathbb{R}^d$ for each $\omega \in \Omega$. Further, if (θ, φ) is the RDS generated by (2.31), then

$$\varphi(t, \omega, x) = T(\theta_t \omega, \psi(t, \omega, T^{-1}(\omega, x))), \quad t > 0, \ x \in \mathbb{R}^d, \ \omega \in \Omega \, ,$$

where ψ is the cocycle corresponding to the random equation

$$\frac{dy}{dt} = [D_x T(\theta_t \omega, y)]^{-1} \left(f(T(\theta_t \omega, y)) + \mu \sum_{j=1}^m \sigma_j(T(\theta_t \omega, y)) z_j(\theta_t \omega) \right) \, .$$

Example 2.5.3. Let us assume that $m = d$ and $\sigma_{ij}(x) = \delta_{ij} \cdot \sigma_i(x_i)$, i.e. we consider the following system of Stratonovich SDEs

$$dx_i = f_i(x_1, \dots, x_d)dt + \sigma_i(x_i) \circ dW_t^i, \quad i = 1, \dots, d . \tag{2.55}$$

Simple calculation shows that condition (2.53) is satisfied and the equations in (2.54) have the form

$$\frac{\partial u_j}{\partial z_i} = 0, \text{ if } i \neq j \text{ and } \frac{\partial u_i}{\partial z_i} = \sigma_i(u_i); \; u_i(0, x) = x_i . \tag{2.56}$$

Here $i, j = 1, \dots, d$. Suppose that $|\sigma_i(u)| > 0$ for all $u \in \mathbb{R}$ and $i = 1, \dots, d$. Then it is easy to see that

$$u_i(z, x) = H_i^{-1}(z_i + H_i(x_i)), \quad i = 1, \dots, d ,$$

where $H_i(u)$ is a primitive for $\frac{1}{\sigma_i(u)}$. In this case

$$T(\omega, x) = \left(H_1^{-1}(z_1(\omega) + H_1(x_1)), \dots, H_d^{-1}(z_d(\omega) + H_d(x_d)) \right) ,$$

where every $z_i(\omega)$ generates the stationary solution to the Ornstein-Uhlenbeck equation $dz_i + \mu z_i dt = dW_t^i$ via the formula $z_i(\theta_t \omega)$. In this case the cocycle ψ is generated by the random equation

$$\dot{y}_i = \sigma_i(y_i)g_i(\theta_t \omega, y_1, \dots, y_d), \quad i = 1, \dots, d ,$$

where

$$g_i(\omega, y) = \frac{f_i\left(H_1^{-1}(z_1(\omega) + H_1(y_1)), \dots, H_d^{-1}(z_d(\omega) + H_d(y_d)) \right)}{\sigma_i(H_i^{-1}(z_i(\omega) + H_i(y_i)))} + \mu z_i(\omega)$$

for $i = 1, \dots, d$.

Equations (2.56) can be also solved in the case when $\sigma_i(u) = \sigma_i \cdot u$. In this case we have $u_i(z, x) = x_i \exp\{\sigma_i z_i\}$ for $i = 1, \dots, d$. In particular, this observation means that the SDE (2.36) which arises in a stochastic binary biochemical model is conjugate with the RDE

$$\dot{y}_1 = e_1(\theta_t \omega)^{-1} g(e_2(\theta_t \omega)y_2) - (\alpha_1 - \mu \sigma_1 z_1(\theta_t \omega))y_1 ,$$

$$\dot{y}_2 = e_2(\theta_t \omega)^{-1} e_1(\theta_t \omega)y_1 - (\alpha_2 - \mu \sigma_2 z_2(\theta_t \omega))y_2 ,$$

where $e_i(\omega) = \exp\{\sigma_i z_i(\omega)\}$ for $i = 1, 2$.

Other examples of the conjugacy of SDE and RDE can be found in KELLER/SCHMALFUSS [63], IMKELLER/SCHMALFUSS [59] and in Chap. 6 below.

3. Order-Preserving Random Dynamical Systems

In this chapter we first consider properties of partially ordered Banach spaces and prove some auxiliary results concerning random sets in these spaces. In Sect. 3.3 we introduce a general concept of order-preserving (monotone) random dynamical systems and consider several examples. We also define sub- and super-equilibria and prove a theorem on the existence of an equilibrium between them. On the one hand this theorem is a random version of the well-known deterministic assertion (see, e.g., HIRSCH [54] and SMITH [102]). On the other hand it generalizes statements on the existence of random fixed points in comparison with the theorems presented in SCHMALFUSS [94] and ARNOLD/SCHMALFUSS [11]. We consider the simplest examples and we give a counterexample that shows that ω-limit sets for monotone random dynamical systems can contain a non-trivial ordering subset of elements. This phenomenon does not take place in the deterministic case. We also prove that a global attractor of a monotone random dynamical system must be between two equilibria. We conclude this chapter with discussion of a comparison principle for order-preserving RDS in Sect. 3.7.

3.1 Partially Ordered Banach Spaces

In this section we describe some well-known results concerning cones and partially ordered spaces. We mainly follow KRASNOSELSKII [68] and KRASNOSELSKII/LIFSHITS/SOBOLEV [71].

Let V be a real Banach space with a closed convex cone $V_+ \subset V$ such that $V_+ \cap (-V_+) = \{0\}$. This cone defines a partial order relation on V via $x \le y$ if $y - x \in V_+$ which is compatible with the vector space structure of V. We write $x < y$ when $x \le y$ and $x \ne y$. If V_+ has nonempty interior $\text{int}V_+$ we say that the cone V_+ is *solid* and V is *strongly ordered*. We write $x \ll y$ if $y - x \in \text{int}V_+$. For any elements a and b from V such that $a \le b$ we define the (conic) *interval* $[a, b]$ as the set of the form

$$[a, b] = \{x \in V : a \le x \le b\} .$$

If the cone V_+ is solid, then any bounded set $B \subset V$ is contained in some interval.

The cones of nonnegative elements in $\mathbb{R}^d$ and in $C(\mathbb{D})$, where $\mathbb{D}$ is compact in $\mathbb{R}^d$, are solid. This cone in $L^p(\mathbb{D})$ is not solid.

Definition 3.1.1 (Upper and Lower Bounds). *An element $v \in V$ is said to be a* upper bound *for a subset $B \subset V$ if $x \leq v$ for each $x \in B$. Similarly, $u \in V$ is called a* lower bound *for a subset $B \subset V$ if $x \geq u$ for each $x \in B$. An upper bound v_0 is said to be the* least upper bound *(or* supremum*) and denoted by $v_0 = \sup B$, if any other upper bound v satisfies $v \geq v_0$. Similarly, a lower bound u_0 is said to be the* greatest lower bound *(or* infimum*) and denoted by $u_0 = \inf B$, if $u_0 \geq u$ for any other lower bound u. If the set B has an upper bound, it is said to be* bounded from above. *If it has a lower bound, it is said to be* bounded from below. *Finally, a set which is bounded from both above and below is said to be* order-bounded.

We note that suprema and infima, if they exist, are unique. Simple examples on the plane $\mathbb{R}^2$ show that $\sup B$ and $\inf B$, if they exist, do not belong to the closure $\overline{B}$ of B in general.

Definition 3.1.2 (Maximal and Minimal Elements). *An element $v \in B$ is said to be* maximal (minimal) *in B if the property $x \geq v$ $(x \leq v)$ for some $x \in B$ implies that $x = v$.*

A maximal element need not be an upper bound, nor a minimal element a lower bound.

Definition 3.1.3 (u-norm). *Let $u \in V_+$. An element $x \in V$ is said to be* u-subordinate *if we have the inequality $-\alpha u \leq x \leq \alpha u$ for some $\alpha \geq 0$. The smallest such α is denoted by $\|x\|_u$ and called the* u-norm *of x.*

It is easy to see that the functional $\|x\|_u$ is a norm on the linear set V_u of all u-subordinate elements from V. If the interval $[-u, u]$ is bounded in the norm of the space V, then $\|x\| \leq R\|x\|_u$ for any $x \in V_u$, where R is the radius of a ball containing $[-u, u]$. Moreover the space V_u is complete with respect to the u-norm if and only if the interval $[-u, u]$ is bounded in the norm of V.

Definition 3.1.4 (Part (Birkhoff) Metric). *(i) The equivalence classes under the equivalence relation defined by $x \sim y$ if there exists $\alpha > 0$ such that $\alpha^{-1}x \leq y \leq \alpha x$ on the cone V_+ are called the* parts *of V_+.*
(ii) Let C be a part of V_+. Then

$$p(x, y) := \inf\{\log \alpha : \alpha^{-1}x \leq y \leq \alpha x\}, \quad x, y \in C,$$

defines a metric on C called the part metric *(or* Birkhoff metric*) of C.*

Clearly intV_+ is a part and every part is also a convex cone in V. For a proof of the fact that p is a metric on C and for other properties of the part metric we refer to KRASNOSELSKII/LIFSHITS/SOBOLEV [71] and BAUER/BEAR [14]. The concept of the part metric plays an important role in the study of sublinear RDS. In Sect.4.1 we prove that these RDS are nonexpansive with respect to p.

Let $u \in V_+ \setminus \{0\}$ and C_u be the part of the cone which contains u. Then it easy to prove the following relations between the part metric and the u-norm

$$\|x - y\|_u \leq \left(e^{p(x,y)} - 1 \right) \cdot \max \left\{ e^{p(x,u)}, e^{p(y,u)} \right\}, \quad x, y \in C_u, \tag{3.1}$$

and

$$p(x, y) \leq \log \left\{ 1 + \|x - y\|_u \cdot (\|u\|_x + \|u\|_y) \right\}, \quad x, y \in C_u. \tag{3.2}$$

Therefore C_u is a complete space with respect to the part metric provided the interval $[-u, u]$ is bounded in the norm of V. If the cone V_+ is solid, then

$$p(x, y) \leq \log \left\{ 1 + r^{-1} \cdot \|x - y\| \right\}, \quad x, y \in \mathrm{int} V_+, \tag{3.3}$$

where $r = \min \{ \mathrm{dist}(x, \partial V_+), \mathrm{dist}(y, \partial V_+) \}$ (see KRASNOSELSKII/BURD/KO-LESOV [70, p.136] or KRAUSE/NUSSBAUM [72, Lemma 2.3]). We also note that in $\mathbb{R}^d$ with the standard cone

$$\mathbb{R}^d_+ = \{ x = (x_1, \dots, x_d) \in \mathbb{R}^d : x_i \geq 0, \ i = 1, \dots, d \}$$

every set

$$\{ x = (x_1, \dots, x_d) \in \mathbb{R}^d_+ : x_i > 0, \ i \in I; \ x_i = 0, \ i \notin I; \ I \subset \{1, \dots, d\} \}$$

is a part and the part metric in $\mathrm{int}\mathbb{R}^d_+$ has the form

$$p(x, y) = \log \max \left\{ \frac{x_i}{y_i}, \frac{y_i}{x_i} : i = 1, \dots, d \right\} = \max_i \left| \log \frac{x_i}{y_i} \right|. \tag{3.4}$$

In particular this formula shows that the part metric is not strictly convex, i.e. for some points a and b from $\mathrm{int}\mathbb{R}^d_+$ the set

$$\{ x : p(a, x) \leq \alpha p(a, b) \} \cap \{ x : p(b, x) \leq (1 - \alpha) p(a, b) \}, \quad 0 < \alpha < 1,$$

may consist of more than one point.

Below we also use the following monotonicity property of the part metric.

Lemma 3.1.1. *Assume that $a \leq b$ are elements from a part C of the cone V_+. Then $[a, b] \subset C$ and $p(v_1, v_2) \leq p(a, b)$ for any $v_1, v_2 \in [a, b]$.*

Proof. The relation $a \leq b \leq \lambda a$ with $\lambda \geq 1$ implies that $v_1 \leq b \leq \lambda a \leq \lambda v_2$ and $v_2 \leq b \leq \lambda a \leq \lambda v_1$ for any $v_1, v_2 \in [a, b]$. Thus $\lambda^{-1} v_1 \leq v_2 \leq \lambda v_1$ and therefore $p(v_1, v_2) \leq p(a, b)$. $\square$

The following concept of a normal cone is also important in applications.

Definition 3.1.5 (Normal Cone). *Let V be a real Banach space. A cone V_+ is said to be* normal *if the norm $\| \cdot \|$ in V is semi-monotone, i.e., there exist a constant c such that the property $0 \leq x \leq y$ implies that $\|x\| \leq c \cdot \|y\|$.*

The following proposition (see, e.g., KRASNOSELSKII/LIFSHITS/SOBOLEV [71]) characterizes the normality property of cones.

Proposition 3.1.1. *The cone V_+ is normal if and only if one of the following assertions is valid:*

(i)   *the relations $u_n \leq x_n \leq v_n$ for all $n \in \mathbb{Z}_+$ and convergences $u_n \to z$ and $v_n \to z$ imply that $x_n \to z$;*

(ii) *the original norm $\| \cdot \|$ in V is equivalent to the norm*

$$\|x\|_* = \max\left\{ \inf_{u \leq x} \|u\|, \ \inf_{u \geq x} \|u\| \right\} ; \tag{3.5}$$

(iii) *every interval $[a, b] = \{x \in V : a \leq x \leq b\}$ is bounded in the norm of V;*

(iv) *for any $u \in V_+$ the space V_u of all u-subordinate elements from V is complete with respect to u-norm;*

(iv) *every part of the cone V_+ is a complete space with respect to the part metric.*

Remark 3.1.1. Let V_+ be a normal cone. Then it is easy to see that the norm $\| \cdot \|_*$ defined by (3.5) is monotone, i.e. the relation $0 \leq x \leq y$ implies that $\|x\|_* \leq \|y\|_*$. For any monotone norm $\| \cdot \|$ we have the inequality

$$\|x - y\| \leq \left(2e^{p(x,y)} - e^{-p(x,y)} - 1 \right) \cdot \min\left\{ \|x\|, \|y\| \right\}, \quad x, y \in V_+ \setminus \{0\},$$

where $p(x, y)$ is the part metric and we suppose $p(x, y) = \infty$ if x and y belong to different parts (the proof see in KRAUSE/NUSSBAUM [72]).

In general the monotonicity and boundedness of a sequence does not imply its convergence. Counterexamples can be easily constructed in the space $C[0, 1]$ of continuous functions on the interval $[0, 1] \subset \mathbb{R}$ with the cone of nonnegative functions. Therefore the following concept is useful in applications.

Definition 3.1.6 (Regular Cone). *Let V be a real Banach space. A cone V_+ is said to be* regular *if every monotone sequence*

$$x_1 \leq x_2 \leq \ldots \leq x_n \leq \ldots$$

which is bounded from above, converges in the norm of the space V.

Every cone in $\mathbb{R}^d$ is regular. The cone of nonnegative functions in $L^p(Q)$ is regular, $1 \leq p < \infty$. The same is true for the space ℓ^p of sequences $\{a_i\}_{i=1}^{\infty}$ of real numbers with the property $\sum |a_i|^p < \infty$, $1 \leq p < \infty$. One can prove (see, e.g., KRASNOSELSKII/LIFSHITS/SOBOLEV [71]) that every regular cone is normal.

Definition 3.1.7 (Minihedral Cone). *A cone V_+ is said to be* minihedral *if every finite set M in V which is order-bounded has a supremum. A cone V_+ is called* strongly minihedral *if every set M in V which is order-bounded has a supremum.*

Minihedrality means that the nonempty intersection of any finite number of sets $x_i + V_+$ has the form $u + V_+$. The cones of nonnegative elements in $\mathbb{R}^d$ and in $C(\mathbb{D})$, where $\mathbb{D}$ is compact in $\mathbb{R}^d$, are minihedral. Moreover they are solid and normal. The cones of nonnegative elements in $L^p(\mathbb{D})$ is strongly minihedral. This cone in $C^1(\mathbb{D})$ is not minihedral.

The following theorem shows a relation between minihedrality and regularity of the cone.

Theorem 3.1.1. *A solid minihedral cone cannot be regular in an infinite-dimensional space. In a separable Banach space every regular minihedral cone is strongly minihedral.*

Below we use the following result on the existence of suprema for compact sets.

Theorem 3.1.2. *Let V_+ be a solid normal minihedral cone in the real Banach space V. Then every compact set $B \subset V$ has a supremum.*

The next assertion on the representation of spaces with cones is also important in our subsequent considerations.

Theorem 3.1.3. *Let V_+ be a solid normal minihedral cone in the real Banach space V. Then there exists a linear homeomorphism of V onto the space $C(Q)$ of continuous functions on some compact topological Hausdorff space such that the image of the cone V_+ is the cone of nonnegative functions in $C(Q)$.*

For the proofs of Theorems 3.1.1, 3.1.2 and 3.1.3 and comments we refer to the monograph KRASNOSELSKII/LIFSHITS/SOBOLEV [71].

Theorem 3.1.3 allow us to establish the following properties of normal solid minihedral cones.

Corollary 3.1.1. *Let V be a real Banach space with a solid normal minihedral cone V_+. Suppose that Φ is the linear homeomorphism of V onto the space $C(Q)$ given by Theorem 3.1.3. Then*

(i) for any $u, v \in V$ the supremum $u \vee v := \sup\{u, v\}$ and the infimum $u \wedge v := \inf\{u, v\}$ exist and

$$\Phi(u \vee v) = \Phi(u) \vee \Phi(v), \quad \Phi(u \wedge v) = \Phi(u) \wedge \Phi(v) ;$$

(ii) the functions $\{u, v\} \mapsto u \vee v$ and $\{u, v\} \mapsto u \wedge v$ are continuous mappings from $V \times V$ into V.

Proof. (i) It is easy to check that the elements

$$z_{\text{sup}} = \Phi^{-1}(\Phi(u) \vee \Phi(v)) \quad \text{and} \quad z_{\text{inf}} = \Phi^{-1}(\Phi(u) \wedge \Phi(v))$$

are the supremum and infimum for the pair u and v.

(ii) Simple calculation gives that

$$u \vee v = \frac{1}{2} \cdot (u + v + (u - v)_+ + (u - v)_-) \tag{3.6}$$

and

$$u \wedge v = \frac{1}{2} \cdot (u + v - (u - v)_+ - (u - v)_-), \tag{3.7}$$

where $w_+ = w \vee 0$ and $w_- = -(w \wedge 0) = (-w)_+$. Therefore it is sufficient to prove that $w \mapsto w_+$ is a continuous mapping from V into V. We obviously have

$$\|w_+ - w_+^*\| \leq C_1 \|\Phi(w_+) - \Phi(w_+^*)\|_{C(Q)}$$

$$\tag{3.8}$$

$$\leq C_1 \|\Phi(w) - \Phi(w^*)\|_{C(Q)} \leq C_1 C_2 \|w - w^*\|$$

for any w and w^* from V, where C_1 and C_2 are the operator norms of Φ^{-1} and Φ. $\square$

3.2 Random Sets in Partially Ordered Spaces

In this section we assume that V is a real separable Banach space with a closed convex cone V_+ and establish several properties of random sets in V. Below a mapping $\omega \mapsto v(\omega)$ from Ω into V is called a random variable in V if it is measurable with respect to the Borel σ-algebra $\mathcal{B}(V)$ generated by the open sets of V. We also note that the sets V_+, $V_+ \setminus \{0\}$ and $\text{int}V_+$ are elements of $\mathcal{B}(V)$.

Proposition 3.2.1. *Let $a(\omega)$ and $b(\omega)$ be random variables in V. Then the semiintervals*

$$\mathcal{I}_a(\omega) = \{x : x \geq a(\omega)\} \equiv a(\omega) + V_+$$

and

$$\mathcal{I}^b(\omega) = \{x : x \leq b(\omega)\} \equiv b(\omega) - V_+$$

are random closed sets. If the cone V_+ is solid and $a(\omega) \ll b(\omega)$ for all $\omega \in \Omega$, then the interval

$$[a, b](\omega) = \{x : a(\omega) \leq x \leq b(\omega)\}$$

is a random closed set. If $a(\omega) \leq b(\omega)$ for all $\omega \in \Omega$, then $[a, b](\omega)$ is a random closed set provided that either V_+ is a solid normal minihedral cone or V is a finite-dimensional space.

Proof. Since $\text{dist}(y, \mathcal{I}_a(\omega)) = \text{dist}(-y + a(\omega), V_+)$ and $\text{dist}(z, V_+)$ is a continuous function with respect to z, we have that $\text{dist}(y, \mathcal{I}_a(\omega))$ is measurable for any $y \in V$. Therefore by Definition 1.3.1 the semiinterval $\mathcal{I}_a(\omega))$ is a random closed set. The same is true for $\mathcal{I}^b(\omega)$.

Since $[a, b] = a + [0, b - a]$, by Proposition 1.3.2 it is sufficient to consider the case $a(\omega) \equiv 0$.

We first assume that the cone V_+ is solid and $b(\omega) \gg 0$. In this case for any open set $U \subset V$ we have

$$U \cap \{x : 0 \ll x \le b(\omega)\} = (U \cap \text{int} V_+) \cap \mathcal{I}^b(\omega)$$

Hence by Proposition 1.3.1(i)

$$\{\omega : U \cap \{x : 0 \ll x \le b(\omega)\} \ne \emptyset\} \in \mathcal{F}$$

for all open sets $U \subset V$ and therefore $\overline{\{x : 0 \ll x \le b(\omega)\}}$ is a random set. By Proposition 1.3.1(ii) $[0, b](\omega) = \overline{\{x : 0 \ll x \le b(\omega)\}}$ is a random closed set.

Assume now that V_+ is a solid normal minihedral cone and $b(\omega) \ge 0$. Since $\mathcal{I}^b(\omega)$ is a random closed set, by Proposition 1.3.2 there exists a sequence $\{v_n(\omega)\} \subset \mathcal{I}^b(\omega)$ of random variables such that $\overline{\{v_n(\omega)\}} = \mathcal{I}^b(\omega)$ for every $\omega \in \Omega$. Let $v_n^+(\omega) = 0 \vee v_n(\omega)$. Corollary 3.1.1(ii) implies that $\overline{\{v_n^+(\omega)\}} = [0, b](\omega)$. Thus by Proposition 1.3.2 $[0, b](\omega)$ is a random closed set.

If V is finite-dimensional, then we use the representation

$$[0, b](\omega) = \bigcup_{n \in \mathbb{N}} D_n(\omega) \quad \text{with} \quad D_n(\omega) = \mathcal{I}^b(\omega) \cap \{x \in V_+ : \|x\| \le n\}$$

and Proposition 1.3.1(iv,v). $\qquad\square$

Proposition 3.2.2. *Let V_+ be a solid normal cone. Then a random set $\{D(\omega)\}$ is bounded if and only if there exists a random variable $v(\omega) \gg 0$ such that*

$$D(\omega) \subset [-v(\omega), v(\omega)] \quad \text{for all} \quad \omega \in \Omega . \tag{3.9}$$

The set $\{D(\omega)\}$ is tempered (see Definition 1.3.3) if and only if the random variable $v(\omega)$ in (3.9) is tempered.

Proof. Let $D(\omega)$ be bounded. Then there exists a random variable $r(\omega) \in \mathbb{R}_+$ such that

$$D(\omega) \subset \{x \in V : \|x\| \le r(\omega)\} \quad \text{for all} \quad \omega \in \Omega . \tag{3.10}$$

Since V_+ is solid, there exists $u \in \text{int} V_+$ such that the interval $[-u, u]$ contains the unit ball of V. Therefore we have (3.9) with $v(\omega) = r(\omega) \cdot u$. If $\{D(\omega)\}$ is tempered, then $r(\omega)$ is a tempered random variable. Therefore $v(\omega) = r(\omega) \cdot u$ is a tempered element in V_+.

Assume that (3.9) is valid with $v(\omega) \geq 0$. Then for any $x \in D(\omega)$ we have $0 \leq x + v(\omega) \leq 2v(\omega)$. Therefore the normality of the cone implies that

$$\|x\| \leq \|v(\omega)\| + \|x + v(\omega)\| \leq (1 + 2c)\|v(\omega)\|, \quad x \in D(\omega) .$$

Thus we have (3.10) with $r(\omega) = (1 + 2c)\|v(\omega)\|$. Here above c is the constant from Definition 3.1.5 and $r(\omega)$ is tempered if $v(\omega)$ is a tempered element in V_+.   □

Lemma 3.2.1. *Let $a(\omega)$ and $b(\omega)$ be random variables in V. If V_+ is a solid normal minihedral cone, then*

$$u(\omega) = a(\omega) \vee b(\omega) \quad and \quad v(\omega) = a(\omega) \wedge b(\omega)$$

are random variables in V. If we additionally assume that $a(\omega)$ and $b(\omega)$ are tempered random variables in V, then the same property is true for $u(\omega)$ and $v(\omega)$.

Proof. The first part follows easily from Corollary 3.1.1(ii). For the second part, due to (3.6) and (3.7) it is sufficient to prove that $w_+(\omega) = w(\omega) \vee 0$ is tempered for any tempered variable $w(\omega) \in V$. Using (3.8) with $w^* = 0$ we have $\|w_+(\omega)\| \leq C_1 C_2 \|w(\omega)\|$. Therefore the temperedness of $w(\omega)$ implies the temperedness of $w_+(\omega)$.   □

Theorem 3.2.1. *Assume that the cone V_+ is a solid normal minihedral cone. Let $\{D(\omega)\}$ be a random compact set in V. Then $\sup D(\omega)$ and $\inf D(\omega)$ are random variables in V. If we assume additionally that $D(\omega)$ is tempered, then $\sup D(\omega)$ and $\inf D(\omega)$ are tempered random elements in V.*

Proof. Since $\inf D = -\sup\{-D\}$, it is sufficient to consider $\sup D(\omega)$ only. Proposition 1.3.2 implies that there exists a sequence $\{v_n(\omega) : n \in \mathbb{N}\}$ of random variables in V such that

$$v_n(\omega) \in D(\omega) \quad and \quad D(\omega) = \overline{\{v_n(\omega), n \in \mathbb{N}\}} \quad for \ all \quad \omega \in \Omega.$$

Let

$$w_N(\omega) = \sup\{v_1(\omega), \ldots, v_N(\omega)\}, \quad N = 1, 2, \ldots .$$

Lemma 3.2.1 implies that $w_N(\omega)$ is a random variable in V for every $N = 1, 2, \ldots$. It is also clear that

$$w_1(\omega) \leq w_2(\omega) \leq \ldots \leq w_N(\omega) \leq \ldots \tag{3.11}$$

Let us prove that for every fixed $\omega \in \Omega$ the limit

$$\overline{w}(\omega) = \lim_{N \to \infty} w_N(\omega) \tag{3.12}$$

exists. Due to property (3.11) it is sufficient to prove that there exists a convergent subsequence $w_{N_k}(\omega)$. Since $D(\omega)$ is a compact set, for any $k \in \mathbb{N}$ there exists $N_k \in \mathbb{N}$ such that

$$D(\omega) \subset \cup_{n=1}^{N_k}(v_n(\omega) + 2^{-k} \cdot B), \quad k = 1, 2, \ldots , \tag{3.13}$$

where B is the unit ball of V. Since V_+ is solid, there exists an interval $[-u, u]$ which contains the ball B. It follows from (3.13) that for any $m \in \mathbb{N}$ there exists $n_m \leq N_k$ such that

$$v_m(\omega) \in v_{n_m}(\omega) + 2^{-k} \cdot B \subset [v_{n_m}(\omega) - 2^{-k}u, v_{n_m}(\omega) + 2^{-k}u]$$

and therefore

$$v_m(\omega) \leq v_{n_m}(\omega) + 2^{-k}u \leq w_{N_k} + 2^{-k}u, \quad m \in \mathbb{N}.$$

Consequently

$$w_{N_k}(\omega) \leq w_{N_{k+1}}(\omega) \leq w_{N_k}(\omega) + 2^{-k}u, \quad k \in \mathbb{N}.$$

The normality of the cone V_+ implies that

$$\|w_{N_{k+1}}(\omega) - w_{N_k}(\omega)\| \leq C \cdot 2^{-k}\|u\|, \quad k \in \mathbb{N}.$$

Therefore the sequence $\{w_{N_k}\}$ is a Cauchy sequence. This implies the existence of the limit in (3.12). It is clear that $\overline{w}(\omega) = \sup D(\omega)$. The temperedness of $\sup D(\omega)$ and $\inf D(\omega)$ for tempered $D(\omega)$ follows from Proposition 3.2.2. □

The following uniform randomization of the definition of the part of a cone will turn out to be useful.

Definition 3.2.1. *For every random variable $v : \Omega \to V_+$ we define the* part *C_v of v, or generated by v in the cone V_+ to be the collection of random variables $w : \Omega \to V_+$ possessing the property*

$$\alpha^{-1}v(\omega) \leq w(\omega) \leq \alpha v(\omega) \quad for \; all \; \; \omega \in \Omega$$

for some nonrandom number $\alpha \geq 1$.

Note that $w \in C_v$ if and only if there is a *deterministic r* such that $w(\omega) \in B_r(v(\omega))$ for all $\omega \in \Omega$, where $B_r(v)$ is the sphere in the part metric centered at v with radius r. Hence C_v consists of those random variables which can be included in some sphere around v with deterministic radius. Note that for any $w \in C_v$ we have $C_w = C_v$.

If α in the above definition were allowed to depend on ω, C_v would just be a part-valued random variable, assigning to each ω the part of $v(\omega)$.

Proposition 3.2.3. *Let V_+ be a normal cone and $v : \Omega \mapsto V_+ \setminus \{0\}$ be a random variable. Then the part C_v generated by v is a complete metric space with respect to the metric*

$$\varrho(u, w) = \sup_{\omega \in \Omega} p(u(\omega), w(\omega)), \quad u, w \in C_v , \tag{3.14}$$

where p is the part metric (see Definition 3.1.4).

Proof. We only need to prove the completeness of C_v with respect to the metric (3.14). Let $\{u_m(\omega)\}$ be a Cauchy sequence in C_v. Then $\varrho(u_m, v)$ is bounded and therefore there exists $\lambda > 1$ such that

$$\lambda^{-1} v(\omega) \leq u_m(\omega) \leq \lambda v(\omega) \quad \text{for all} \quad m \in \mathbb{Z}_+, \, \omega \in \Omega .$$

Using (3.1) we have

$$\lim_{n,m \to \infty} \sup_{\omega \in \Omega} \|u_n(\omega) - u_m(\omega)\|_{v(\omega)} = 0 ,$$

where $\| \cdot \|_v$ is the v-norm. Since the cone V_+ is normal, we have that $\|w\| \leq C_\lambda(\omega)\|w\|_{v(\omega)}$ for any $w \in [-\lambda v(\omega), \lambda v(\omega)]$. Therefore $\{u_m(\omega)\}$ is a Cauchy sequence in V for each $\omega \in \Omega$. Thus there exists a random variable $u(\omega)$ such that

$$\lim_{m \to \infty} \|u_n(\omega) - u(\omega)\| = 0 \quad \text{for all} \quad \omega \in \Omega \tag{3.15}$$

and

$$\lambda^{-1} v(\omega) \leq u(\omega) \leq \lambda v(\omega), \quad \omega \in \Omega .$$

In particular $u(\omega) \in C_v$. Since $\{u_m(\omega)\}$ is a Cauchy sequence in C_v, for any $\varepsilon > 0$ there exists N_ε such that

$$p(u_m(\omega), u_n(\omega)) \leq \varepsilon \quad \text{for all} \quad m, n \geq N_\varepsilon, \, \omega \in \Omega .$$

Therefore by Definition 3.1.4

$$e^{-2\varepsilon} u_m(\omega) \leq u_n(\omega) \leq e^{2\varepsilon} u_m(\omega) \quad \text{for all} \quad m, n \geq N_\varepsilon, \, \omega \in \Omega .$$

If we let $n \to \infty$, then using (3.15) we obtain the inequality

$$e^{-2\varepsilon} u_m(\omega) \leq u(\omega) \leq e^{2\varepsilon} u_m(\omega) \quad \text{for all} \quad m \geq N_\varepsilon, \, \omega \in \Omega .$$

This implies that $\varrho(u, u_m) \leq 2\varepsilon$ for $m \geq N_\varepsilon$. Thus $\varrho(u, u_m) \to 0$ as $m \to \infty$. $\quad \square$

Below we also need the following property of the part metric $p(u, v)$.

Proposition 3.2.4. *Let $u(\omega)$ and $v(\omega)$ be random variables in V_+ such that $p(u(\omega), v(\omega))$ exists for every $\omega \in \Omega$. Then the function $\omega \mapsto p(u(\omega), v(\omega))$ is a random variable.*

Proof. From Definition 3.1.4 we have

$$A_c := \{\omega : p(u(\omega), w(\omega)) < c\} = \{\omega : e^{-c}u(\omega) < v(\omega) < e^c u(\omega)\}$$

$$= \{\omega : v(\omega) - e^{-c}u(\omega) > 0\} \cap \{\omega : e^c u(\omega) - v(\omega) > 0\}$$

for every $c > 0$. Since $V_+ \setminus \{0\}$ is a Borel set, we have that $A_c \in \mathcal{F}$.   □

3.3 Definition of Order-Preserving RDS

Let X be a subset of a real separable Banach space V with a closed convex cone V_+.

Definition 3.3.1. *An RDS (θ, φ) with phase space X is said to be*
(i) order-preserving if

$$x \leq y \quad \text{implies} \quad \varphi(t, \omega)x \leq \varphi(t, \omega)y \quad \text{for all} \quad t \geq 0 \quad \text{and} \quad \omega \in \Omega;$$

(ii) strictly order-preserving if it is order-preserving and

$$x < y \quad \text{implies} \quad \varphi(t, \omega)x < \varphi(t, \omega)y \quad \text{for all} \quad t \geq 0 \quad \text{and} \quad \omega \in \Omega;$$

(iii) strongly order-preserving if it is order-preserving and

$$x < y \quad \text{implies} \quad \varphi(t, \omega)x \ll \varphi(t, \omega)y \quad \text{for all} \quad t \geq 0 \quad \text{and} \quad \omega \in \Omega.$$

We now give several simple examples. For more complicated examples of order-preserving RDS we refer to Chaps. 5 and 6 below.

Example 3.3.1 (Markov Chain). Let $(\Omega_0, \mathcal{F}_0, \mathbb{P}_0)$ be a probability space. Assume that $f(\alpha, x)$ is a measurable mapping from $\Omega_0 \times \mathbb{R}$ into $\mathbb{R}$ which is continuous and nondecreasing with respect to x for every fixed $\alpha \in \Omega_0$. Then the RDS (θ, φ) constructed in Example 1.2.1 with $X = \mathbb{R}$ is order-preserving. If $f(\alpha, x)$ is an increasing function for every α, then (θ, φ) is a strongly order-preserving RDS.

Example 3.3.2 (Kick Model). Let V be a Banach space with a cone V_+ and $g : V \mapsto V$ be an order-preserving (deterministic) mapping. Consider the RDS (θ, φ) generated by the difference equation

$$x_{n+1} = g(x_n) + \xi(\theta_{n+1}\omega), \quad n \in \mathbb{Z}_+ ,$$

over a metric dynamical system $\theta = (\Omega, \mathcal{F}, \mathbb{P}, \{\theta_n, n \in \mathbb{Z}\})$, where ξ is a random variable in V. It is clear that (θ, φ) is an order-preserving RDS. If $g(x)$ is a strictly (strongly) order-preserving mapping, then (θ, φ) is a strictly (strongly) order-preserving RDS.

Example 3.3.3 (1D Random Equation). Let $\theta = (\Omega, \mathcal{F}, \mathbb{P}, \{\theta_t, t \in \mathbb{R}\})$ be a metric dynamical system. Consider the random ordinary differential equation

$$\dot{x}(t) = f(\theta_t \omega, x(t)). \tag{3.16}$$

Assume that the function $f : \Omega \times \mathbb{R} \mapsto \mathbb{R}$ possesses properties (2.1), (2.2) and (2.10) which guarantee a well-posedness of the Cauchy problem for (3.16). Then Corollary 2.1.1 implies that equation (3.16) generates an RDS with state space $\mathbb{R}$. For any two solutions to equation (3.16) we obviously have the relation

$$x(t) - y(t) = (x(0) - y(0)) \cdot \exp\left\{ \int_0^t \xi(\tau, \omega) \, d\tau \right\},$$

where the function

$$\xi(t, \omega) = \frac{f(\theta_t \omega, x(t)) - f(\theta_t \omega, y(t))}{x(t) - y(t)}$$

is locally integrable with respect to t for every fixed $\omega \in \Omega$. Therefore the RDS generated by (3.16) is strongly order-preserving with respect to the cone $\mathbb{R}_+$.

Example 3.3.4 (1D Stochastic Equation). Let $\{W_t\}$ be the one-dimensional Wiener process (see Example 1.1.7 and Section 2.3), and take two scalar functions $h(x)$ and $\sigma(x)$ which belong to $C_b^{0,1}(\mathbb{R})$ (see Definition 2.4.1). Then Theorem 2.4.1 implies that the Itô stochastic differential equation

$$dx(t) = h(x(t))dt + \sigma(x(t))dW_t \tag{3.17}$$

generates an RDS in $\mathbb{R}$. Due to the comparison principle (see IKEDA/WATANABE [57]) we can assert that this RDS is strictly order-preserving with the cone $\mathbb{R}_+$. Of course, the same conclusion remains true if we understand the stochastic equation (3.17) in the Stratonovich sense and assume that $h(x) \in C_b^{1,\delta}(\mathbb{R})$, $\sigma(x) \in C_b^{2,\delta}(\mathbb{R})$ and $\sigma(x) \cdot \sigma'(x) \in C_b^{1,\delta}(\mathbb{R})$ for some $\delta \in (0,1]$ (see Theorem 2.4.3).

Example 3.3.5 (Binary Biochemical Model). It follows from results presented in Chaps. 5 and 6 that both random (2.12) and stochastic (2.36) equations generate order-preserving random dynamical systems in $\mathbb{R}_+^2$ provided that $g(0) \geq 0$ and $g'(x) \geq 0$ for $x > 0$.

Example 3.3.6 (Affine Order-Preserving RDS). Let V be a Banach space with a cone V_+. Let us consider an affine RDS (θ, φ) with state space V (see Definition 1.2.3). This system is an order-preserving RDS if and only if in the representation (cf. (1.4))

$$\varphi(t, \omega)x = \Phi(t, \omega)x + \psi(t, \omega), \quad x \in V, \ \omega \in \Omega,$$

the linear operator $\Phi(t, \omega)$ maps V_+ into itself for any $t \geq 0$ and $\omega \in \Omega$. We obtain particular cases of affine order-preserving RDS if in Examples 3.3.1 – 3.3.4 we additionally assume that $f(\alpha, x)$, $f(\omega, x)$, $g(x, h(x)$ and $\sigma(x)$ are linear functions with respect to x. We study properties of affine order-preserving RDS in detail in Sect. 4.6.

3.4 Sub-Equilibria and Super-Equilibria

The following concepts of sub- and super-equilibria turn out to be of prime importance for the study of order-preserving RDS. They are the stochastic analog of the corresponding deterministic notions (see, e.g., HIRSCH [54] and SMITH [102]).

Let (θ, φ) be an order-preserving RDS on a subset X of a real separable Banach space V with a closed convex cone V_+.

Definition 3.4.1. *A random variable $u : \Omega \mapsto X$ is said to be*
(i) a sub-equilibrium *if*

$$\varphi(t, \omega) u(\omega) \geq u(\theta_t \omega) \quad \text{for all} \quad t \geq 0 \quad \text{and all} \quad \omega \in \Omega ; \tag{3.18}$$

(ii) a super-equilibrium *if*

$$\varphi(t, \omega) u(\omega) \leq u(\theta_t \omega) \quad \text{for all} \quad t \geq 0 \quad \text{and all} \quad \omega \in \Omega . \tag{3.19}$$

It is clear that any equilibrium (see Definition 1.7.1) is both a sub- and super-equilibrium. Below we will refer to sub- and super-equilibria as semi-equilibria.

We note that inequality (3.18) is equivalent to

$$\varphi(t - s, \theta_s \omega) u(\theta_s \omega) \geq u(\theta_t \omega) \quad \text{for all} \quad t \geq s \quad \text{and} \quad \omega \in \Omega . \tag{3.20}$$

Indeed, if we let $s = 0$ in (3.20) we obtain (3.18). On the other hand (3.18) implies that

$$\varphi(\tau, \theta_s \omega) u(\theta_s \omega) \geq u(\theta_{\tau+s} \omega) \quad \text{for any} \quad \tau \geq 0 \quad \text{and} \quad s \in \mathbb{T} .$$

Therefore after substituting $\tau = t - s$ we have (3.20). In the same way the inequality (3.19) is equivalent to

$$\varphi(t - s, \theta_s \omega) u(\theta_s \omega) \leq u(\theta_t \omega) \quad \text{for all} \quad t \geq s \quad \text{and} \quad \omega \in \Omega . \tag{3.21}$$

Remark 3.4.1. Assume that (θ, φ) is an order-preserving RDS with state space $X = V$. Then for any sub-equilibrium $a(\omega)$ and for any super-equilibrium $b(\omega)$ the random semiintervals

$$\mathcal{I}_a(\omega) = \{x \,:\, x \geq a(\omega)\} \equiv a(\omega) + V_+$$

and

$$\mathcal{I}^b(\omega) = \{x \,:\, x \leq b(\omega)\} \equiv b(\omega) - V_+$$

are forward invariant (see Definition 1.3.4) random closed sets. This follows from Proposition 3.2.1, Definitions (3.18) and (3.19) and from the order-preserving property of φ. In particular, if $a(\omega)$ and $b(\omega)$ are sub- and super-equilibria such that $a(\omega) \leq b(\omega)$, then the random interval

$$[a, b](\omega) = \{x \,:\, a(\omega) \leq x \leq b(\omega)\} \tag{3.22}$$

is forward invariant. Moreover an interval of the type (3.22) is forward invariant if and only if $a(\omega)$ is a sub-equilibrium and $b(\omega)$ is a super-equilibrium. A similar fact is also valid for the semiintervals $\mathcal{I}_a$ and $\mathcal{I}^b$. However we emphasize that in general the interval (3.22) is not backward invariant even for the case when $a(\omega)$ and $b(\omega)$ are equilibria. Indeed, let us consider the linear mapping $T_\lambda : \mathbb{R}^2 \mapsto \mathbb{R}^2$ given by the formula

$$T_\lambda(x_1, x_2) = \left(\frac{1+\lambda}{2}x_1 + \frac{1-\lambda}{2}x_2, \frac{1-\lambda}{2}x_1 + \frac{1+\lambda}{2}x_2 \right)$$

with $\lambda \in (0, 1)$. This mapping is a contraction along the line $l = \{(s, -s) \,:\, s \in \mathbb{R}\}$ and it is order-preserving with respect to $\mathbb{R}^2_+$ with equilibria $a = (-1, -1)$ and $b = (1, 1)$. A simple calculation shows that any element $(\alpha, -\alpha)$ with $\alpha \in (\lambda, 1]$ belongs to $[a, b]$ and it does not belong to $T_\lambda[a, b]$.

Example 3.4.1 (Semi-equilibria for 1D Random Equation). Let (θ, φ) be the order-preserving RDS generated in $\mathbb{R}$ by random differential equation (3.16) under the conditions given in Example 3.3.3. Assume that there exists $a \in \mathbb{R}$ such that $f(\omega, a) \geq 0$ for all $\omega \in \Omega$. Then $a(\omega) \equiv a$ is a sub-equilibrium for (θ, φ). Indeed, Corollary 2.2.1 implies that $[a, +\infty)$ is a deterministic forward invariant set for the RDS (θ, φ). Thus (see Remark 3.4.1) $a(\omega) \equiv a$ is a sub-equilibrium. The same argument gives that $b(\omega) \equiv b$ is a super-equilibrium provided $b \in \mathbb{R}$ satisfies the inequality $f(\omega, b) \leq 0$ for all $\omega \in \Omega$.

Assume now that the RDS (θ, φ) generated by (3.16) has a (random) sub-equilibrium $u(\omega)$. Since

$$\varphi(t - s, \theta_s\omega)x = x + \int_s^t f(\theta_\tau\omega, \varphi(\tau - s, \theta_s\omega)x)\, d\tau$$

for any $t > s$ and $x \in \mathbb{R}$, it follows from (3.20) that

$$u(\theta_t\omega) \leq u(\theta_s\omega) + \int_s^t f(\theta_\tau\omega, \varphi(\tau - s, \theta_s\omega)u(\theta_s\omega))\, d\tau, \quad t > s, \ \omega \in \Omega \,.$$

If we assume that the mapping $(t, x) \mapsto f(\theta_t\omega, x)$ is continuous, then we obtain

$$D^+ u(\theta_t \omega) := \limsup_{h \to +0} \frac{u(\theta_{t+h}\omega) - u(\theta_t \omega)}{h} \leq f(\theta_t \omega, u(\theta_t \omega)), \quad t \in \mathbb{R}, \ \omega \in \Omega \ .$$

Thus the sub-equilibrium $u(\omega)$ is a random variable such that the stationary process $u(t) := u(\theta_t \omega)$ solves (in some sense) the differential inequality

$$\dot{u}(t) \leq f(\theta_t \omega, u(t)) \quad \text{for all} \quad t \in \mathbb{R}, \ \omega \in \Omega \ .$$

Similarly, if $u(\omega)$ is a super-equilibrium for (θ, φ), then the process $u(t) = u(\theta_t \omega)$ solves the inequality

$$\dot{u}(t) \geq f(\theta_t \omega, u(t)) \quad \text{for all} \quad t \in \mathbb{R}, \ \omega \in \Omega \ .$$

Moreover using the comparison principle (cf. Theorem 5.3.1 below) one can prove that a random variable $u(\omega)$ is a semi-equilibrium for (θ, φ) if and only if the process $u(t) = u(\theta_t \omega)$ solves one of these differential inequalities. In particular a number c is a sub-equilibrium (resp. super-equilibrium) if and only if $f(\omega, c) \geq 0$ (resp. $f(\omega, c) \leq 0$) for all $\omega \in \Omega$. Similar results remain true for order-preserving RDS generated by systems of random or stochastic differential equations.

Example 3.4.2 (Semi-equilibria for Binary Biochemical Model). Consider the RDS presented in Example 2.1.1. If $g(0) \geq 0$ and $g'(x) \geq 0$ for $x > 0$, then $a = 0$ is a sub-equilibrium (cf. Example 3.3.5). Assume that $\alpha_i(\omega) \geq \alpha_i^0 > 0$ for $i = 1, 2$ and $\omega \in \Omega$. If there exists $r > 0$ such that $g(r) - \alpha_1^0 \alpha_2^0 r \leq 0$, then $b = (\alpha_2^0 r, r)$ is a super-equilibrium. To prove these results we can use Corollary 2.2.1.

Example 3.4.3 (Semi-equilibria for 1D Stochastic Equation). Let (θ, φ) be the order-preserving RDS on $\mathbb{R}$ constructed in Example 3.3.4. Assume that there exists $a \in \mathbb{R}$ such that $h(a) \geq 0$ and $\sigma(a) = 0$. Then Corollary 2.5.1 implies that there exists a θ-invariant set $\Omega^* \subset \Omega$ of full measure such that

$$\varphi(t, \omega)[a, +\infty) \subset [a, +\infty) \quad \text{for} \quad \omega \in \Omega^*.$$

Therefore there exists a version of the cocycle φ (see Remark 1.2.1(ii)) such that $\varphi(t, \omega)a \geq a$ for all $\omega \in \Omega$ and $t \geq 0$. Thus $a(\omega) \equiv a$ is a sub-equilibrium for (θ, φ). The same argument gives that $b(\omega) \equiv b$ is a super-equilibrium provided $b \in \mathbb{R}$ satisfies the conditions $h(b) \leq 0$ and $\sigma(b) = 0$.

The following assertion contains some monotonicity properties of sub- and super-equilibria.

Proposition 3.4.1. *Let $a(\omega)$ be a sub-equilibrium of the order-preserving RDS (θ, φ) on $X \subset V$. Then*

$$a_s(\omega) := \varphi(s, \theta_{-s}\omega)a(\theta_{-s}\omega) \tag{3.23}$$

is a sub-equilibrium for any $s > 0$. These sub-equilibria possess the property

$$a_s(\omega) \geq a_\sigma(\omega) \geq a(\omega) \qquad (3.24)$$

for any $s \geq \sigma \geq 0$ and $\omega \in \Omega$. The same assertion is valid for super-equilibria with the reversed inequality signs in (3.24).

Proof. The cocycle property gives

$$\varphi(t, \omega)a_s(\omega) = \varphi(t, \omega)\varphi(s, \theta_{-s}\omega)a(\theta_{-s}\omega) = \varphi(t + s, \theta_{-s}\omega)a(\theta_{-s}\omega)$$

$$\qquad (3.25)$$

$$= \varphi(s, \theta_{t-s}\omega)\varphi(t, \theta_{-s}\omega)a(\theta_{-s}\omega) .$$

Using the inequality (cf. (3.18))

$$\varphi(t, \theta_{-s}\omega)a(\theta_{-s}\omega) \geq a(\theta_{t-s}\omega) \qquad (3.26)$$

and the order-preserving property we have

$$\varphi(s, \theta_{t-s}\omega)\varphi(t, \theta_{-s}\omega)a(\theta_{-s}\omega) \geq \varphi(s, \theta_{t-s}\omega)a(\theta_{t-s}\omega) = a_s(\theta_t\omega) .$$

Consequently (3.25) implies that $a_s(\omega)$ is a sub-equilibrium.

It follows from (3.26) with $t = s$ that $a_s(\omega) \geq a(\omega)$ for any $s \geq 0$ and $\omega \in \Omega$. Since $a_s(\omega)$ is a sub-equilibrium, the last inequality gives

$$(a_s)_\sigma(\omega) = \varphi(\sigma, \theta_{-\sigma}\omega)a_s(\theta_{-\sigma}\omega) \geq a_s(\omega) \quad \text{for any} \quad s, \sigma \geq 0 . \qquad (3.27)$$

From the definition of $a_s(\omega)$ we have

$$(a_s)_\sigma(\omega) = \varphi(\sigma, \theta_{-\sigma}\omega)a_s(\theta_{-\sigma}\omega) = \varphi(\sigma, \theta_{-\sigma}\omega)\varphi(s, \theta_{-s-\sigma}\omega)a(\theta_{-s-\sigma}\omega) .$$

Therefore the cocycle property gives

$$(a_s)_\sigma(\omega) = \varphi(s + \sigma, \theta_{-s-\sigma}\omega)a(\theta_{-s-\sigma}\omega) = a_{s+\sigma}(\omega), \quad s, \sigma \geq 0 . \qquad (3.28)$$

Consequently inequality (3.27) implies (3.24).

The assertion for super-equilibria can be proved in a similar way. $\qquad \square$

For minihedral cones we also have the following property of semi-equilibria.

Proposition 3.4.2. *Let the cone V_+ be minihedral. Then $\sup\{u_1(\omega), u_2(\omega)\}$ is a sub-equilibrium provided that $u_1(\omega)$ and $u_2(\omega)$ are sub-equilibria and also*

$$\sup\{u_1(\omega), u_2(\omega)\} \in X \quad \text{for all} \quad \omega \in \Omega .$$

Similarly, if $u_1(\omega)$ and $u_2(\omega)$ are super-equilibria and $\inf\{u_1(\omega), u_2(\omega)\} \in X$ for all $\omega \in \Omega$, then $\inf\{u_1(\omega), u_2(\omega)\}$ is a super-equilibrium.

Proof. For any two sub-equilibria $u_1(\omega)$ and $u_2(\omega)$ we have

$$\varphi(t,\omega)(\sup\{u_1(\omega),u_2(\omega)\}) \geq \varphi(t,\omega)u_j(\omega) \geq u_j(\theta_t\omega), \quad j = 1,2\,.$$

Therefore $\varphi(t,\omega)(\sup\{u_1(\omega),u_2(\omega)\}) \geq \sup\{u_1(\theta_t\omega),u_2(\theta_t\omega)\}$. The same argument applies for super-equilibria.   □

The following simple assertion on the existence of sub- and super-equilibria is useful in what follows.

Lemma 3.4.1. *Let V_+ be a normal solid minihedral cone. Assume that an order-preserving RDS (θ,φ) possesses a backward invariant random compact set $A(\omega) \subset X$, i.e. $\varphi(t,\omega)A(\omega) \supseteq A(\theta_t\omega)$. Then $b(\omega) := \sup A(\omega)$ is a sub-equilibrium and $a(\omega) := \inf A(\omega)$ is a super-equilibrium for the RDS (θ,φ) provided that $a(\omega)$ and $b(\omega)$ belong to X for all $\omega \in \Omega$.*

Proof. Since (θ,φ) is order-preserving, the equation

$$a(\omega) \leq w(\omega) \leq b(\omega) \quad \text{for all} \quad w(\omega) \in A(\omega)$$

implies that

$$\varphi(t,\omega)a(\omega) \leq \varphi(t,\omega)w(\omega) \leq \varphi(t,\omega)b(\omega) \quad \text{for all} \quad w(\omega) \in A(\omega)\,.$$

The invariance property $\varphi(t,\omega)A(\omega) \supseteq A(\theta_t\omega)$ now gives

$$\varphi(t,\theta_{-t}\omega)a(\theta_{-t}\omega) \leq w(\omega) \leq \varphi(t,\theta_{-t}\omega)b(\theta_{-t}\omega) \quad \text{for all} \quad w(\omega) \in A(\omega)\,.$$

Since $a(\omega) = \inf A(\omega)$ and $b(\omega) = \sup A(\omega)$, the last relation implies that

$$\varphi(t,\theta_{-t}\omega)a(\theta_{-t}\omega) \leq a(\omega) \quad \text{and} \quad \varphi(t,\theta_{-t}\omega)b(\theta_{-t}\omega) \geq b(\omega)$$

for all $t \geq 0$ and $\omega \in \Omega$.   □

Remark 3.4.2. (i) If the cone V_+ is normal, solid and miniedral and if for an RDS (θ,φ) on $X = V$ there exists a random element $x \in V$ such that the closure $\overline{\gamma_x^\tau}(\omega)$ of the orbit

$$\gamma_x^\tau(\omega) = \bigcup_{t \geq \tau} \varphi(t,\theta_{-t}\omega)x(\theta_{-t}\omega)$$

emanating from $\varphi(\tau,\theta_{-\tau}\omega)x(\theta_{-\tau}\omega)$ is a random compact set for some $\tau > 0$, then there exist a sub-equilibrium $b(\omega)$ and a super-equilibrium $a(\omega)$ for (θ,φ) such that $a(\omega) \leq b(\omega)$. In fact the compactness of $\overline{\gamma_x^\tau}(\omega)$ implies (see Sect.1.6) that the corresponding omega-limit set

$$\Gamma_x(\omega) = \bigcap_{t>0} \overline{\bigcup_{\tau \geq t} \varphi(\tau,\theta_{-\tau}\omega)x(\theta_{-\tau}\omega)}$$

is an invariant random compact set. Therefore we can apply Lemma 3.4.1 with $A(\omega) = \Gamma_x(\omega)$.

(ii) Assume in addition to the hypotheses of Lemma 3.4.1 that $A(\omega)$ is an invariant set and $b(\omega) = \sup A(\omega) \in A(\omega)$ (resp. $a(\omega) = \inf A(\omega) \in A(\omega)$) for all $\omega \in \Omega$. The second assumption is always true for one-dimensional RDS. Then it is easy to see that $b(\omega)$ (resp. $a(\omega)$) is an equilibrium. This property is not true in general without the assumption $\sup A(\omega) \in A(\omega)$. As an example we can consider the following two-dimensional discrete deterministic system with $X = \mathbb{R}_+^2$ and

$$\varphi(1; x_1, x_2) = \left(\sqrt{x_1} + x_1 x_2, \sqrt{x_2} + x_1 x_2\right).$$

It is clear that the set $A = \{(1,0), (0,1)\}$ is invariant and $\sup A = (1,1)$ is a strict sub-equilibrium.

3.5 Equilibria

The monotonicity properties of semi-equilibria given by Proposition 3.4.1 allow us to establish a result on the existence of equilibria.

Theorem 3.5.1. *Let (θ, φ) be an order-preserving RDS. Assume that there exist a sub-equilibrium $a(\omega)$ and a super-equilibrium $b(\omega)$ such that $a(\omega) \leq b(\omega)$ and the interval $[a, b](\omega)$ defined by (3.22) belongs to X. Assume also that the set $\phi(t_0, \omega)[a, b](\omega)$ is relatively compact in X for some $t_0 > 0$ and for every $\omega \in \Omega$. Then the limits*

$$\underline{u}(\omega) = \lim_{s \to +\infty} a_s(\omega) = \sup_{s>0} a_s(\omega) \tag{3.29}$$

and

$$\bar{u}(\omega) = \lim_{s \to +\infty} b_s(\omega) = \inf_{s>0} b_s(\omega) \tag{3.30}$$

exist, where $a_s(\omega)$ and $b_s(\omega)$ are defined as in (3.23). These limits are equilibria of (θ, φ) such that

$$a(\omega) \leq \underline{u}(\omega) \leq \bar{u}(\omega) \leq b(\omega). \tag{3.31}$$

Proof. Let us consider the discrete ($\mathbb{T} = \mathbb{Z}$) RDS $(\hat{\theta}, \hat{\varphi})$, where $\hat{\theta}_n = \theta_{nt_0}$ and $\hat{\varphi}(n, \omega) = \varphi(nt_0, \omega)$. It is clear that $a(\omega)$ and $b(\omega)$ are sub- and super-equilibria of $\hat{\varphi}$. Let $\hat{a}_n(\omega)$ and $\hat{b}_n(\omega)$ be defined for $(\hat{\theta}, \hat{\varphi})$ as in (3.23). From Proposition 3.4.1 one can see that

$$a(\omega) \leq \hat{a}_n(\omega) \leq \hat{a}_{n'}(\omega) \leq \hat{b}_{m'}(\omega) \leq \hat{b}_m(\omega) \leq b(\omega) \tag{3.32}$$

for any $n, n', m, m' \in \mathbb{Z}_+$ such that $n \leq n'$ and $m \leq m'$. From (3.28) we have

$$\hat{a}_{n+1}(\omega) = (\hat{a}_n)_1 (\omega) = \hat{\varphi}(1, \hat{\theta}_{-1}\omega)\hat{a}_n(\hat{\theta}_{-1}\omega) = \varphi(t_0, \theta_{-t_0}\omega)a_{nt_0}(\theta_{-t_0}\omega) .$$

Consequently $\hat{a}_{n+1}(\omega) \in \varphi(t_0, \theta_{-t_0}\omega)[a, b](\theta_{-t_0}\omega)$ for any $n \in \mathbb{Z}_+$. Thus the sequence $\{\hat{a}_n(\omega)\}$ is relatively compact for every ω. The monotonicity property (3.32) implies that this sequence has a unique limit point. Indeed, let us assume that there exist two points $v(\omega)$ and $w(\omega)$ such that

$$v(\omega) = \lim_{k \to \infty} \hat{a}_{n_k}(\omega), \quad \hat{a}_{n_k}(\omega) \leq v(\omega), \; k = 1, 2, \ldots$$

and

$$w(\omega) = \lim_{k \to \infty} \hat{a}_{m_k}(\omega), \quad \hat{a}_{m_k}(\omega) \leq w(\omega), \; k = 1, 2, \ldots$$

for some sequences $\{n_k\}$ and $\{m_k\}$. However for any k there exists l such that $n_k \leq m_l$ and, therefore, $\hat{a}_{n_k}(\omega) \leq \hat{a}_{m_l}(\omega) \leq w(\omega)$. Hence

$$v(\omega) = \lim_{k \to \infty} \hat{a}_{n_k}(\omega) \leq w(\omega) .$$

In the same way we have $w(\omega) \leq v(\omega)$. Hence $v(\omega) = w(\omega)$. Thus the limit

$$\underline{u}(\omega) = \lim_{n \to \infty} \hat{a}_n(\omega) \tag{3.33}$$

exists. Since for any $s \in \mathbb{T}_+$ there exists $n \in \mathbb{Z}_+$ such that $(n-1)t_0 < s \leq nt_0$ we have $\hat{a}_n(\omega) \leq a_s(\omega) \leq \hat{a}_{n+1}(\omega)$. Therefore (3.33) implies that the element $\underline{u}(\omega)$ possesses property (3.29).

It remains to prove that $\underline{u}(\omega)$ is an equilibrium. The continuity of the cocycle φ and the structure of $a_s(\omega)$ imply that

$$\varphi(t, \theta_{-t}\omega)\underline{u}(\theta_{-t}\omega) = \lim_{s \to \infty} \varphi(t, \theta_{-t}\omega)\phi(s, \theta_{-s-t}\omega)a(\theta_{-s-t}\omega) .$$

Therefore the cocycle property gives

$$\varphi(t, \theta_{-t}\omega)\underline{u}(\theta_{-t}\omega) = \lim_{s \to \infty} \varphi(t + s, \theta_{-s-t}\omega)a(\theta_{-s-t}\omega)$$

$$= \lim_{s \to \infty} a_{s+t}(\omega) = \underline{u}(\omega) .$$

This relation means that $\underline{u}(\omega)$ is an equilibrium. In the same way one can prove the existence of an equilibrium $\bar{u}(\omega)$ possessing property (3.30). The inequalities (3.32) imply (3.31). □

Remark 3.5.1. For regular cones (see Definition 3.1.6) Theorem 3.5.1 is valid without the assumption concerning the relative compactness of the set $\phi(t_0, \omega)[a, b](\omega)$.

Corollary 3.5.1. *Assume that the hypotheses of Theorem 3.5.1 hold. If $a(\omega)$ is measurable with respect to the past σ-algebra $\mathcal{F}_-$ (see the definition in Sect.1.10), then the equilibrium $\underline{u}(\omega)$ is also $\mathcal{F}_-$-measurable and the random Dirac measure $\delta_{\underline{u}(\omega)}$ is a disintegration of the invariant Markov measure μ on $(\Omega \times X, \mathcal{F} \times \mathcal{B}(X))$ which has the form*

$$\mu(\mathcal{A}) = \mathbb{P}\left\{\omega \ : \ (\omega, \underline{u}(\omega)) \in \mathcal{A}\right\}, \quad \mathcal{A} \in \mathcal{F} \times \mathcal{B}(X). \qquad (3.34)$$

The same is true concerning $b(\omega)$ and $\bar{u}(\omega)$.

Proof. Since

$$\mathcal{F}_- = \sigma\{\omega \mapsto \varphi(\tau, \theta_{-t}\omega) \ : \ 0 \leq \tau \leq t\},$$

the mapping $\omega \mapsto \varphi(s, \theta_{-s}\omega)x$ is $\mathcal{F}_-$-measurable for any $x \in X$. If $a(\omega)$ is $\mathcal{F}_-$-measurable, then $a(\theta_{-s}\omega)$ is also $\mathcal{F}_-$-measurable for $s \geq 0$. Therefore $a_s(\omega) = \varphi(s, \theta_{-s}\omega)a(\theta_{-s}\omega)$ possesses the same property and (3.29) implies that the equilibrium $\underline{u}(\omega)$ is $\mathcal{F}_-$-measurable. It is clear that the measure defined by (3.34) is invariant for the RDS (θ, φ) (see Example 1.10.1). Its disintegration μ_ω has the form $\mu_\omega(B) = \chi_B(\underline{u}(\omega))$, where $\chi_B(x) = 1$ for $x \in B$ and $\chi_B(x) = 0$ if $x \notin B$. Therefore $\omega \mapsto \mu_\omega(B)$ is $\mathcal{F}_-$-measurable for any $B \in \mathcal{B}$. Thus μ is a Markov measure by Definition 1.10.2.   □

Example 3.5.1 (Equilibria for 1D Random Equations). Let (θ, φ) be the order-preserving RDS on $\mathbb{R}$ constructed in Example 3.3.3. Assume that there exist $a < b$ such that $f(\omega, a) \geq 0$ and $f(\omega, b) \leq 0$ for all $\omega \in \Omega$. Then $a(\omega) \equiv a$ is a sub-equilibrium and $b(\omega) \equiv b$ is a super-equilibrium for (θ, φ) (see Example 3.4.1). Thus Theorem 3.5.1 implies the existence of equilibria $\underline{u}(\omega)$ and $\bar{u}(\omega)$ such that

$$a \leq \underline{u}(\omega) \leq \bar{u}(\omega) \leq b.$$

By Corollary 3.5.1 these equilibria are $\mathcal{F}_-$-measurable and therefore they generate invariant Markov measures for the RDS (θ, φ) connected with equation (3.16).

Example 3.5.2 (Equilibria for Binary Biochemical Model). Consider the situation described in Example 3.4.2. Theorem 3.5.1 and Corollary 3.5.1 imply that the RDS (θ, φ) generated in $\mathbb{R}_+^2$ by equations (2.12) possesses $\mathcal{F}_-$-measurable equilibria $\underline{u}(\omega)$ and $\bar{u}(\omega)$ such that

$$0 \leq \underline{u}(\omega) \leq \bar{u}(\omega) \leq (\alpha_2^0 r, r), \quad \omega \in \Omega.$$

It is clear that $\underline{u}(\omega) > 0$ provided that $g(0) > 0$.

The following assertion shows that for a one-dimensional order-preserving RDS every ergodic invariant measure is generated by some equilibrium.

Proposition 3.5.1. *Let (θ, φ) be an order-preserving RDS whose state space $X = [c_1, c_2]$ is an interval (which need not be finite) in $\mathbb{R}$ and let μ be an invariant measure for (θ, φ). Assume that μ possesses a disintegration μ_ω such that $\varphi(t, \omega)\mu_\omega = \mu_{\theta_t \omega}$ for all $t \geq 0$ and $\omega \in \Omega$ (cf. Remark 1.10.1). Then (θ, φ) has at least one equilibrium $u(\omega) \in X$ with*

$$\mu_\omega\{(c_1, u(\omega)]\} \geq 1/2 \quad \text{and} \quad \mu_\omega\{[u(\omega), c_2)\} \geq 1/2 . \tag{3.35}$$

If (θ, φ) is strictly order-preserving and μ is a φ-ergodic measure, then its disintegration μ_ω is a random Dirac measure, $\mu_\omega = \delta_{u(\omega)}$, where $u(\omega)$ is an equilibrium.

Proof. The main idea of the proof is due to Hans Crauel (see ARNOLD [3, Sect.1.8], where the second assertion of this proposition is proved for one-dimensional RDS with continuous two-sided time). For the sake of simplicity we consider the case $X = \mathbb{R}$ only. The proofs for other cases are similar.

We start with some preliminary observations. Let ν be a probability measure on $\mathbb{R}$ and

$$I_- = \{a : \nu\{(-\infty, a]\} \geq 1/2\} \quad \text{and} \quad I_+ = \{b : \nu\{[b, \infty)\} \geq 1/2\} .$$

Since $g_-(x) := \nu\{(-\infty, x]\}$ is a right continuous function and $g_+(x) := \nu\{[x, \infty)\}$ is a left continuous function, the sets I_- and I_+ have the form $I_- = [\alpha, \infty)$ and $I_+ = (-\infty, \beta]$ for some $\alpha, \beta \in \mathbb{R}$. It is also easy to see that $I_- \cap I_+ \neq \emptyset$, i.e. $\beta \geq \alpha$. Thus for any probability measure ν on $\mathbb{R}$ we have the relation

$$\nu\{(-\infty, x]\} \geq 1/2 \quad \text{and} \quad \nu\{[x, \infty)\} \geq 1/2$$

for all $x \in [\alpha, \beta]$, where

$$\alpha = \min\{a : \nu\{(-\infty, a]\} \geq 1/2\} \quad \text{and} \quad \beta = \max\{b : \nu\{[b, \infty)\} \geq 1/2\} .$$

Let μ_ω be the disintegration of μ and

$$\alpha(\omega) = \min\{a : \mu_\omega\{(-\infty, a]\} \geq 1/2\} ,$$

$$\beta(\omega) = \max\{b : \mu_\omega\{[b, \infty)\} \geq 1/2\} .$$

Since

$$\{\omega : \alpha(\omega) > c\} = \{\omega : \mu_\omega\{(-\infty, c]\} < 1/2\}$$

and

$$\{\omega : \beta(\omega) < c\} = \{\omega : \mu_\omega\{[c, \infty)\} < 1/2\} ,$$

the values $\alpha(\omega)$ and $\beta(\omega)$ are random variables. Since φ is order-preserving, we have

$$(-\infty, x] \subseteq \varphi(t, \omega)^{-1}(-\infty, \varphi(t, \omega)x] \ .$$

Thus by the invariance of μ_ω, we obtain the relation

$$\mu_\omega\{(-\infty, x]\} \leq \mu_\omega\{\varphi(t, \omega)^{-1}(-\infty, \varphi(t, \omega)x]\} = \mu_{\theta_t\omega}\{(-\infty, \varphi(t, \omega)x]\} \ .$$

Similarly,

$$\mu_\omega\{[x, \infty)\} \leq \mu_{\theta_t\omega}\{[\varphi(t, \omega)x, \infty)\} \ .$$

These properties imply that $\alpha(\omega)$ is a sub-equilibrium and $\beta(\omega)$ is a super-equilibrium for (θ, φ) such that $\alpha(\omega) \leq \beta(\omega)$. Therefore we can apply Theorem 3.5.1 to conclude that there exist at least one equilibrium $u(\omega) \in [\alpha(\omega), \beta(\omega)]$ with properties (3.35).

The random semiinterval $\mathcal{I}^u(\omega) := (-\infty, u(\omega)]$ is a forward invariant random closed set (see Remark 3.4.1), i.e. $\mathcal{I}^u(\omega) \subset \varphi(t, \omega)^{-1}\mathcal{I}^u(\theta_t\omega)$. If for some $\omega \in \Omega$ there exists $x \in \varphi(t, \omega)^{-1}\mathcal{I}^u(\theta_t\omega)$ such that $x \notin \mathcal{I}^u(\omega)$, then $\varphi(t, \omega)x \leq u(\theta_t\omega)$ and $x > u(\omega)$ which is impossible if φ is strictly order-preserving. Thus $\mathcal{I}^u(\omega) = \varphi(t, \omega)^{-1}\mathcal{I}^u(\theta_t\omega)$. This relation implies that the set

$$M_- = \{(\omega, x) \ : \ x \leq u(\omega)\} \in \mathcal{F} \times \mathcal{B}(\mathbb{R})$$

satisfies $\pi_t^{-1}M_- = M_-$, where π_t is the skew-product semiflow corresponding to (θ, φ) (see (1.6)). We also have

$$\mu(M_-) = \int_\Omega \mu_\omega\{\mathcal{I}^u(\omega)\}\mathbb{P}(d\omega) \geq 1/2 \ .$$

Thus the φ-ergodicity of μ implies that $\mu(M_-) = 1$. Therefore

$$\mu_\omega\{(-\infty, u(\omega)]\} = 1 \quad \text{almost surely} \ .$$

In a similar way we obtain that $\mu_\omega\{[u(\omega), \infty)\} = 1$ almost surely. Consequently $\mu_\omega\{\{u(\omega)\}\} = 1$ for almost all $\omega \in \Omega$. This implies that $\delta_{u(\omega)}$ is a disintegration for μ.   □

The following proposition shows that the pull back omega-limit set (see Definition 1.6.1) emanating from a semi-equilibrium consists of a single equilibrium.

Proposition 3.5.2. *Assume c is either a sub- or a super-equilibrium and for any $\omega \in \Omega$ there exists $t_0 = t_0(\omega)$ such that the closure $\overline{\gamma_c^{t_0(\omega)}}(\omega)$ of the tail of the orbit $\gamma_c^0(\omega)$ emanating from c,*

$$\gamma_c^{t_0(\omega)}(\omega) = \bigcup_{t \geq t_0(\omega)} \varphi(t, \theta_{-t}\omega)c(\theta_{-t}\omega) \,,$$

is a compact set in X. Then the omega-limit set of c,

$$\Gamma_c(\omega) = \bigcap_{t>0} \overline{\bigcup_{\tau \geq t} \varphi(\tau, \theta_{-\tau}\omega)c(\theta_{-\tau}\omega)} \,,$$

consists of a single equilibrium u and

$$\lim_{t\to\infty} \varphi(t, \theta_{-t}\omega)c(\theta_{-t}\omega) = u(\omega) \quad \text{for all} \quad \omega \in \Omega \text{ monotonically}.$$

Proof. Using Proposition 3.4.1 we obviously have that for any $\omega \in \Omega$ the sequence $c_n(\omega) = \varphi(n, \theta_{-n}\omega)c(\theta_{-n}\omega)$ is monotone and relatively compact for each ω. Therefore we can repeat the argument given in the proof of Theorem 3.5.1. ☐

Below we also need the following assertion.

Lemma 3.5.1. *Let (θ, φ) be an order-preserving RDS over an ergodic metric dynamical system θ with phase space $X = V_+$. Assume that (θ, φ) is strongly positive, i.e. $\varphi(t, \omega)(V_+ \setminus \{0\}) \subset \operatorname{int}V_+$. Then for any equilibrium $u(\omega)$ there exists a θ-invariant subset $\mathcal{B} \in \mathcal{F}$ of full $\mathbb{P}$-measure such that either $u(\omega) = 0$ for all $\omega \in \mathcal{B}$ or $u(\omega) \gg 0$ for all $\omega \in \mathcal{B}$.*

Proof. Let $\mathcal{B}_0 = \{\omega : u(\omega) > 0\}$, where $u(\omega)$ is an equilibrium. The set $\mathcal{B}_0$ is $\mathcal{F}$-measurable because $V_+ \setminus \{0\}$ is a Borel set in V. Since $u(\theta_t\omega) = \varphi(t, \omega)u(\omega)$ for all $t \geq 0$, the strong positivity assumption implies that $\theta_t\mathcal{B}_0 \subset \mathcal{B}_0$ for all $t \geq 0$. It is clear that $\mathcal{B} := \cap_{n \in \mathbb{Z}_+} \theta_n \mathcal{B}_0 \subset \mathcal{B}_0$ is invariant with respect to θ, i.e. $\theta_t\mathcal{B} = \mathcal{B}$ for any $t \in \mathbb{R}$, and $\mathbb{P}(\mathcal{B}) = \mathbb{P}(\mathcal{B}_0)$. Moreover $u(\omega) \gg 0$ for $\omega \in \mathcal{B}$. The ergodicity implies that $\mathbb{P}(\mathcal{B})$ is equal either 1 or 0. If $\mathbb{P}(\mathcal{B}) = 1$, then the lemma is proved. If $\mathbb{P}(\mathcal{B}) = 0$ then we have $\mathbb{P}(\mathcal{B}_0) = 0$ and $u(\omega) = 0$ for $\omega \in \mathcal{A}_0 := \Omega \setminus \mathcal{B}_0$, where $\mathbb{P}(\mathcal{A}_0) = 1$. Since $u(\omega)$ is an equilibrium, it is easy to see that $\mathcal{A}_0 \subset \theta_t\mathcal{A}_0$ for any $t \geq 0$. This implies that $\mathcal{A} = \cap_{n \in \mathbb{Z}_+} \theta_{-n}\mathcal{A}_0 \subset \mathcal{A}_0$ possesses the properties (a) $\mathbb{P}(\mathcal{A}) = \mathbb{P}(\mathcal{A}_0) = 1$, (b) $\mathcal{A}$ is invariant with respect to θ, (c) $u(\omega) = 0$ for $\omega \in \mathcal{A}$. ☐

3.6 Properties of Invariant Sets of Order-Preserving RDS

In this section we prove a theorem on the structure of the random pull back attractor (see Definition 1.8.1) of an order-preserving RDS. We obtain this result as a corollary of the following general assertion which can also be useful to prove the existence of equilibria. We consider an order-preserving RDS (θ, φ) on a subset X of a real separable Banach space V with a closed convex cone V_+ such that $V_+ \cap \{-V_+\} = \{0\}$. We do not assume any additional properties of the cone V_+ here.

Below for an element $v \in X$ and a subset $A \subset X$ we write $v \geq A$ if $v \geq a$ for any $a \in A$. We understand the relation $v \leq A$ similarly.

Theorem 3.6.1. *Let $A(\omega)$ be an invariant random compact set for an order-preserving RDS (θ, φ). Then the following assertions are valid:*

(i) *if there exists a random variable $v(\omega)$ such that $v(\omega) \geq A(\omega)$ for all $\omega \in \Omega$ and*

$$\lim_{t \to +\infty} \mathrm{dist}_X \left(\varphi(t, \theta_{-t}\omega)v(\theta_{-t}\omega), A(\omega) \right) = 0 , \quad \omega \in \Omega , \tag{3.36}$$

then there exists an equilibrium $\overline{u}(\omega) \in A(\omega)$ such that $\overline{u}(\omega) = \sup A(\omega)$ and

$$\lim_{t \to +\infty} \varphi(t, \theta_{-t}\omega)v(\theta_{-t}\omega) = \overline{u}(\omega) \quad \text{for all} \quad \omega \in \Omega ; \tag{3.37}$$

(ii) *if there exists a random variable $v(\omega)$ satisfying (3.36) and such that $v(\omega) \leq A(\omega)$ for all $\omega \in \Omega$, then there exists an equilibrium $\underline{u}(\omega) \in A(\omega)$ such that $\underline{u}(\omega) = \inf A(\omega)$ and*

$$\lim_{t \to +\infty} \varphi(t, \theta_{-t}\omega)v(\theta_{-t}\omega) = \underline{u}(\omega) \quad \text{for all} \quad \omega \in \Omega . \tag{3.38}$$

Proof. We prove assertion (i) only. Since $A(\omega) \leq v(\omega)$, the invariance property of $A(\omega)$ implies that

$$a(\omega) \leq \varphi(t, \theta_{-t}\omega)v(\theta_{-t}\omega) \quad \text{for any} \quad a(\omega) \in A(\omega) . \tag{3.39}$$

The compactness of $A(\omega)$ and property (3.36) imply that for any $\omega \in \Omega$ there exist a sequence $t_n = t_n(\omega) \to \infty$ and an element $\overline{u}(\omega) \in A(\omega)$ such that

$$\varphi(t_n, \theta_{-t_n}\omega)v(\theta_{-t_n}\omega) \to \overline{u}(\omega) \quad \text{when} \quad n \to \infty .$$

From (3.39) we have that $a(\omega) \leq \overline{u}(\omega)$ for any $a(\omega) \in A(\omega)$, i.e. $\overline{u}(\omega)$ is the least upper bound for $A(\omega)$. Let us prove (3.37). Assume that this relation is not true for some $\omega \in \Omega$. Then there exists a sequence $\tau_k \to \infty$ such that

$$\|\varphi(\tau_k, \theta_{-\tau_k}\omega)v(\theta_{-\tau_k}\omega) - \overline{u}(\omega)\| \geq \delta \quad \text{for all} \quad k \in \mathbb{Z}_+ \tag{3.40}$$

with some positive δ. As above the compactness of $A(\omega)$ and property (3.36) allow us to extract a subsequence $\{\tau_{k_m}\}$ and to find an element $\overline{u}^* \in A(\omega)$ such that

$$\varphi(\tau_{k_m}, \theta_{-\tau_{k_m}}\omega)v(\theta_{-\tau_{k_m}}\omega) \to \overline{u}^* \quad \text{when} \quad m \to \infty .$$

It is also clear that $\overline{u}^* \geq A(\omega)$ and therefore $\overline{u}^* = \sup A(\omega)$. Consequently $\overline{u}^* = \overline{u}(\omega)$ which contradicts (3.40). Thus we have (3.37). Property (3.37) implies that $\bar{u}(\omega)$ is a random variable in X. Since $\bar{u}(\omega) = \sup A(\omega)$, the invariance of $A(\omega)$ implies

$$a(\omega) \leq \varphi(t, \theta_{-t}\omega)\bar{u}(\theta_{-t}\omega) \quad \text{for any} \quad a(\omega) \in A(\omega), \ t \geq 0 .$$

Therefore

$$\varphi(t, \theta_{-t}\omega)\bar{u}(\theta_{-t}\omega) \geq \bar{u}(\omega) \quad \text{for all} \quad \omega \in \Omega, \ t \geq 0 .$$

Hence using the property $\bar{u}(\omega) \in A(\omega)$ we find that

$$\varphi(t, \theta_{-t}\omega)\bar{u}(\theta_{-t}\omega) = \bar{u}(\omega) ,$$

i.e. $\bar{u}(\omega)$ is an equilibrium.   □

The main corollary of Theorem 3.6.1 is the following result concerning the structure of the global attractor for an order-preserving RDS.

Theorem 3.6.2. *Assume that the order-preserving RDS (θ, φ) on X has a random compact pull back attractor $A(\omega)$ in some universe $\mathcal{D}$ and that this attractor is order-bounded in the following sense: there exists a random interval*

$$[b, c](\omega) = \{x \ : \ b(\omega) \leq x \leq c(\omega)\}$$

such that $\{b(\omega)\}, \{c(\omega)\} \in \mathcal{D}$ and $A(\omega) \subset [b, c](\omega)$. Then there exist two equilibria $\underline{u}(\omega)$ and $\bar{u}(\omega)$ in $A(\omega)$ such that $\underline{u}(\omega) \leq \bar{u}(\omega)$ and the random attractor $A(\omega)$ belongs to the interval $[\underline{u}, \bar{u}](\omega)$, i.e.

$$\underline{u}(\omega) \leq a(\omega) \leq \bar{u}(\omega) \quad \text{for any} \quad a(\omega) \in A(\omega) . \tag{3.41}$$

These equilibria $\underline{u}(\omega)$ and $\bar{u}(\omega)$ are globally asymptotically stable from below and from above respectively, i.e.

$$\lim_{t \to +\infty} \varphi(t, \theta_{-t}\omega)w(\theta_{-t}\omega) = \underline{u}(\omega) \tag{3.42}$$

and

$$\lim_{t \to +\infty} \varphi(t, \theta_{-t}\omega)v(\theta_{-t}\omega) = \bar{u}(\omega) \tag{3.43}$$

for any $w(\omega) \leq \underline{u}(\omega)$ and for any $v(\omega) \geq \bar{u}(\omega)$ such that $\{w(\omega)\}$ and $\{v(\omega)\}$ belong to $\mathcal{D}$.

Proof. The application of Theorem 3.6.1 gives us the existence of the equilibria $\underline{u}(\omega)$ and $\bar{u}(\omega)$. To prove (3.42) we note that

$$\varphi(t, \theta_{-t}\omega)w(\theta_{-t}\omega) \leq \underline{u}(\omega) \quad \text{and} \quad \varphi(t, \theta_{-t}\omega)w(\theta_{-t}\omega) \to A(\omega)$$

for any $w(\omega) \leq \underline{u}(\omega)$ such that $\{w(\omega)\} \in \mathcal{D}$. Now the compactness of $A(\omega)$ and an argument similar to that used in the proof of Theorem 3.6.1 give (3.42). The same argument can be applied to prove (3.43).   □

Remark 3.6.1. The theorem on the existence of a random attractor (see Sect. 1.8) implies that the conditions of Theorem 3.6.2 hold, for example, if we assume that the order-preserving RDS (θ, φ) is asymptotically compact and possesses an absorbing interval in $\mathcal{D}$, i.e. there exists an interval $[b, c](\omega) \in \mathcal{D}$ with the property: for every $D \in \mathcal{D}$ there exists a time $t_0(\omega, D) > 0$ such that

$$\phi(t, \theta_{-t}\omega)D(\theta_{-t}\omega) \in [b, c](\omega) \quad \text{for all} \quad t \geq t_0(\omega, D) .$$

This remark allow us to derive from Theorem 3.6.2 the following corollary.

Corollary 3.6.1. *Let the order-preserving RDS (θ, φ) be asymptotically compact in some universe $\mathcal{D}$. Assume that $\mathcal{D}$ contains an absorbing interval for this RDS. If (θ, φ) has a unique equilibrium point $u(\omega)$ in $\mathcal{D}$, then $\{u(\omega)\}$ is a random attractor for this RDS in $\mathcal{D}$.*

In the connection with Remark 3.6.1 and Corollary 3.6.1 it is convenient to introduce the concept of an absorbing semi-equilibrium.

Definition 3.6.1. *A super-equilibrium $u(\omega)$ is said to be* absorbing *in the universe $\mathcal{D}$ if for any $B \in \mathcal{D}$ there exists $t_B(\omega) > 0$ such that*

$$\varphi(t, \theta_{-t}\omega)B(\theta_{-t}\omega) \subset \mathcal{I}^u(\omega) = u(\omega) - V_+, \quad \omega \in \Omega ,$$

for all $t \geq t_B(\omega)$, i.e. we have

$$\varphi(t, \theta_{-t}\omega)v(\theta_{-t}\omega) \leq u(\omega), \quad t \geq t_B(\omega), \quad \omega \in \Omega , \tag{3.44}$$

for all $v(\omega) \in B(\omega)$. Similarly a sub-equilibrium $w(\omega)$ is said to be absorbing *in the universe $\mathcal{D}$ if instead of (3.44) we have*

$$\varphi(t, \theta_{-t}\omega)v(\theta_{-t}\omega) \geq w(\omega), \quad t \geq t_B(\omega), \quad \omega \in \Omega ,$$

for all $v(\omega) \in B(\omega)$.

Proposition 3.6.1. *Assume that an order-preserving RDS (θ, φ) possesses an absorbing super-equilibrium $u(\omega)$ and an absorbing sub-equilibrium $w(\omega)$ in some universe $\mathcal{D}$ such that $\{u(\omega)\}, \{w(\omega)\} \in \mathcal{D}$. Then $w(\omega) \leq u(\omega)$ and the interval $[w(\omega), u(\omega)]$ is absorbing and forward invariant for RDS (θ, φ) in the universe $\mathcal{D}$.*

Proof. From Proposition 3.4.1 we have that $w(\omega) \leq \varphi(t, \theta_{-t}\omega)w(\theta_{-t}\omega)$ for all $t > 0$. Therefore (3.44) implies that $w(\omega) \leq u(\omega)$ and the interval $[w(\omega), u(\omega)]$ is absorbing by Definition 3.6.1. It is forward invariant by Remark 3.4.1. $\square$

We conclude this section with the following example which demonstrates a phenomenon which is impossible in deterministic order-preserving dynamical systems.

Example 3.6.1. Let us consider the following scalar Stratonovich equation

$$dx(t) = g(x(t)) \circ dW_t, \quad t > 0, \tag{3.45}$$

where $g(x)$ is a smooth function on $\mathbb{R}$ possessing the properties $g(u_0) = g(u_1) = 0$ and $g(x) > 0$ for $x \in (u_0, u_1)$. Here $u_0 < u_1$ are real numbers. It is easy to see that the cocycle for RDS generated by (3.45) in $[u_0, u_1]$ has the form

$$\varphi(t, \omega)x = G^{-1}(G(x) + W_t), \quad t > 0, \quad x \in (u_0, u_1),$$

where $G(x)$ is a primitive for $\frac{1}{g(u)}$ on the interval (u_0, u_1) and G^{-1} is the inverse mapping for $G : (u_0, u_1) \mapsto \mathbb{R}$. It is clear that this RDS is strongly order-preserving. A simple calculation shows that for this case there are no equilibria except u_0 and u_1 and that ω-limit set for any point x from (u_0, u_1) coincides with the whole interval $[u_0, u_1]$: a non-trivial completely ordered set. This phenomenon does not take place in deterministic (autonomous or periodic) strongly order-preserving systems (cf. SMITH [102]). Furthermore, in this example all the trajectories oscillate between the two equilibria u_0 and u_1 and there is no equilibrium inside the interval.

3.7 Comparison Principle

All the above considerations give ample proof of the crucial role played by sub- and super-equilibria in the study of qualitative properties of order-preserving RDS. One of the methods of proving their existence relies on the *comparison principle*. Let V be a Banach space with a cone V_+ and let X_1 and X_2 be subsets of V. Let (θ, φ_1) and (θ, φ_2) be two RDS on X_1 and X_2 over the same metric dynamical system θ and take $Y \subset X_1 \cap X_2$. The system (θ, φ_2) is said to *dominate* (θ, φ_1) *from above* on Y (or (θ, φ_1) dominates (θ, φ_2) from below on Y) if

$$\varphi_1(t, \omega, x) \leq \varphi_2(t, \omega, x) \quad \text{for any} \quad t > 0, \quad x \in Y, \quad \omega \in \Omega. \tag{3.46}$$

Clearly (3.46) implies that any super-equilibrium $v(\omega)$ for (θ, φ_2) such that $v(\omega) \in Y$ for all $\omega \in \Omega$ is a super-equilibrium for (θ, φ_1) and any sub-equilibrium $w(\omega)$ for (θ, φ_1) with the property $w(\omega) \in Y$ for all $\omega \in \Omega$ is a sub-equilibrium for (θ, φ_2).

In many applications, e.g. for the construction of random attractors (see Chaps.5 and 6 below), a nonlinear RDS can be shown to be dominated by an affine RDS (see Definition 1.2.3), whose equilibrium then becomes a sub- or super-equilibrium of the corresponding nonlinear RDS.

As an example of an application of the comparison principle we prove the following assertion.

Proposition 3.7.1. *Let (θ, φ) be an order-preserving RDS in a cone V_+ of a separable Banach space V. Assume that the system (θ, φ) is dominated from above on the cone V_+ by an affine RDS $(\theta, \varphi_{\text{aff}})$. Suppose that RDS $(\theta, \varphi_{\text{aff}})$ satisfies the hypotheses of Proposition 1.9.2, i.e. $(\theta, \varphi_{\text{aff}})$ is asymptotically compact in a universe $\mathcal{D}$ with the properties (a) $\{0\} \in \mathcal{D}$, (b) for any $D \in \mathcal{D}$ and $\lambda > 0$ the set $\omega \mapsto \lambda D(\omega)$ belongs to $\mathcal{D}$ and (c) an attracting random compact set B_0 belongs to $\mathcal{D}$. Let $u(\omega)$ be the unique equilibrium for $(\theta, \varphi_{\text{aff}})$ in $\mathcal{D}$. Then $u(\omega) \geq 0$ and for any $\mu \geq 1$ the random variable $v_\mu(\omega) := \mu u(\omega)$ is a super-equilibrium for (θ, φ). If the cone V_+ is solid, then the interval $[0, e(\omega) + u(\omega)]$ with arbitrary $e(\omega) \in \text{int} V_+$ is absorbing for (θ, φ) in the universe*

$$\mathcal{D}^* = \{D \in \mathcal{D} : D(\omega) \subset V_+ \text{ for all } \omega \in \Omega\}.$$

If $u(\omega) \gg 0$, then $v_\mu(\omega)$ is an absorbing super-equilibrium for (θ, φ) in $\mathcal{D}^*$ for any $\mu > 1$. In this case, if V is a finite-dimensional space, then the RDS (θ, φ) possesses a random attractor in the universe $\tilde{\mathcal{D}}$ consisting of all random closed sets $\{B(\omega)\}$ such that $B(\omega) \subset [0, \alpha u(\omega)]$ for some $\alpha > 0$ and the conclusions of Theorem 3.6.2 hold.

Proof. The cocycle φ_{aff} has the form

$$\varphi_{\text{aff}}(t, \omega)x = \Phi(t, \omega)x + \psi(t, \omega), \quad x \in V.$$

Since $\varphi(t, \omega)0 \geq 0$ and $\varphi(t, \omega)x \leq \varphi_{\text{aff}}(t, \omega)x$ for all $x \in V_+$, we have

$$\psi(t, \omega) = \varphi_{\text{aff}}(t, \omega)0 \geq \varphi(t, \omega)0 \geq 0.$$

Therefore (1.51) implies that $u(\omega) \geq 0$. We also have

$$\varphi(t, \omega)v_\mu(\omega) \leq \mu \Phi(t, \omega)u(\omega) + \psi(t, \omega)$$

$$= \mu \varphi_{\text{aff}}(t, \omega)u(\omega) + (1 - \mu)\psi(t, \omega) = \mu u(\theta_t \omega) + (1 - \mu)\psi(t, \omega).$$

Hence $v_\mu(\omega)$ is a super-equilibrium for (θ, φ) for any $\mu \geq 1$. Relation (1.52) implies that $[0, e(\omega) + u(\omega)]$ is an absorbing interval for (θ, φ). If $u(\omega) \gg 0$ we can choose $e(\omega) = (\mu - 1) \cdot u(\omega)$ and therefore $v_\mu(\omega)$ is an absorbing super-equilibrium, $\mu > 1$. Thus $[0, \mu u(\omega)]$ is an absorbing set for (θ, φ) in $\tilde{\mathcal{D}}$. Hence (θ, φ) is dissipative. If V is finite-dimensional, then Corollary 1.8.1 implies that a random attractor exists in the universe $\tilde{\mathcal{D}}$ and we can apply Theorem 3.6.2. □

Example 3.7.1 (Binary Biochemical Model). Consider the random differential equations

$$\dot{x}_1 = g(x_2) - \alpha_1(\theta_t \omega)x_1,$$
$$\dot{x}_2 = x_1 - \alpha_2(\theta_t \omega)x_2, \tag{3.47}$$

over an ergodic metric dynamical system θ. Assume that $g(x)$ is a C^1 function such that

$$g(0) \geq 0, \quad 0 \leq g(x) \leq g_0, \quad g'(x) \geq 0 \quad \text{for all} \quad x \in \mathbb{R}_+ ,$$

where g_0 is a constant. Let $\alpha_i(\omega) \in L^1(\Omega, \mathfrak{F}, \mathbb{P})$ be a random variable such that $\alpha_i(\theta_t\omega) \in L^1_{\text{loc}}(\mathbb{R})$ and $\mathbb{E}\alpha_i > 0$ for $i = 1, 2$. Equations (3.47) generate a strictly order-preserving RDS in $\mathbb{R}^2_+$ (see Example 3.3.5). It is easy to see that (θ, φ) is dominated from above by the affine RDS $(\theta, \varphi_{\text{aff}})$ generated by the equations

$$\dot{x}_1 = g_0 - \alpha_1(\theta_t\omega)x_1 ,$$
$$\dot{x}_2 = x_1 - \alpha_2(\theta_t\omega)x_2 .$$

Let $u(\omega) = (u_1(\omega), u_2(\omega))$, where

$$u_1(\omega) = g_0 \int_{-\infty}^0 \exp\left\{ - \int_s^0 \alpha_1(\theta_\tau\omega)d\tau \right\} ds$$

and

$$u_2(\omega) = \int_{-\infty}^0 u_1(\theta_s\omega) \exp\left\{ - \int_s^0 \alpha_2(\theta_\tau\omega)d\tau \right\} ds .$$

A simple calculation shows that $u(\omega) \gg 0$ is an equilibrium for $(\theta, \varphi_{\text{aff}})$. Thus $u(\omega)$ is a super-equilibrium for (θ, φ). Proposition 3.7.1 can be applied here with the universe $\mathcal{D}$ consisting of all tempered sets from $\mathbb{R}^2$.

4. Sublinear Random Dynamical Systems

In this chapter we consider a class of order-preserving RDS which possess certain concavity properties. The deterministic versions of these properties play an important role in many studies and applications, see KRASNOSELSKII [68, 69], KRAUSE ET AL. [72, 73], SMITH [101], TAKÁČ [103] and the references therein. For the sake of simplicity we assume that the state space X is equal to a solid cone V_+ of a real Banach space V,

$$X = V_+ = \{x \in V : x \geq 0\} ,$$

i.e. we consider random dynamical systems (θ, φ) which possess the positivity property: $\varphi(t, \omega)V_+ \subset V_+$ for all t and ω. For order-preserving systems this property is equivalent to the relation $\varphi(t, \omega)0 \geq 0$. Our main result in this chapter is a random limit set trichotomy which describes all possible types of long-time behaviour in sublinear random systems. This result is a clear manifestation of the general experience that monotonicity, and even more so sublinearity, drastically simplifies the possible long-term behaviour of a dynamical system.

4.1 Sublinear and Concave RDS

We start with the most general concavity property which we call *sublinearity* (sometimes also named subhomogeneity). Sublinearity means concavity for the particular case in which one of the reference points is 0, hence asks less (and is thus more general) than classical concavity.

Definition 4.1.1 (Sublinear RDS). *An order-preserving RDS (θ, φ) on $X = V_+$ is said to be* sublinear *if for any $x \in V_+$ and for any $\lambda \in (0,1)$ we have*

$$\lambda\varphi(t, \omega, x) \leq \varphi(t, \omega, \lambda x) \quad \text{for all} \quad t > 0 \quad \text{and} \quad \omega \in \Omega . \tag{4.1}$$

The RDS is said to be (i) strictly sublinear *if we have in addition for any $x \in \text{int}V_+$ the strict inequality*

$$\lambda\varphi(t, \omega, x) < \varphi(t, \omega, \lambda x) \quad \text{for all} \quad t > 0 \quad \text{and} \quad \omega \in \Omega , \tag{4.2}$$

and (ii) strongly sublinear if in addition to (4.1) we have

$$\lambda\varphi(t,\omega,x) \ll \varphi(t,\omega,\lambda x) \quad \text{for all} \quad t > 0, \ x \in \text{int}V_+, \quad \text{and} \quad \omega \in \Omega, \quad (4.3)$$

i.e. $\varphi(t,\omega,\lambda x) - \lambda\varphi(t,\omega,x) \in \text{int}V_+$.

Equation (4.1) holds automatically for $t = 0$ and for $\lambda = 0$ and 1. In one-dimensional case the properties of strict sublinearity (4.2) and strong sublinearity (4.3) coincide. Property (4.1) can be equivalently rewritten as follows: For any $x \in V_+$ and for any $\lambda > 1$ we have

$$\varphi(t,\omega,\lambda x) \leq \lambda\varphi(t,\omega,x) \quad \text{for all} \quad t > 0 \quad \text{and} \quad \omega \in \Omega. \quad (4.4)$$

Similarly for (4.2) and (4.3).

Using conditions (4.1) and (4.4) it is easy to see that if $u \geq 0$ is

(i)   a sub-equilibrium, then $\lambda u(\cdot)$ is a sub-equilibrium for any $\lambda \in [0,1]$;
(ii) a super-equilibrium, then $\lambda u(\cdot)$ is a super-equilibrium for any $\lambda \geq 1$;
(iii) an equilibrium, then $\lambda u(\cdot)$ is a sub-equilibrium for any $\lambda \in [0,1]$ and a super-equilibrium for any $\lambda \geq 1$.

Example 4.1.1 (Binary Biochemical Model). Let (θ,φ) be the RDS in $\mathbb{R}_+^2$ generated by the equations

$$\dot{x}_1 = g(x_2) - \alpha_1(\theta_t\omega)x_1 \,,$$
$$\dot{x}_2 = x_1 - \alpha_2(\theta_t\omega)x_2 \,,$$

over a metric dynamical system θ. We assume that the function $g(x)$ and the random variables $\alpha_1(\omega)$ and $\alpha_2(\omega)$ satisfy the assumptions of Examples 2.1.1 and 3.3.5. In Sect.5.7 we prove that (θ,φ) is a strictly sublinear RDS if $g(x)$ is a sublinear mapping from $\mathbb{R}_+$ into $\mathbb{R}$, i.e. if $\lambda g(x) \leq g(\lambda x)$ for all $x \geq 0$ and $0 < \lambda < 1$. This system is strongly sublinear if we assume additionally that $g'(x) > 0$ for $x > 0$. Similar result remains true for the stochastic case (cf. Example 2.4.3 and Sect.6.8). We note that sublinear functions $g(x)$ appear in the Griffith ($g(x) = x \cdot (1 + x)^{-1}$ for $x > 0$) and in the Othmer-Tyson ($g(x) = (1 + x) \cdot (k + x)^{-1}$, $x > 0$, $k > 1$) biochemical models. We refer to SELGRADE [96] for a discussion and for the references.

We note that an order-preserving affine (see Definition 1.2.3 and Example 3.3.6) RDS which maps V_+ into itself is automatically sublinear. It is strictly (resp. strongly) sublinear if $\psi(t,\omega) > 0$ (resp. $\psi(t,\omega) \gg 0$) in the representation (1.4) for $t > 0$. It is also easy to see that a scalar function $g : \mathbb{R}_+ \mapsto \mathbb{R}$ is sublinear if and only if $g(u)/u$ is nonincreasing.

Definition 4.1.2 (Concave RDS). *An order-preserving RDS (θ,φ) on $X = V_+$ is said to be concave if for any $0 \leq x \leq y$ and for any $\lambda \in (0,1)$ we have*

$$\lambda\varphi(t,\omega,x) + (1 - \lambda)\varphi(t,\omega,y) \leq \varphi(t,\omega,\lambda x + (1 - \lambda)y) \quad (4.5)$$

for all $t > 0$ and $\omega \in \Omega$. The RDS is said to be strictly concave *if in addition we have strict inequality in (4.5) for all $0 \ll x \ll y$ and it is* strongly concave *if*

$$\lambda\varphi(t,\omega,x) + (1-\lambda)\varphi(t,\omega,y) \ll \varphi(t,\omega,\lambda x + (1-\lambda)y), \quad 0 \ll x \ll y .$$

It is clear that (strict, strong) concavity implies (strict, strong) sublinearity. A simple one-dimensional example $f(x) = (1+x)^{-1}$, $x \in V_+ = \mathbb{R}_+$ shows that the converse is not valid.

If (θ, φ) is a C^1-smooth RDS we can establish the following necessary and sufficient conditions for sublinearity and concavity. Below we denote by D_x the Frechet derivative with respect to x.

Proposition 4.1.1. *Assume that (θ, φ) is a C^1-smooth order-preserving RDS in V_+. Then*
(i) it is sublinear if and only if

$$D_x\varphi(t,\omega,x)x \leq \varphi(t,\omega,x) \quad \text{for all} \quad t \geq 0, \ \omega \in \Omega, \ x \in V_+ \setminus \{0\} , \quad (4.6)$$

and it is strictly (strongly) sublinear provided that in (4.6) we have strict (strong) inequality for $x \in \mathrm{int}V_+$;
(ii) the system (θ, φ) is concave if and only if for any $x, z \in V_+ \setminus \{0\}$ we have

$$D_x\varphi(t,\omega,x+z)z \leq D_x\varphi(t,\omega,x)z \quad \text{for all} \quad t > 0, \ \text{and} \ \omega \in \Omega , \quad (4.7)$$

and it is strictly (strongly) concave if in (4.7) the inequality is strict (strong) for all x and z from $\mathrm{int}V_+$.

Proof. Since

$$\frac{d}{d\lambda}\left\{\frac{1}{\lambda} \cdot \varphi(t,\omega,\lambda x)\right\} = -\frac{1}{\lambda^2}\left\{\varphi(t,\omega,\lambda x) - D_x\varphi(t,\omega,\lambda x)\lambda x\right\} ,$$

we have

$$\frac{1}{\nu}\varphi(t,\omega,\nu x) - \frac{1}{\lambda} \cdot \varphi(t,\omega,\lambda x) = -\int_\lambda^\nu \frac{1}{\mu^2}\left\{\varphi(t,\omega,\mu x) - D_x\varphi(t,\omega,\mu x)\mu x\right\} d\mu ,$$

for any $0 < \lambda < \nu \leq 1$, $\omega \in \Omega$ and $x \in V_+ \setminus \{0\}$. This implies (i).
    To prove (ii) we first note that

$$\frac{d}{d\lambda}\left\{\frac{1}{\lambda} \cdot [\varphi(t,\omega,x+\lambda z) - \varphi(t,\omega,x)]\right\}$$

$$= -\frac{1}{\lambda^2}\left\{\int_0^\lambda [D_x\varphi(t,\omega,x+\mu z) - D_x\varphi(t,\omega,x+\lambda z)] d\mu\right\} z$$

for any x and z from $V_+ \setminus \{0\}$. Therefore for any $\lambda_2 > \lambda_1 > 0$ we have

$$\varphi(t, \omega, x + \lambda_1 z) - \frac{\lambda_1}{\lambda_2} \varphi(t, \omega, x + \lambda_2 z) - \left(1 - \frac{\lambda_1}{\lambda_2}\right) \varphi(t, \omega, x)$$

$$= \lambda_1 \int_{\lambda_1}^{\lambda_2} \frac{1}{\lambda^2} \left\{ \int_0^\lambda [D_x \varphi(t, \omega, x + \mu z) z - D_x \varphi(t, \omega, x + \lambda z) z]\, d\mu \right\} d\lambda.$$

$$(4.8)$$

Consequently the (strict, strong) concavity in the differential form (4.7) implies (strict, strong) concavity in the sense of Definition 4.1.2. It is clear from (4.5) that

$$D_x \varphi(t, \omega, x + z) z \le \varphi(t, \omega, x + z) - \varphi(t, \omega, x) \le D_x \varphi(t, \omega, x) z \qquad (4.9)$$

for all x and z from $V_+ \setminus \{0\}$. Consequently (4.5) implies (4.7). $\qquad \square$

Remark 4.1.1. A simple example of the C^1-mapping

$$f(x) = \begin{cases} 2x + x(x-1)^2 & \text{if } 0 \le x \le 1, \\ 2 + x - \frac{1}{x} & \text{if } x > 1, \end{cases}$$

from $\mathbb{R}_+$ into itself shows that strict (strong) sublinearity does not imply strict (strong) inequality in (4.6).

Below we also use the following concept of concavity for a C^1-smooth RDS which was introduced by SMITH [101] in the deterministic case.

Definition 4.1.3 (S-Concave RDS). *A C^1-smooth order-preserving RDS (θ, φ) on $X = V_+$ is said to be s-concave if for any $0 \ll x \ll y$ and $z \in \mathrm{int} V_+$ we have*

$$D_x \varphi(t, \omega, y) z < D_x \varphi(t, \omega, x) z \quad \text{for all} \quad t > 0, \quad \text{and } \omega \in \Omega. \qquad (4.10)$$

It is clear from Proposition 4.1.1 that any s-concave RDS is strictly concave.

4.2 Equilibria and Semi-Equilibria for Sublinear RDS

In this section we prove a uniqueness theorem for equilibria of strongly sublinear RDS and study their stability properties.

We start with the following important lemma. We recall (see Definition 3.1.4) that any equivalence class C under the equivalence relation on the cone V_+ defined by

$$\{x \sim y\} \quad \Longleftrightarrow \quad \{\exists \alpha \in \mathbb{R}_+ \setminus \{0\}, \ \alpha^{-1} x \le y \le \alpha x\} \qquad (4.11)$$

is called a part of V_+ and every part C is a metric space with respect to the part (Birkhoff) metric defined by

$$p(x,y) := \inf\{\log\alpha : \alpha^{-1}x \leq y \leq \alpha x\}, \quad x,y \in C . \tag{4.12}$$

Lemma 4.2.1. *Let (θ,φ) be a sublinear order-preserving RDS on V_+. Then*

(i) φ preserves the equivalence relation (4.11) and is nonexpansive under the part metric on every part C of V_+, i.e. for all $x,y \in C$

$$p(\varphi(t,\omega)x, \varphi(t,\omega)y) \leq p(x,y) \quad \text{for all} \quad t \geq 0 \quad \text{and} \quad \omega \in \Omega .$$

(ii) (θ,φ) is strongly sublinear if and only if it is contractive under the part metric, i.e. for all $x,y \in \mathrm{int}V_+$, $x \neq y$,

$$p(\varphi(t,\omega)x, \varphi(t,\omega)y) < p(x,y) \quad \text{for all} \quad t > 0, \quad \omega \in \Omega , \tag{4.13}$$

and $\varphi(t,\omega)\mathrm{int}V_+ \subset \mathrm{int}V_+$ for $t > 0$ and $\omega \in \Omega$.

Proof. (i) It follows from (4.1) and (4.4) that if for $x,y \in V_+$ and some $\alpha \geq 1$

$$\alpha^{-1}x \leq y \leq \alpha x$$

then also

$$\alpha^{-1}\varphi(t,\omega)x \leq \varphi(t,\omega)y \leq \alpha\varphi(t,\omega)x \quad \text{for all} \quad t \geq 0 \quad \text{and} \quad \omega \in \Omega$$

and hence by (4.12) we have $p(\varphi(t,\omega)x, \varphi(t,\omega)y) \leq p(x,y)$ for all $t \geq 0$ and $\omega \in \Omega$, proving (i).

(ii) Assume that $x,y \in \mathrm{int}V_+$ and there is no $\lambda > 0$ such that $y = \lambda x$. In this case $p(x,y) > 0$ and

$$e^{-p(x,y)}x < y < xe^{p(x,y)} .$$

Thus (4.3) implies that

$$e^{-p(x,y)}\varphi(t,\omega)x \ll \varphi(t,\omega)y \ll e^{p(x,y)}\varphi(t,\omega)x \quad \text{for} \quad t > 0 \quad \text{and} \quad \omega \in \Omega .$$

It is clear that for every $t > 0$ and $\omega \in \Omega$ there exists $\mu := \mu(t,\omega,x,y) > 0$ such that

$$e^{\mu}e^{-p(x,y)}\varphi(t,\omega)x \ll \varphi(t,\omega)y \ll e^{-\mu}e^{p(x,y)}\varphi(t,\omega)x .$$

Therefore

$$p(\varphi(t,\omega)x, \varphi(t,\omega)y) \leq p(x,y) - \mu < p(x,y) .$$

Thus we obtain (4.13) for these x and y. If $y = \lambda x$ for some $\lambda > 1$, then $p(x, y) = \log \lambda$ and

$$\varphi(t, \omega)x \leq \varphi(t, \omega)y \ll \lambda\varphi(t, \omega)x \quad \text{for} \quad t > 0 \quad \text{and} \quad \omega \in \Omega. \tag{4.14}$$

As above this implies (4.13). The case $y = \lambda x$ with $0 < \lambda < 1$ is similar. Thus a strongly sublinear RDS possesses property (4.13). It is also clear from (4.14) that $\varphi(t, \omega)\text{int}V_+ \subset \text{int}V_+$.

If (4.13) holds for some order-preserving RDS, then for any $x \in \text{int}V_+$ and $0 < \lambda < 1$ we have

$$p(\varphi(t, \omega)[\lambda x], \varphi(t, \omega)x) \leq \log\frac{1}{\lambda} - \mu$$

with some positive μ. Hence $e^\mu \lambda\varphi(t, \omega)x \leq \varphi(t, \omega)[\lambda x]$. This property and the invariance of $\text{int}V_+$ imply (4.3). □

Lemma 4.2.1 is a motivation for the following definition.

Definition 4.2.1. *A sublinear order-preserving RDS (θ, φ) is said to be strongly sublinear on a part C of V_+ if $\varphi(t, \omega)C \subset C$ for $t > 0$ and $\omega \in \Omega$ and it is contractive under the part metric, i.e. (4.13) holds for all $x, y \in C$, $x \neq y$.*

Theorem 4.2.1 (Uniqueness of Equilibrium). *If a sublinear order-preserving RDS (θ, φ) is strictly sublinear on some part C of the cone V_+, then any two equilibria in C are equal on a set of full measure in Ω which is invariant with respect to θ.*

Proof. Consider the function $V(\omega, u, v) = p(u, v)$ on $\Omega \times C \times C$, where p is the part metric. Proposition 3.2.4 implies that the function $\omega \mapsto V(\omega, u(\omega), v(\omega))$ is measurable for any random variables $u(\omega)$ and $v(\omega)$ from C. Lemma 4.2.1 gives that for any u and v from C we have

$$V(\theta_t\omega, \varphi(t, \omega)u, \varphi(t, \omega)v) \leq V(\omega, u, v) \quad \text{for all} \quad t > 0, \ \omega \in \Omega,$$

with strict inequality, if $u \neq v$. Thus the function $V(\omega, u, v) = p(u, v)$ satisfies the hypotheses of Proposition 1.7.1 which gives the assertion. □

Remark 4.2.1. (i) Theorem 4.2.1 is wrong without the assumption of *strong* sublinearity. Consider for example $\dot{x} = a(\theta_t\omega)x$ on $X = \mathbb{R}_+$, where $a(\theta_t\omega) = db(\theta_t\omega)/dt$ is the derivative of a stationary process $t \mapsto b(\theta_t\omega)$ with absolutely continuous trajectories. Then the sublinear (but not strongly or strictly sublinear) solution is

$$\varphi(t, \omega)x = xe^{-b(\omega)}e^{b(\theta_t\omega)},$$

meaning that φ is a coboundary, i.e. is a cocycle which is cohomologous to the trivial cocycle $\psi(t, \omega) \equiv 1$ (ARNOLD [3, Chap.5]), and any $x(\omega) = ce^{b(\omega)}$,

$c \in \mathbb{R}_+$, is an equilibrium. It is also easy to see that we cannot replace the strong sublinearity by property (4.2). The deterministic mapping $f(x_1, x_2) = (\sqrt{x_1}, x_2)$ of $\mathbb{R}_+^2$ into itself provides an example.

(ii) If two equilibria coincide on a set of full measure in Ω, then they generate the same φ-invariant measure on $\Omega \times V_+$ by equation (1.59). Thus Theorem 4.2.1 means that for any part C of the cone V_+ a strongly sublinear RDS (θ, φ) has at most one invariant measure generated by a random Dirac measure supported by C. We also note that every part of the cone can contain a positive equilibrium which is stable in its part (see Remark 4.5.1(ii) below).

Proposition 4.2.1. *Let* $v : \Omega \to V_+$. *Let the order-preserving RDS* (θ, φ) *be strongly sublinear on the part* C_v *generated by* $v(\omega)$ *(see Definition 3.2.1). Then any two equilibria in* C_v *are equal on a* θ-*invariant set in* Ω *of full measure.*

The idea of the proof is the same as for Theorem 4.2.1 and Proposition 1.7.1.

We next make a statement about the global asymptotic stability of an equilibrium u in its own part C_u.

Theorem 4.2.2. *Let* (θ, φ) *be a strongly sublinear order-preserving RDS in* V_+. *Assume that it has an equilibrium* $u : \Omega \to \mathrm{int}V_+$. *Suppose that there exists a constant* $\mu_0 \geq 1$ *such that for all* $\mu > \mu_0$ *the orbits emanating from* μu *and* $\mu^{-1} u$ *are relatively compact in* V_+. *Then* u *is globally asymptotically stable in* C_u, *i.e. there exists a* θ-*invariant set* $\Omega^* \in \mathcal{F}$ *of full measure such that for any* $w \in C_u$

$$\lim_{t \to +\infty} \varphi(t, \theta_{-t}\omega) w(\theta_{-t}\omega) = u(\omega) \quad \text{for all} \quad \omega \in \Omega^* . \tag{4.15}$$

Proof. Let $w \in C_u$. Then there exists an integer $\mu = \mu(w) > \mu_0 \geq 1$ such that

$$\mu^{-1} u(\omega) \leq w(\omega) \leq \mu u(\omega) . \tag{4.16}$$

Since μu is a super-equilibrium, Proposition 3.5.2 and our assumption ensure the existence of an equilibrium $\overline{w}_\mu$ such that

$$\lim_{t \to +\infty} \varphi(t, \theta_{-t}\omega) \mu u(\theta_{-t}\omega) = \overline{w}_\mu(\omega) \quad \text{for all} \quad \omega \in \Omega . \tag{4.17}$$

Similarly, there exists an equilibrium $\underline{w}_\mu$ such that

$$\lim_{t \to +\infty} \varphi(t, \theta_{-t}\omega) \mu^{-1} u(\theta_{-t}\omega) = \underline{w}_\mu(\omega) \quad \text{for all} \quad \omega \in \Omega . \tag{4.18}$$

By Theorem 3.5.1

$$\mu^{-1} u(\omega) \leq \underline{w}_\mu(\omega) \leq \overline{w}_\mu(\omega) \leq \mu u(\omega).$$

Hence $\underline{w}_\mu, \overline{w}_\mu \in C_u$, and by Proposition 4.2.1 there exists a θ-invariant set $\Omega^*_\mu \in \mathcal{F}$ of full measure such that

$$\underline{w}_\mu(\omega) = \overline{w}_\mu(\omega) = u(\omega) \quad \text{for all} \quad \omega \in \Omega^*_\mu, \; \mu \in \mathbb{N}, \; \mu > \mu_0 .$$

Therefore

$$\underline{w}_\mu(\omega) = \overline{w}_\mu(\omega) = u(\omega) \quad \text{for all} \quad \omega \in \Omega^* := \cap_{\mu \in \mathbb{N}, \mu > \mu_0} \Omega^*_\mu . \tag{4.19}$$

It is clear that Ω^* is a θ-invariant set of full measure. By sublinearity, (4.16) implies

$$\varphi(t, \theta_{-t}\omega)\mu^{-1}u(\theta_{-t}\omega) \le \varphi(t, \theta_{-t}\omega)w(\theta_{-t}\omega) \le \varphi(t, \theta_{-t}\omega)\mu u(\theta_{-t}\omega) .$$

Consequently (4.17) to (4.19) imply (4.15). □

We next present a criterion for the existence and half-sided attractivity of an equilibrium.

Theorem 4.2.3. *Let (θ, φ) be a strongly sublinear order-preserving RDS in V_+.*

(i) Let $a : \Omega \to \operatorname{int}V_+$ be a sub-equilibrium such that the orbits emanating from elements λa are relatively compact in V_+ for all $0 < \lambda \le 1$ and $\omega \in \Omega$. Then there exists an equilibrium u such that $u(\omega) \ge a(\omega)$ for all $\omega \in \Omega$ and

$$\lim_{t \to +\infty} \varphi(t, \theta_{-t}\omega)w(\theta_{-t}\omega) = u(\omega) \tag{4.20}$$

on a θ-invariant set in Ω of full measure for any w possessing the property

$$\alpha^{-1}a(\omega) \le w(\omega) \le u(\omega) \quad \text{for some number } \alpha \ge 1 . \tag{4.21}$$

(ii) Let $b : \Omega \to \operatorname{int}V_+$ be a super-equilibrium such that the orbits emanating from elements λb are relatively compact in V_+ for all $\lambda \ge 1$ and $\omega \in \Omega$. Then there exists an equilibrium v such that $v(\omega) \le b(\omega)$ for all $\omega \in \Omega$. If $v(\omega) \gg 0$ for all $\omega \in \Omega$, then

$$\lim_{t \to +\infty} \varphi(t, \theta_{-t}\omega)w(\theta_{-t}\omega) = v(\omega) \tag{4.22}$$

on a θ-invariant set in Ω of full measure for any w possessing the property

$$v(\omega) \le w(\omega) \le \beta b(\omega) \quad \text{for some number } \beta \ge 1 .$$

Proof. We only prove (i). Since λa is a sub-equilibrium for every $0 < \lambda \le 1$, by Proposition 3.5.2 there exists an equilibrium u_λ such that

$$\lim_{t \to +\infty} \varphi(t, \theta_{-t}\omega)\lambda a(\theta_{-t}\omega) = u_\lambda(\omega).$$

It is clear that $u_\lambda(\omega) \leq u_1(\omega)$ for $0 < \lambda \leq 1$. By (4.1)

$$\lambda\varphi(t,\theta_{-t}\omega)a(\theta_{-t}\omega) \leq \varphi(t,\theta_{-t}\omega)\lambda a(\theta_{-t}\omega) \quad \text{for all} \quad t > 0 \text{ and } \omega \in \Omega \,,$$

hence

$$\lambda u_1(\omega) \leq u_\lambda(\omega) \leq u_1(\omega) \quad \text{for all} \quad 0 < \lambda \leq 1 \,.$$

This means that $u_\lambda \in C_{u_1}$ for all $\lambda \in (0,1]$, thus by Proposition 4.2.1, $u_\lambda(\omega) = u_1(\omega)$ on a θ-invariant set Ω^* of full measure. As in the proof of Theorem 4.2.2 we can choose Ω^* independent of λ. By Theorem 3.5.1, $u_1(\omega) \geq a(\omega)$.

For all w satisfying (4.21) with $u(\omega) := u_1(\omega)$ clearly

$$\varphi(t,\theta_{-t}\omega)\alpha^{-1}a(\omega) \leq \varphi(t,\theta_{-t}\omega)w(\omega) \leq \varphi(t,\theta_{-t}\omega)u_1(\omega) = u_1(\omega) \,.$$

Letting $t \to \infty$ gives (4.20) with $u(\omega) = u_1(\omega)$. $\qquad\qquad\square$

Using Uniqueness Theorem 4.2.1 it is easy to derive from Theorem 4.2.3 the following assertion.

Corollary 4.2.1. *Let (θ,φ) be a strongly sublinear order-preserving RDS in V_+. Assume that there exist a sub-equilibrium $a(\omega)$ and a super-equilibrium $b(\omega)$ such that $a(\omega) \leq b(\omega)$ and the hypotheses of Theorem 4.2.3 concerning a and b hold. Then there exists an equilibrium $u(\omega) \in [a(\omega),b(\omega)]$ such that (4.20) holds for any $w(\omega)$ satisfying*

$$\beta^{-1}a(\omega) \leq w(\omega) \leq \beta b(\omega) \quad \text{for some} \quad \beta \geq 1 \,.$$

Remark 4.2.2. The compactness assumptions in Theorems 4.2.2 and 4.2.3(ii) can be omitted if the cone V_+ is regular (see Definition 3.1.6). Moreover, in this case relation (4.22) holds for all $w(\omega)$ satisfying

$$\beta^{-1}v(\omega) \leq w(\omega) \leq \beta b(\omega) \quad \text{for some} \quad \beta \geq 1 \,.$$

As for the first statement of Theorem 4.2.3, the regularity of the cone V_+ and the compactness of the trajectory γ_a imply (4.20) for all $w(\omega)$ satisfying

$$\alpha^{-1}a(\omega) \leq w(\omega) \leq \alpha u(\omega) \quad \text{for some} \quad \alpha \geq 1 \,.$$

In the case when 0 is an equilibrium we have the following assertion for concave systems.

Proposition 4.2.2. *Let (θ,φ) be a concave order-preserving C^1 RDS in a normal cone V_+ and $v \equiv 0$ be an equilibrium. Let (θ,Φ) be the linearization of (θ,φ) at 0. If the top Lyapunov exponent λ of (θ,Φ) is negative, then there exist a θ-invariant set Ω^* of full measure such that*

$$\lim_{t\to\infty}\left\{e^{\gamma t}\|\varphi(t,\theta_{-t}\omega)x\|\right\} = 0, \quad x \in V_+, \; \omega \in \Omega^* \,, \qquad (4.23)$$

for every $\gamma < -\lambda$.

Proof. From (4.7) we have $D_x\varphi(t,\omega,z)z \le D_x\varphi(t,\omega,0)z$ for any $z > 0$. Therefore from

$$\varphi(t,\omega)x = \int_0^1 D_x\varphi(t,\omega,sx)x\,ds$$

we get that

$$\varphi(t,\omega)x \le D_x\varphi(t,\omega,0)x \equiv \Phi(t,\omega)x, \quad x \in V_+.$$

Therefore (4.23) follows from Definition 1.9.1 of the top Lyapunov exponent.

□

4.3 Almost Equilibria

In this section we introduce the notion of an almost equilibrium and prove a theorem which gives a description of the long-time behaviour of strongly sublinear RDS with a strongly positive sub-equilibrium.

Definition 4.3.1. *A random variable* $u(\omega)$ *in* V_+ *is said to be an* almost equilibrium *of an RDS* (θ,φ) *if it is invariant under* φ *for almost all* $\omega \in \Omega$, *i.e. if there exists a set* $\Omega^* \in \mathcal{F}$ *such that* $\mathbb{P}(\Omega^*) = 1$ *and*

$$\varphi(t,\omega)u(\omega) = u(\theta_t\omega) \quad \text{for all} \quad t \ge 0 \quad \text{and all} \quad \omega \in \Omega^*. \tag{4.24}$$

The following assertion shows that we can choose the set Ω^* in (4.24) to be θ-invariant.

Proposition 4.3.1. *If* $u(\omega) \ge 0$ *is an almost equilibrium of an RDS* (θ,φ), *then there exists a* θ-*invariant set* $\Omega^{**} \in \bar{\mathcal{F}}^{\mathbb{P}}$ *of full measure such that* (4.24) *holds.*

Proof. Let

$$\tilde{\Omega} := \{\omega \ : \ \varphi(t,\omega)u(\omega) = u(\theta_t\omega) \text{ for all } t \ge 0\}.$$

Since $\Omega^* \subseteq \tilde{\Omega}$ and $\mathbb{P}(\Omega^*) = 1$, we have that $\theta_s\tilde{\Omega} \in \bar{\mathcal{F}}^{\mathbb{P}}$ and $\bar{\mathbb{P}}(\theta_s\tilde{\Omega}) = 1$ for every fixed $s \in \mathbb{R}$. Here $\bar{\mathbb{P}}$ is the extension of $\mathbb{P}$ on $\bar{\mathcal{F}}^{\mathbb{P}}$. Using the cocycle property we get

$$\varphi(t,\theta_s\omega)u(\theta_s\omega) = \varphi(t,\theta_s\omega)\varphi(s,\omega)u(\omega) = \varphi(t+s,\omega)u(\omega) = u(\theta_{t+s}\omega)$$

for all $t,s \ge 0$ and $\omega \in \tilde{\Omega}$. Hence $\theta_s\tilde{\Omega} \subset \tilde{\Omega}$ for all $s \ge 0$. Let $\Omega^{**} := \cap_{n\ge0}\theta_n\tilde{\Omega}$. It is clear that $\theta_t\Omega^{**} \subset \Omega^{**}$ for $t \ge 0$, $\bar{\mathbb{P}}(\Omega^{**}) = 1$ and (4.24) holds for $\omega \in \Omega^{**}$. Let $k - 1 \le t < k$ for $k \in \mathbb{N}$. Then

$$\theta_{-t}\Omega^{**} \subset \bigcap_{n\ge0}\theta_{n-t}\tilde{\Omega} = \bigcap_{n\ge0}\theta_{n-k}\theta_{k-t}\tilde{\Omega} \subset \bigcap_{m\ge0}\theta_m\theta_{k-t}\tilde{\Omega}$$

Since $\theta_{k-t}\tilde{\Omega} \subset \tilde{\Omega}$, we obtain $\theta_{-t}\Omega^{**} \subset \Omega^{**}$. Thus Ω^{**} is a θ-invariant set. □

Remark 4.3.1. If (θ, φ) is an RDS with discrete time, then in the proof of Proposition 4.3.1 we have $\tilde{\Omega} \in \mathcal{F}$ and therefore $\Omega^{**} \in \mathcal{F}$. Under this condition it is possible (cf. Remark 1.2.1(ii)) to find a version $\tilde{\varphi}$ of the cocycle φ such that $u(\omega)$ is an equilibrium for $(\theta, \tilde{\varphi})$. We also refer to SCHEUTZOW [90], where the perfection problem of crudely invariant elements is discussed for invertible cocycles.

For the main result of this section we need the following definitions.

Definition 4.3.2. *Let* $U \in \mathcal{F}$. *The orbit* $\gamma_a(\omega) = \cup_{t \geq 0} \varphi(t, \theta_{-t}\omega) a(\theta_{-t}\omega)$ *of the RDS* (θ, φ) *in* $X = V_+$ *emanating from* a *is said to be* bounded *on* U *if there exists a random variable* C *on* U *such that*

$$\|\varphi(t, \theta_{-t}\omega) a(\theta_{-t}\omega)\| \leq C(\omega) \quad \text{for all} \quad t \geq 0 \text{ and } \omega \in U .$$

The orbit γ_a *is said to be* bounded, *if it is bounded on the whole* Ω. *We say that the orbit* γ_a *is* unbounded *if it is not bounded.*

Definition 4.3.3. *An RDS* (θ, φ) *in* V_+ *is said to be* conditionally compact *if for any* $U \in \mathcal{F}$ *and for any orbit* $\gamma_a(\omega)$ *which is bounded on* U *there exists a family of compact sets* $K(\omega)$ *such that*

$$\lim_{t \to \infty} \text{dist}\, (\varphi(t, \theta_{-t}\omega) a(\theta_{-t}\omega), K(\omega)) = 0 \quad \omega \in U . \qquad (4.25)$$

We note that an RDS (θ, φ) in V_+ is conditionally compact if any orbit $\gamma_a(\omega)$ which is bounded on some $U \in \mathcal{F}$ is a relatively compact set for any $\omega \in U$.

Theorem 4.3.1. *Let* V *be a separable Banach space with a normal solid cone* V_+. *Assume that* (θ, φ) *is a strongly sublinear conditionally compact order-preserving RDS over an ergodic metric dynamical system* θ. *Suppose that there exists a sub-equilibrium* $a(\omega) \in \text{int}\, V_+$. *Then either*

(i) we have $\|\varphi(t, \theta_{-t}\omega) v(\theta_{-t}\omega)\| \to \infty$ *almost surely as* $t \to \infty$ *for every* $v(\omega) \in \text{int}\, V_+$ *such that* $v(\omega) \geq \alpha a(\omega)$ *for some nonrandom* $\alpha > 0$ *and for every* $\omega \in \Omega$ *or*

(ii) there exists a unique almost equilibrium $u(\omega) \gg 0$ *defined on a* θ-*invariant set* $\Omega^* \in \mathcal{F}$ *of full measure such that*

$$\lim_{t \to +\infty} \varphi(t, \theta_{-t}\omega) v(\theta_{-t}\omega) = u(\omega), \quad \omega \in \Omega^* , \qquad (4.26)$$

for any random variable $v(\omega)$ *possessing the property* $\alpha a(\omega) \leq v(\omega) \leq \lambda u(\omega)$ *for all* $\omega \in \Omega^*$ *and for some nonrandom positive* α *and* λ.

Proof. From Proposition 3.4.1 we get that $\{a_s(\omega) := \varphi(s, \theta_{-s}\omega) a(\theta_{-s}\omega), \, s > 0\}$ is a monotone family of sub-equilibria. Since the cone V_+ is normal, there exists an equivalent norm $\|\cdot\|_*$ on V such that $s \mapsto \|a_s(\omega)\|_*$ is a monotone nondecreasing function for every $\omega \in \Omega$ (see Remark 3.1.1) and therefore the limit $\lim_{s \to \infty} \|a_s(\omega)\|_*$ exists (finite or infinite). Thus if (i) is not true, then there exists $v(\omega) \in V_+$ such that $v(\omega) \geq \alpha a(\omega)$ for some $0 < \alpha < 1$ and $\|\varphi(t, \theta_{-t}\omega)v(\theta_{-t}\omega)\|_* \nrightarrow \infty$ for $\omega \in U$, where $U \in \mathcal{F}$ and $\mathbb{P}(U) > 0$. Therefore for any $\omega \in U$ there exists a sequence $\{t_n(\omega)\}$ such that $t_n(\omega) \to \infty$ as $n \to \infty$ and

$$\sup_n \|\varphi(t_n, \theta_{-t_n}\omega)v(\theta_{-t_n}\omega)\|_* < \infty, \quad \omega \in U .$$

Since (θ, φ) is sublinear, we have $\sup_n \|a_{t_n(\omega)}(\omega)\|_* < \infty$ for $\omega \in U$. This implies that

$$\sup_{s \geq 0} \|a_s(\omega)\|_* < \infty, \quad \omega \in U ,$$

because for any $\omega \in U$ and $s > 0$ there exists $t_n(\omega)$ such that $\|a_s(\omega)\|_* \leq \|a_{t_n(\omega)}(\omega)\|_*$. Consider the set

$$\tilde{U} := \{\omega : \sup_{s \geq 0} \|a_s(\omega)\|_* < \infty\} .$$

The monotonicity of $\|a_s(\omega)\|_*$ implies that

$$\tilde{U} = \{\omega : \sup_{k \in \mathbb{N}} \|a_k(\omega)\|_* < \infty\} = \bigcup_{N \in \mathbb{N}} \bigcap_{k \in \mathbb{N}} \{\omega : \|a_k(\omega)\|_* < N\}.$$

Thus $\tilde{U} \in \mathcal{F}$. Let us prove that $\tilde{U}$ is θ-invariant. Indeed, using the cocycle property for $0 \leq t \leq s$ we have

$$a_s(\theta_t\omega) = \varphi(s, \theta_{-s+t}\omega)a(\theta_{-s+t}\omega)$$

$$= \varphi(t, \omega)\varphi(s - t, \theta_{-s+t}\omega)a(\theta_{-s+t}\omega) = \varphi(t, \omega)a_{s-t}(\omega) .$$

Since $\{a_{s-t}(\omega) : s \geq t\}$ is a bounded set for every $t \geq 0$ and $\omega \in \tilde{U}$, it belongs to some interval $[0, b_t(\omega)]$ for all $\omega \in \tilde{U}$ and $t \geq 0$. Therefore

$$a_s(\theta_t\omega) \in [0, \varphi(t, \omega)b_t(\omega)], \quad \omega \in \tilde{U}.$$

Thus $\sup_{s \geq t} \|a_s(\theta_t\omega)\|_* < \infty$ for $\omega \in \tilde{U}$. Since $a_s(\theta_t\omega) \leq a_t(\theta_t\omega)$ for $0 \leq s \leq t$, we have

$$\sup_{s \geq 0} \|a_s(\theta_t\omega)\|_* < \infty, \quad \text{for all} \quad t \geq 0, \omega \in \tilde{U}.$$

Consequently $\theta_t\tilde{U} \subset \tilde{U}$ for $t \geq 0$ and therefore $U^* := \cap_{t \geq 0} \theta_t\tilde{U} = \cap_{n \in \mathbb{Z}_+} \theta_n\tilde{U}$ is a θ-invariant set such that $\mathbb{P}(U^*) = \mathbb{P}(\tilde{U}) > 0$. By the ergodicity of θ we have $\mathbb{P}(U^*) = 1$. Thus $\sup_{s \geq 0} \|a_s(\omega)\|_* < \infty$ on the θ-invariant set U^* of full measure.

Now we restrict the RDS (θ, φ) to U^*. Since (θ, φ) is conditionally compact, the limit

$$u(\omega) = \lim_{s \to \infty} a_s(\omega), \quad \omega \in U^*,$$

exists, and this is a strongly positive equilibrium by Proposition 3.5.2. Since $nu(\omega)$ is a super-equilibrium for every $n \in \mathbb{N}$, $\varphi(t, \theta_{-t}\omega)[nu(\theta_{-t}\omega)]$ converges to a strongly positive equilibrium which coincides with $u(\omega)$ on a θ-invariant set $\Omega^* \subset U^*$ of full measure (see Theorem 4.2.1). The set Ω^* can be chosen independent of n. Therefore using Theorem 4.2.3 we obtain (4.26).   □

Corollary 4.3.1. *Let V be a separable Banach space with a normal solid cone V_+. Assume that (θ, φ) is a strongly sublinear conditionally compact order-preserving RDS over an ergodic metric dynamical system θ. Suppose that $\varphi(t, \omega)0 \gg 0$ for all $t > 0$ and $\omega \in \Omega$. Then either*
  (i) for any $x \in V_+$ we have $\|\varphi(t, \theta_{-t}\omega)x\| \to \infty$ almost surely as $t \to \infty$
or
  (ii) there exists a unique almost equilibrium $u(\omega) \gg 0$ defined on a θ-invariant set $\Omega^* \in \mathcal{F}$ of full measure such that (4.26) holds for any random variable $v(\omega)$ possessing the property $0 \leq v(\omega) \leq \lambda u(\omega)$ for all $\omega \in \Omega^*$ and for some nonrandom $\lambda > 0$.

Proof. Proposition 3.4.1 implies that $a^\varepsilon(\omega) := \varphi(\varepsilon, \theta_{-\varepsilon}\omega)0 \gg 0$ is a sub-equilibrium for every $\varepsilon > 0$. It is also clear that $\varphi(t, \theta_{-t}\omega)x \geq a^\varepsilon(\omega)$ for all $x \in V_+$, $\omega \in \Omega$ and $t \geq \varepsilon$. Thus we can apply Theorem 4.3.1.   □

Remark 4.3.2. We note that the uniqueness results stated in Theorem 4.2.1 and Propositions 1.7.1 and 4.2.1 remain true for almost equilibria because the proof of Proposition 1.7.1 invokes only monotonicity arguments for scalar measurable functions and properties of probability distributions.

4.4 Limit Set Trichotomy for Sublinear RDS

In this section we prove the limit set trichotomy theorem which describes the only three possible asymptotic scenarios for sublinear systems. We do not assume the existence of a strongly positive sub-equilibrium here.

In the deterministic discrete time case a limit set trichotomy was discovered (and so named) by KRAUSE/RANFT [73] and generalized by KRAUSE/NUSSBAUM [72].

Below we say that a multifunction $\{F(\omega)\}$ belongs to the part C_v generated by a random variable $v(\omega)$ (see Definition 3.2.1) if there exists a nonrandom number $\lambda > 1$ such that $F(\omega) \subset [\lambda^{-1}v(\omega), \lambda v(\omega)]$ for all $\omega \in \Omega$.

Theorem 4.4.1 (Limit Set Trichotomy). *Let V be a separable Banach space with a normal solid minihedral cone V_+. Assume that (θ, φ) is a strongly* *sublinear conditionally compact order-preserving RDS in V_+. Let $v : \Omega \to$ int V_+ be a random variable, and denote by C_v its part in V_+. Assume there exists $a \in C_v$ such that the orbit emanating from a does not leave C_v, i.e.*

$$a_t(\omega) := \varphi(t, \theta_{-t}\omega)a(\theta_{-t}\omega) \in C_v \quad \text{for all} \quad t \geq 0 . \qquad (4.27)$$

Then C_v is a forward invariant set, i.e. (4.27) holds for any $a \in C_v$, and precisely one of the following three cases applies:

(i)   for all $b \in C_v$, the orbit γ_b emanating from b is unbounded;
(ii) for all $b \in C_v$, the orbit γ_b emanating from b is bounded, but the closure of γ_b does not belong to C_v;
(iii) there exists a unique almost equilibrium $u \in C_v$ measurable with respect to the universal σ-algebra $\mathcal{F}^u$, and for all $b \in C_v$ the orbit emanating from b converges to u, i.e.

$$\lim_{t \to +\infty} \varphi(t, \theta_{-t}\omega)b(\theta_{-t}\omega) = u(\omega) \quad \text{for almost all} \quad \omega \in \Omega . \qquad (4.28)$$

The proof of this theorem relies on the following three lemmas.

Lemma 4.4.1. *Let (θ, φ) be a sublinear order-preserving RDS in V_+ and let $v : \Omega \to V_+$.*

(i) Assume that there exist $a \in C_v$ and $t_0 \geq 0$ such that

$$\varphi(t_0, \theta_{-t_0}\omega)a(\theta_{-t_0}\omega) \in C_v . \qquad (4.29)$$

Then for any $b \in C_v$ we have

$$\varphi(t_0, \theta_{-t_0}\omega)b(\theta_{-t_0}\omega) \in C_v . \qquad (4.30)$$

(ii) Assume that there exists $a \in C_v$ for which (4.27) holds. Then for any $b \in C_v$ the orbit emanating from b does not leave C_v, i.e. C_v is forward invariant.

Proof. (i) Since $a, b \in C_v$, there exists a nonrandom number $\lambda \geq 1$ such that

$$\lambda^{-1}a(\omega) \leq b(\omega) \leq \lambda a(\omega) \quad \text{for all} \quad \omega \in \Omega . \qquad (4.31)$$

Therefore sublinearity and monotonicity give the inequality

$$\lambda^{-1}\varphi(t_0, \omega)a(\omega) \leq \varphi(t_0, \omega)b(\omega) \leq \lambda\varphi(t_0, \omega)a(\omega) \quad \text{for all} \quad \omega \in \Omega . \qquad (4.32)$$

Hence (4.29) implies (4.30).
    Assertion (ii) follows immediately from (i). $\square$

Lemma 4.4.2. *Let (θ, φ) be a sublinear order-preserving RDS in V_+ and let $v : \Omega \to V_+$. Assume that for some $a \in C_v$ the orbit γ_a emanating from a does not leave C_v and is bounded. Then for any $b \in C_v$ the orbit γ_b emanating from b is also bounded.*

Proof. If $\gamma_a \subset C_v \subset V_+$ is bounded, by Proposition 3.2.2 there exists a random element $w(\omega) \in \text{int}V_+$ such that

$$0 \leq \varphi(t, \theta_{-t}\omega)a(\theta_{-t}\omega) \leq w(\omega) \quad \text{for all} \quad t > 0, \ \omega \in \Omega .$$

Hence (4.32) implies that

$$0 \leq \varphi(t, \theta_{-t}\omega)b(\theta_{-t}\omega) \leq \lambda w(\omega) \quad \text{for all} \quad t > 0, \ \omega \in \Omega ,$$

where b is an arbitrary element with property (4.31). The normality of the cone V_+ implies that γ_b is bounded. $\qquad\square$

Lemma 4.4.3. *Let (θ, φ) be a sublinear order-preserving RDS in V_+ and let $v : \Omega \to V_+$. Assume that for some $a \in C_v$ the orbit γ_a is bounded and its closure $\overline{\gamma_a(\cdot)}$ belongs to C_v. Then this property is valid for any $b \in C_v$.*

Proof. Let $\bar{b}(\omega) \in \overline{\gamma_b(\omega)}$. Then there exists a sequence $\{t_n(\omega)\}$ such that

$$\varphi(t_n, \theta_{-t_n}\omega)b(\theta_{-t_n}\omega) \to \bar{b}(\omega) .$$

By (4.31) and (4.32)

$$\lambda^{-1}\varphi(t_n, \theta_{-t_n}\omega)a(\theta_{-t_n}\omega) \leq \varphi(t_n, \theta_{-t_n}\omega)b(\theta_{-t_n}\omega) \leq \lambda\varphi(t_n, \theta_{-t_n}\omega)a(\theta_{-t_n}\omega) .$$
$$(4.33)$$

Since $\overline{\gamma_a(\cdot)} \subset C_v$, there exists $\mu > 1$ such that

$$\mu^{-1}v(\omega) \leq \varphi(t, \theta_{-t}\omega)a(\theta_{-t}\omega) \leq \mu v(\omega) \quad \text{for all} \quad t \geq 0, \ \omega \in \Omega .$$

Therefore (4.33) implies that

$$\mu^{-1}\lambda^{-1}v(\omega) \leq \varphi(t_n, \theta_{-t_n}\omega)b(\theta_{-t_n}\omega) \leq \mu\lambda v(\omega) .$$

Consequently $\bar{b} \in [\mu^{-1}\lambda^{-1}v(\omega), \mu\lambda v(\omega)] \subset C_v$. $\qquad\square$

We are now in a position to prove the limit set trichotomy theorem.

Proof of Theorem 4.4.1. By Lemma 4.4.1(ii), C_v is forward invariant.

We now consider the trichotomy. If (i) is not true, then by Lemma 4.4.2 all orbits are bounded. We hence have either (ii), or there exists an orbit whose closure belongs to C_v. If the latter is the case, Lemma 4.4.3 implies that the closure of each orbit belongs to C_v. Therefore the omega-limit set of each element of C_v belongs to C_v. We will now prove that all omega-limit sets coincide with a one-point set consisting of the unique equilibrium $u \in C_v$.

Let Γ_a be the omega-limit set of $a \in C_v$. Since $\Gamma_a = \cap_{n\in\mathbb{Z}_+}\overline{\gamma_a^n(\omega)}$, where $\gamma_a^n(\omega)$ is the tail of the orbit $\gamma_a(\omega)$, we have from (4.25) and Proposition 1.5.1 that Γ_a is a random compact set with respect to the universal σ-algebra $\mathcal{F}^u$

(cf. Remark 1.6.1). Since $\Gamma_a \subset C_v$, there exists a number $\alpha > 1$ such that

$$\alpha^{-1}v(\omega) \leq w(\omega) \leq \alpha v(\omega) \quad \text{for all} \quad w(\omega) \in \Gamma_a(\omega) .$$

Hence by Theorem 3.2.1

$$\overline{w}(\omega) := \sup \Gamma_a(\omega) \gg 0 ,$$

exists and it is an $\mathcal{F}^u$-measurable variable. We also have

$$\alpha^{-1}v(\omega) \leq \overline{w}(\omega) \leq \alpha v(\omega) . \tag{4.34}$$

The invariance of Γ_a, i.e.

$$\Gamma_a(\theta_t\omega) = \varphi(t,\omega)\Gamma_a(\omega) \quad \text{for all} \quad t \geq 0 ,$$

(cf. Lemma 3.4.1) implies that $\overline{w}$ is a sub-equilibrium. It is clear that the multifunction $\omega \mapsto \Gamma_a(\theta_t\omega)$ is a random compact set with respect to $\mathcal{F}^u$ for any fixed $t \in \mathbb{R}_+$. Therefore $\overline{w}(\theta_t\omega)$ is an $\mathcal{F}^u$-measurable variable for any fixed $t \in \mathbb{R}_+$.

Similarly,

$$\underline{w}(\omega) := \inf \Gamma_a(\omega)$$

is an $\mathcal{F}^u$-measurable super-equilibrium such that $\underline{w}(\theta_t\omega)$ is an $\mathcal{F}^u$-measurable variable for any fixed $t \in \mathbb{R}_+$ and

$$\alpha^{-1}v(\omega) \leq \underline{w}(\omega) \leq \alpha v(\omega) . \tag{4.35}$$

It follows from (4.34) and (4.35) that

$$p(\underline{w}(\omega), \overline{w}(\omega)) < 2 \log \alpha \quad \text{for all} \quad \omega \in \Omega ,$$

where $p(\cdot, \cdot)$ stands for the part metric.

Clearly $\underline{w}(\omega) \leq \overline{w}(\omega)$ for all $\omega \in \Omega$. Since $\underline{w}(\omega)$ and $\overline{w}(\omega)$ are super- and sub-equilibria, respectively, we have

$$\varphi(t,\omega)\underline{w}(\omega) \leq \underline{w}(\theta_t\omega) \leq \overline{w}(\theta_t\omega) \leq \varphi(t,\omega)\overline{w}(\omega) \tag{4.36}$$

for all $\omega \in \Omega$ and $t \geq 0$. This inequality and Lemma 3.1.1 imply

$$p(\varphi(t,\omega)\underline{w}(\omega), \varphi(t,\omega)\overline{w}(\omega)) \geq p(\underline{w}(\theta_t\omega), \overline{w}(\theta_t\omega)) \tag{4.37}$$

for all $\omega \in \Omega$ and $t \geq 0$. On the other hand, since $\varphi(t,\omega)$ is sublinear,

$$p(\varphi(t,\omega)\underline{w}(\omega), \varphi(t,\omega)\overline{w}(\omega)) \leq p(\underline{w}(\omega), \overline{w}(\omega)) \quad \text{for all} \quad \omega \in \Omega \quad \text{and} \quad t \geq 0 ,$$

implying

$$p(\underline{w}(\theta_t\omega), \overline{w}(\theta_t\omega)) \leq p(\underline{w}(\omega), \overline{w}(\omega)) \quad \text{for all} \quad \omega \in \Omega \quad \text{and} \quad t \geq 0 .$$

Proposition 3.2.4 implies that $f_t(\omega) := p(\underline{w}(\theta_t\omega), \overline{w}(\theta_t\omega))$ is an $\mathcal{F}^u$-measurable variable for any fixed $t \in \mathbb{R}_+$. Let us prove that f_t has the same distribution for each $t \in \mathbb{R}_+$. Let $U_c^t = \{\omega : f_t(\omega) \leq c\}$. Since $U_c^t \in \mathcal{F}^u$, there exists $\tilde{U}_c^t \in \mathcal{F}$ such that $\tilde{U}_c^t \subseteq U_c^t$ and $\mathbb{P}(\tilde{U}_c^t) = \bar{\mathbb{P}}(U_c^t)$, where $\bar{\mathbb{P}}$ is the extension of $\mathbb{P}$ to $\mathcal{F}^u$. It is clear that $\theta_{-t}\tilde{U}_c^0 \subset U_c^t$. Therefore

$$\bar{\mathbb{P}}(U_c^t) \geq \mathbb{P}(\theta_{-t}\tilde{U}_c^0) = \mathbb{P}(\tilde{U}_c^0) = \bar{\mathbb{P}}(U_c^0) \ .$$

In a similar way the relation $\theta_t \tilde{U}_c^t \subset U_c^0$ implies $\bar{\mathbb{P}}(U_c^0) \geq \bar{\mathbb{P}}(U_c^t)$. Thus all variables f_t have the same distribution.

Suppose now that $\underline{w}(\omega) = \overline{w}(\omega)$ is not true on a set of positive probability, i.e. there exist a measurable set $U \subset \Omega$ with $\mathbb{P}(U) > 0$ such that

$$\underline{w}(\omega) < \overline{w}(\omega) \quad \text{for} \quad \omega \in U \ . \tag{4.38}$$

Property (4.38) and strong sublinearity imply

$$p(\varphi(t,\omega)\underline{w}(\omega), \varphi(t,\omega)\overline{w}(\omega)) < p(\underline{w}(\omega), \overline{w}(\omega)) \quad \text{for} \quad \omega \in U \quad \text{and} \quad t > 0 \ ,$$

hence

$$p(\underline{w}(\theta_t\omega), \overline{w}(\theta_t\omega)) < p(\underline{w}(\omega), \overline{w}(\omega)) \quad \text{for} \quad \omega \in U \quad \text{and} \quad t > 0 \ .$$

However, both sides of the last inequality have the same distribution, leading, as in the proof of Proposition 1.7.1, to a contradiction of the assumption $\mathbb{P}(U) > 0$. Thus $\underline{w}(\omega) = \overline{w}(\omega)$ almost surely, and (4.36) implies that $u(\omega) \equiv \underline{w}(\omega)$ is an almost equilibrium. Moreover $\Gamma_a = \{u(\omega)\}$ almost surely. It finally follows from (4.35) that $u \in C_v$. Proposition 4.2.1 and Remark 4.3.2 imply that this equilibrium is unique almost surely in C_v. In particular $\Gamma_b = \{u(\omega)\}$ almost surely for any $b \in C_v$ which gives the asymptotic stability (4.28). This completes the proof of Theorem 4.4.1.   □

Remark 4.4.1. (i)   It is clear from the proof that if the cases (i) and (ii) of Theorem 4.4.1 do not apply and if there exists an element $a \in C_v$ such that $\overline{\gamma_a(\omega)}$ is a random compact set with respect to $\mathcal{F}$, then case (iii) holds with the equilibrium measurable with respect to $\mathcal{F}$.

(ii) For a discrete RDS ($\mathbb{T} = \mathbb{Z}$) the equilibrium given by Theorem 4.4.1 in case (iii) is measurable with respect to $\mathcal{F}$ because the closure of any trajectory is $\mathcal{F}$-measurable (see Sect. 1.5).

(iii) Theorem 4.4.1 is wrong without the assumption of *strong* sublinearity, see Remark 4.2.1(i).

By slightly strengthening hypothesis (4.27) we can also prove another version of the trichotomy theorem.

Corollary 4.4.1. *Assume that the assumptions of Theorem 4.4.1 hold and property (4.27) is valid in a strengthened form: there exists an $a \in C_v$ such* that the orbit emanating from a does not leave C_v and for any $T > 0$ there exists $\lambda_T > 1$ such that

$$\lambda_T^{-1} v(\omega) \leq \varphi(t, \theta_{-t}\omega) a(\theta_{-t}\omega) \leq \lambda_T v(\omega) \quad \text{for all} \quad t \in [0, T] \,. \qquad (4.39)$$

Then property (4.39) holds for any $a \in C_v$, and precisely one of the following three cases applies:

(i)   for all $b \in C_v$, the orbit γ_b emanating from b is unbounded;
(ii) for all $b \in C_v$, the orbit γ_b emanating from b is bounded, but

$$\limsup_{t \to \infty} \left\{ \sup_{\omega \in \Omega} p(\varphi(t, \theta_{-t}\omega) b(\theta_{-t}\omega), v(\omega)) \right\} = 0 \,; \qquad (4.40)$$

(iii) there exists a unique almost equilibrium $u \in C_v$ measurable with respect to the universal σ-algebra $\mathfrak{F}^u$, and for all $b \in C_v$ the orbit emanating from b converges to u, i.e. (4.28) holds.

Proof. Theorem 4.4.1 is applicable here. We need only prove (4.40) in case (ii). If (4.40) is not true, then (4.39) implies that

$$\lambda_\infty^{-1} v(\omega) \leq \varphi(t, \theta_{-t}\omega) b(\theta_{-t}\omega) \leq \lambda_\infty v(\omega) \quad \text{for all} \quad t \geq 0, \ \omega \in \Omega$$

with some constant $\lambda_\infty > 1$. This implies that

$$\overline{\gamma_b(\omega)} \subset [\lambda_\infty^{-1} v(\omega), \lambda_\infty v(\omega)] \subset C_v$$

which is impossible in case (ii) of Theorem 4.4.1.   □

For one-dimensional sublinear RDS we have the following version of the trichotomy theorem which requires the continuity of the mapping $t \mapsto \varphi(t, \theta_{-t}\omega)x$.

Theorem 4.4.2. *Let (θ, φ) is be a strongly sublinear order-preserving RDS in $\mathbb{R}_+$ over an ergodic metric dynamical system θ. Assume that the function $t \mapsto \varphi(t, \theta_{-t}\omega)x$ is continuous for all $x \in \mathbb{R}_+$ and $\omega \in \Omega$. Then precisely one of the following three cases applies:*

(i) $\limsup_{t \to +\infty} \varphi(t, \theta_{-t}\omega)x = \infty$ *almost surely for all $x > 0$;*
(ii) $\lim_{t \to +\infty} \varphi(t, \theta_{-t}\omega)x = 0$ *almost surely for all $x \geq 0$;*
(iii) *there exists a unique $\mathfrak{F}$-measurable almost equilibrium $u(\omega) > 0$ defined on a θ-invariant set Ω^* of full measure such that*

$$\lim_{t \to +\infty} \varphi(t, \theta_{-t}\omega) b(\theta_{-t}\omega) = u(\omega), \quad \omega \in \Omega^* \,, \qquad (4.41)$$

for any $b(\omega)$ with the property $\lambda^{-1} u(\omega) \leq b(\omega) \leq \lambda u(\omega)$ for all $\omega \in \Omega^*$ and for some $\lambda > 1$.

Proof. If (i) is not true, then there exist $x_0 > 0$ and a set $U \in \mathcal{F}$ such that $\mathbb{P}(U) > 0$ and $\sup_{t \in \mathbb{R}_+} \varphi(t, \theta_{-t}\omega)x_0 < \infty$ for $\omega \in U$. Let

$$\tilde{U} := \{\omega : \sup_{t \in \mathbb{R}_+} \varphi(t, \theta_{-t}\omega)x_0 < \infty\} .$$

Since

$$\tilde{U} = \{\omega : \sup_{t \in \mathbb{Q} \cap \mathbb{R}_+} \varphi(t, \theta_{-t}\omega)x_0 < \infty\} = \bigcup_{N \in \mathbb{N}} \bigcap_{t \in \mathbb{Q} \cap \mathbb{R}_+} \{\omega : \varphi(t, \theta_{-t}\omega)x_0 < N\} ,$$

the set $\tilde{U}$ is measurable. Thus as in the proof of Theorem 4.3.1 we can obtain that there exists a θ-invariant set Ω^* of full measure such that

$$\limsup_{t \to \infty} \varphi(t, \theta_{-t}\omega)x_0 < \infty \quad \text{for all} \quad \omega \in \Omega^* . \tag{4.42}$$

Therefore by Remark 1.6.1 and Proposition 1.6.4 the omega-limit set $\Gamma_{x_0}(\omega)$ emanating from x_0 is a nonempty invariant compact random set measurable with respect to the σ-algebra $\mathcal{F}$. Since $\sup B \in B$ for any compact set $B \subset \mathbb{R}_+$, Lemma 3.4.1 and Remark 3.4.2(ii) imply that $u(\omega) := \sup \Gamma_{x_0}(\omega) \geq 0$ is an $\mathcal{F}$-measurable equilibrium on Ω^*. By Lemma 3.5.1 we have either $u(\omega) = 0$ or $u(\omega) > 0$ almost surely.

If $u(\omega) = 0$ almost surely, then $\varphi(t, \theta_{-t}\omega)x \to 0$ almost surely for all $0 \leq x \leq x_0$. The sublinearity implies that

$$\varphi(t, \omega)x = \varphi(t, \omega)\left[\frac{x}{x_0} \cdot x_0\right] \leq \frac{x}{x_0} \cdot \varphi(t, \omega)x_0 \quad \text{for all} \quad x \geq x_0 . \tag{4.43}$$

Thus $\varphi(t, \theta_{-t}\omega)x \to 0$ almost surely for all $x \in \mathbb{R}_+$.

If $u(\omega) > 0$ almost surely, then from (4.42) and (4.43) we have

$$\limsup_{t \to \infty} \varphi(t, \theta_{-t}\omega)x < \infty \quad \text{for all} \quad x \in \mathbb{R}_+, \ \omega \in \Omega^* .$$

Therefore by the same argument $u_x(\omega) := \sup \Gamma_x(\omega)$ is an $\mathcal{F}$-measurable positive equilibrium for every $x > 0$. By Theorem 4.2.1 we have that $u_x(\omega) = u(\omega)$ on a θ-invariant set of full measure. Thus Theorem 4.2.2 implies (4.41). $\square$

The following two simple examples of discrete systems show that all three cases of the limit set trichotomy can actually occur. The corresponding examples of RDS with continuous time are discussed in Chaps.5 and 6.

We start with the deterministic case.

Example 4.4.1. Let us consider the scalar function $f_\alpha(x) = \alpha x + \frac{x}{1+x}$ on $\mathbb{R}_+$. It is easy to see that for every $\alpha \in \mathbb{R}_+$ the mapping $x \mapsto f_\alpha(x)$ generates a strongly sublinear dynamical system in $\mathbb{R}_+$. The hypotheses of Theorem 4.4.1 hold for this system with $v = 1$. If $\alpha \geq 1$, then $f_\alpha^n(x) \to \infty$ for any $x > 0$, i.e.

any orbit γ_x emanating from $x > 0$ is unbounded. If $\alpha = 0$, then $f_\alpha^n(x) \to 0$ for any $x \geq 0$, i.e. any orbit γ_x is bounded, but the closure of γ_x contains elements (namely 0) which do not belong to any part $C_v \subset \text{int}\mathbb{R}_+$. Finally for $\alpha \in (0,1)$ there exists a unique globally asymptotically stable positive equilibrium. To produce more complicated limit behaviour we can consider the mapping $f = (f_{\alpha_1}, \ldots, f_{\alpha_d})$ from $\mathbb{R}_+^d$ into itself with appropriate choices of α_i.

Now using the properties of the functions f_α we can can easily construct a random example.

Example 4.4.2. Let us consider the RDS on $\mathbb{R}_+$ constructed in the example given in the Introduction with $f_0(x) = \alpha_0 x + \frac{x}{1+x}$ and $f_1(x) = \alpha_1 x + \frac{x}{2+2x}$, where $0 \leq \alpha_0 \leq \alpha_1$. As in the previous example it is clear that these two mappings generate a strongly sublinear RDS in $\mathbb{R}_+$. Since $\alpha_0 < f_i(1) \leq \alpha_1 + 1$ for $i = 1, 2$, the random part C_v generated by $v(\omega) \equiv 1$ is forward invariant. Therefore the trichotomy theorem applies. As in the previous example it is easy to see that (i) if $\alpha_1 \geq \alpha_0 \geq 1$, then any orbit γ_x emanating from $x > 0$ is unbounded; (ii) if $\alpha_0 = \alpha_1 = 0$, then any orbit γ_x is bounded, but the closure of γ_x contains elements (namely 0) which do not belong to the part C_v; (iii) if $\alpha_0, \alpha_1 \in (0,1)$, then there exists a unique globally asymptotically stable positive equilibrium. As above, using these properties we can produce more complicated limit behaviour.

4.5 Random Mappings

In this section we consider a sublinear order-preserving RDS generated by random mappings in $\mathbb{R}_+^d$.

Let $\theta = (\Omega, \mathcal{F}, \mathbb{P}, \{\theta_n, n \in \mathbb{Z}\})$ be a metric dynamical system with discrete time $\mathbb{T} = \mathbb{Z}$. Assume that the function $f : \Omega \times \mathbb{R}_+^d \to \mathbb{R}_+^d$ is measurable and has the following properties:

(i) $f(\omega, \cdot)$ is continuous for every $\omega \in \Omega$,
(ii) $f(\omega, \cdot)$ is order-preserving, i.e. $f(\omega, x) \leq f(\omega, y)$ for all $0 \leq x \leq y$ and all $\omega \in \Omega$,
(iii) $f(\omega, \cdot)$ is sublinear, i.e. $\lambda f(\omega, x) \leq f(\omega, \lambda x)$ for $0 < \lambda < 1$, all $x \in \mathbb{R}_+^d$ and $\omega \in \Omega$.

Under (i) to (iii) the random difference equation

$$x_{n+1} = f(\theta_n \omega, x_n) \tag{4.44}$$

generates a sublinear order-preserving RDS in $\mathbb{R}_+^d$.

We note that assumptions (i) to (iii) are fulfilled, for instance, for the function

$$f(\omega, x) = \sum_{k=1}^{N} A_k(\omega) x^{\alpha_k} + b(\omega) , \qquad (4.45)$$

where $A_k(\omega)$ are $d \times d$ matrices with nonnegative entries, $b(\omega) \in \mathbb{R}_+^d$, $\alpha_k = (\alpha_k^1, \dots, \alpha_k^d)$ are multi-indices with $0 < \alpha_k^j \leq 1$, and $x^{\alpha_k} := (x_1^{\alpha_k^1}, \dots, x_d^{\alpha_k^d})$. It can be easily checked that the sublinearity condition (iii) is valid in the form

$$\lambda^\alpha f(\omega, x) \leq f(\omega, \lambda x), \quad 0 < \lambda < 1, \quad x \in \mathbb{R}_+^d, \quad \omega \in \Omega ,$$

where $\alpha = \max_{j,k} \alpha_k^j \leq 1$. Consequently

$$p(f(\omega, x), f(\omega, y)) \leq \alpha p(x, y), \quad x, y \in C \subset \mathbb{R}_+^d ,$$

where p is the part metric and C is any part of $\mathbb{R}_+^d$.

Hence if $\alpha < 1$ then f is uniformly contractive with respect to p. This makes it possible to use standard fixed point methods to prove the existence of equilibria for this case.

Proposition 4.5.1. *Assume that $f(\omega, x)$ has the form (4.45) with the parameters α_k^j possessing the property $\alpha := \max_{j,k} \alpha_k^j < 1$. Let $v(\omega) > 0$ for all $\omega \in \Omega$. If the part C_v generated by v is invariant for the RDS (θ, φ) defined by (4.44), then there exists a unique equilibrium $u(\omega)$ in C_v for (θ, φ) and*

$$\sup_{\omega \in \Omega} p(\varphi(n, \theta_{-n}\omega)w(\theta_{-n}\omega), u(\omega)) \leq \alpha^n \sup_{\omega \in \Omega} p(w(\omega), u(\omega)) \qquad (4.46)$$

for all $w \in C_v$ and $n \in \mathbb{Z}_+$.

Proof. We define the mapping $T : C_v \mapsto C_v$ by the formula

$$(Tw)(\omega) := \varphi(1, \theta_{-1}\omega)w(\theta_{-1}\omega) = f(\theta_{-1}\omega, w(\theta_{-1}\omega)), \quad \omega \in \Omega .$$

It is easy to see that

$$\varrho(Tw_1, Tw_2) \leq \alpha \varrho(w_1, w_2) \quad \text{for all} \quad w_1, w_2 \in C_v , \qquad (4.47)$$

where $\varrho(w_1, w_2) = \sup_{\omega \in \Omega} p(w_1(\omega), w_2(\omega))$. By Proposition 3.2.3 C_v is a complete metric space with respect to ϱ. Therefore we can apply the contraction principle and conclude that the mapping T has unique stationary point $u(\omega)$ in C_v. Relation (4.46) easily follows from (4.47). □

The following assertion gives a sufficient condition for the existence of an invariant part C_v.

Corollary 4.5.1. *Assume that the entries of the matrices $A_k(\omega)$ are bounded from above by a nonrandom constant and $\alpha = \max_{j,k} \alpha_k^j < 1$. Let $b(\omega) = b_0(\omega) \cdot v$, where $v \in \mathrm{int}\mathbb{R}_+^d$ and $b_0(\omega) > 0$ is a scalar random variable such that*

$$0 < \beta_0 \le \beta_1 b_0(\omega) \le b_0(\theta_1 \omega) \le \beta_2 b_0(\omega), \quad \omega \in \Omega,$$

for some nonrandom β_i. Then the part C_b generated by b is invariant for the RDS (θ, φ) generated by (4.44) with f given by (4.45) and the conclusions of Proposition 4.44 hold.

Proof. A simple calculation shows that $b(\omega) \le f(\omega, b(\omega)) \le Cb(\omega)$ for some constant $C > 0$. This implies that the part C_b is invariant and therefore we can apply Proposition 4.5.1.   □

Remark 4.5.1. (i) In the situation of Corollary 4.5.1 the equilibrium $u(\omega)$ is globally stable not only in C_b. It is easy to see that

$$p(\varphi(n,\omega)w(\omega), u(\theta_n\omega)) \le \alpha^n p(w(\omega), u(\omega)) \quad \text{for all} \quad \omega \in \Omega \qquad (4.48)$$

provided that $p(w(\omega), u(\omega))$ is finite for each $\omega \in \Omega$. Therefore for each $\omega \in \Omega$ we have that

$$p(\varphi(n, \theta_{-n}\omega)w(\theta_{-n}\omega), u(\omega)) \to 0 \quad \text{as} \quad n \to \infty$$

with exponential rate provided that $p(w(\omega), u(\omega))$ is a tempered random variable. We also note that relation (4.48) means that $u_n(\omega) := u(\theta_n\omega)$ is a forward exponentially attracting stationary process.

(ii) The deterministic example $f(x) = (\sqrt{x_1}, \dots, \sqrt{x_d})$ shows that every part of the cone $\mathbb{R}_+^d$ can contain an equilibrium which is exponentially stable in this part.

(iii) Assertions similar to Proposition 4.5.1 and Corollary 4.5.1 can be proved for more general mappings. Assume that $f : \Omega \times \mathbb{R}_+^d \to \mathbb{R}_+^d$ is a measurable function such that $f(\omega, \cdot) \in C^1(\mathrm{int}\mathbb{R}_+^d)$ for every $\omega \in \Omega$. Then the property

$$\sum_{j=1}^{d} x_j \left| \frac{\partial f_i(\omega, x)}{\partial x_j} \right| < \alpha(\omega) f_i(\omega, x), \quad i = 1, \dots, d, \ x \in \mathrm{int}\mathbb{R}_+^d, \ \omega \in \Omega,$$

where $\alpha(\omega)$ is a positive random variable, implies that

$$p(f(\omega, x), f(\omega, y)) < \alpha(\omega) p(x, y), \quad x, y \in \mathrm{int}\mathbb{R}_+^d, \ \omega \in \Omega,$$

where p is the part metric (see KRAUSE/NUSSBAUM [72, Theorem 4.1]). Thus under the condition $\alpha(\omega) \le \alpha_0 < 1$ we can obtain the same results as for the mapping (4.45).

The following assertion deals with another class of mappings and is an application of the limit set trichotomy theorem.

Proposition 4.5.2. *Assume that the measurable mapping* $f : \Omega \times \mathbb{R}_+^d \to \mathbb{R}_+^d$ *possesses properties (i) and (ii) and also*

(iii*) for each $\omega \in \Omega$ *the function* $f(\omega, \cdot)$ *is strongly sublinear, i.e.*

$$\lambda f(\omega, x) \ll f(\omega, \lambda x) \quad \text{for all} \quad 0 < \lambda < 1 \quad \text{and} \quad x \in \text{int}\mathbb{R}_+^d \ ;$$

(iv)   there exist points $\underline{x}$ *and* $\overline{x}$ *in* $\text{int}\mathbb{R}_+^d$ *such that*

$$f(\omega, \underline{x}) \geq \underline{x} \quad \text{and} \quad f(\omega, \overline{x}) \leq \overline{x} \quad \text{for all} \quad \omega \in \Omega \ . \tag{4.49}$$

Then there exists a unique strongly positive equilibrium $u(\omega)$ *for the RDS generated by (4.44). This equilibrium is uniformly separated from 0 and from* ∞*, i.e. there exist positive constants* α *and* β *such that* $\alpha \mathbf{e} \leq u(\omega) \leq \beta \mathbf{e}$ *for all* $\omega \in \Omega$*, where* $\mathbf{e} = (1, \dots, 1)$*. Moreover the equilibrium* $u(\omega)$ *is globally asymptotically stable in* $\text{int}\mathbb{R}_+^d$*, i.e. for every* $x \in \text{int}\mathbb{R}_+^d$ *we have*

$$\lim_{n \to +\infty} \varphi(n, \theta_{-n}\omega)x = u(\omega) \quad \text{for almost all} \quad \omega \in \Omega. \tag{4.50}$$

Proof. Relations (4.49) imply that $\underline{x}^m := m^{-1}\underline{x}$ is a super-equilibrium and $\overline{x}^m := m\overline{x}$ is a sub-equilibrium for each $m \in \mathbb{N}$. Therefore every deterministic interval $[\underline{x}^m, \overline{x}^m]$ with m large enough is an invariant set (see Remark 3.4.1). Hence the part $C_{\mathbf{e}}$ is invariant and option (iii) of Theorem 4.4.1 is the only possible one.   □

Proposition 4.5.2 allow as to obtain the following assertion which slightly strengthens a result by BHATTACHARYA/LEE [16, Sect.4] concerning asymptotic behaviour of a class of Markov chains generated by families of random mappings as described in Example 1.2.1.

Let $[a, b]$ be an interval in $\text{int}\mathbb{R}_+^d$. In a similar way to BHATTACHARYA/LEE [16] we introduce the class $\mathcal{A}_{a,b}$ of sets in $[a, b]$ of the form $A_c = \{x \ : \ h(x) \leq c\}$, where h varies over the class of all continuous nondecreasing functions from $[a, b]$ into itself and we define the semidistances

$$d_{a,b}(\nu_1, \nu_2) = \sup\{|\nu_1(A) - \nu_2(A)| \ : \ A \in \mathcal{A}_{a,b}\}$$

on the space all probability measures on $(\text{int}\mathbb{R}_+^d, \mathcal{B}(\text{int}\mathbb{R}_+^d))$. The function $d_{a,b}(\cdot, \cdot)$ is a distance if we restrict ourselves to measures with support in $[a, b]$.

Theorem 4.5.1. *Assume that the hypotheses of Proposition 4.5.2 hold and that the random mappings* $f(\theta_n\omega, \cdot)$*,* $n \in \mathbb{Z}$*, are independent and identically distributed (i.i.d.). Let* $u(\omega)$ *be the equilibrium given by Proposition 4.5.2 for the RDS* (θ, φ) *generated by (4.44). Then*

(i)   the family of the random sequences

$$\{\Phi_n^x := \varphi(n,\omega)x \ : \ n \in \mathbb{Z}_+, \ x \in \mathbb{R}_+^d\}$$

is a homogeneous Markov chain with state space $\mathbb{R}_+^d$ and transition probability

$$P(x,B) := \mathbb{P}\{\Phi_{n+1} \in B \,|\, \Phi_n = x\} = \mathbb{P}\{\omega \ : \ f(\omega,x) \in B\}, \quad B \in \mathcal{B}(\mathbb{R}_+^d) \, ;$$

(ii)   the measure ν on $(\mathbb{R}_+^d, \mathcal{B}(\mathbb{R}_+^d))$ defined by the formula

$$\nu(A) := \mathbb{P}\{\omega \ : \ u(\omega) \in A\}, \quad A \in \mathcal{B}(\mathbb{R}_+^d) \, ,$$

has compact support in $\mathrm{int}\mathbb{R}_+^d$ and is an invariant probability measure for the Markov chain $\{\Phi_n^x\}$, i.e

$$\nu(A) = (P^*\nu)(A) := \int_{\mathbb{R}_+^d} P(x,A)\nu(dx), \quad \text{for all} \quad A \in \mathcal{B}(\mathbb{R}_+^d) \, ;$$

(iii) for every compact set $K \subset \mathrm{int}\mathbb{R}_+^d$ we have that

$$\lim_{n\to\infty} \sup_{x\in K} |p^{(n)}(x,[a,b]) - \nu([a,b])| = 0 \qquad (4.51)$$

for any interval $[a,b] \subset \mathrm{int}\mathbb{R}_+^d$, where $p^{(n)}(x,A) = \mathbb{P}\{\omega \ : \ \varphi(n,\omega)x \in A\}$ for $A \in \mathcal{B}(\mathbb{R}_+^d)$;

(iv) $(P^{*n}\lambda)(A) \to \nu(A)$ as $n \to \infty$ for all $A \in \mathcal{B}(\mathbb{R}_+^d)$ and for any probability measure λ on $(\mathbb{R}_+^d, \mathcal{B}(\mathbb{R}_+^d))$ with compact support in $\mathrm{int}\mathbb{R}_+^d$.

If the equilibrium $u(\omega)$ possesses the property

$$\mathbb{P}\{\omega \ : \ u(\omega) \le a\} > 0 \quad \text{and} \quad \mathbb{P}\{\omega \ : \ u(\omega) \ge a\} > 0 \qquad (4.52)$$

for some $a \in \mathrm{int}\mathbb{R}_+^d$, then for any $m \in \mathbb{N}$ large enough we have

$$\sup\left\{ d_{\underline{x}^m,\overline{x}^m}\left(p^{(n)}(x,\cdot),\nu\right) \ : \ x \in [\underline{x}^m,\overline{x}^m]\right\} \to 0 \qquad (4.53)$$

exponentially fast as $n \to \infty$. Here $\underline{x}^m = m^{-1}\underline{x}$ and $\overline{x}^m = m\overline{x}$, where $\underline{x}$ and $\overline{x}$ satisfy (4.49).

Proof. Items (i) and (ii) follow from the general assertion proved by ARNOLD [3, Theorem 2.1.4] (see also the discussion in Sect.1.10). The support of ν is a compact set in $\mathrm{int}\mathbb{R}_+^d$ because $u(\omega) \in [\alpha e, \beta e]$ for all $\omega \in \Omega$.

(iii) Any compact K belongs to the interval $[\underline{x}^m, \overline{x}^m]$ with m large enough. Therefore the relation

$$\varphi(n,\omega)\underline{x}^m \le \varphi(n,\omega)x \le \varphi(n,\omega)\overline{x}^m, \quad x \in K \, , \qquad (4.54)$$

implies

$$\mathbb{P}\{\omega \,:\, \varphi(n,\omega)\overline{x}^m \le b\} \le p^{(n)}(x,[0,b]) \le \mathbb{P}\{\omega \,:\, \varphi(n,\omega)\underline{x}^m \le b\}$$

for every $x \in K \subset [\underline{x}^m, \overline{x}^m]$. From (4.50) we have

$$\mathbb{P}\{\omega \,:\, \varphi(n,\omega)z \le b\} = \mathbb{P}\{\omega \,:\, \varphi(n,\theta_{-n}\omega)z \le b\} \to \mathbb{P}\{\omega \,:\, u(\omega) \le b\}$$

as $n \to \infty$ for any $z \in \mathrm{int}\mathbb{R}_+^d$. Hence

$$p^{(n)}(x,[0,b]) \to \nu([0,b]), \quad n \to \infty, \quad b \in \mathrm{int}\mathbb{R}_+^d \,,$$

uniformly with respect to $x \in K$. This implies (4.51).

(iv) Since $(P^{*n}\lambda)(A) = \int_{\mathbb{R}_+^d} p^{(n)}(x,A)\lambda(dx)$ and $\mathrm{supp}\lambda \subset [\underline{x}^m, \overline{x}^m]$ for some m, it follows from (4.51) that $(P^{*n}\lambda)([a,b]) \to \nu([a,b])$ for any interval $[a,b] \subset \mathbb{R}_+^d$. This implies the weak convergence of $(P^{*n}\lambda)$ to ν as $n \to \infty$.

To prove (4.53) under condition (4.52) we use a result from BHAT-TACHARYA/LEE [16]. Relation (4.51) implies that

$$\mathbb{P}\{\omega \,:\, \varphi(n,\omega)\overline{x}^m \le a\} \to \mathbb{P}\{\omega \,:\, u(\omega) \le a\}, \quad n \to \infty \,,$$

and

$$\mathbb{P}\{\omega \,:\, \varphi(n,\omega)\underline{x}^m \ge a\} \to \mathbb{P}\{\omega \,:\, u(\omega) \ge a\}, \quad n \to \infty \,,$$

for any fixed m. Therefore it follows from (4.52) that there exists $n_0 = n_0(m)$ with m large enough such that

$$\mathbb{P}\{\omega \,:\, \varphi(n_0,\omega)\overline{x}^m \le a\} > 0 \quad \text{and} \quad \mathbb{P}\{\omega \,:\, \varphi(n_0,\omega)\underline{x}^m \ge a\} > 0 \,.$$

Hence using (4.54) we have

$$\mathbb{P}\{\omega \,:\, \varphi(n_0,\omega)x \le a, \,\forall x \in [\underline{x}^m, \overline{x}^m]\} \ge \mathbb{P}\{\omega \,:\, \varphi(n_0,\omega)\overline{x}^m \le a\} > 0 \quad (4.55)$$

and

$$\mathbb{P}\{\omega \,:\, \varphi(n_0,\omega)x \ge a, \,\forall x \in [\underline{x}^m, \overline{x}^m]\} \ge \mathbb{P}\{\omega \,:\, \varphi(n_0,\omega)\underline{x}^m \ge a\} > 0 \,. \tag{4.56}$$

Since the interval $[\underline{x}^m, \overline{x}^m]$ is forward invariant with respect to $\varphi(n,\omega)$, we can apply Theorem 2.1 from BHATTACHARYA/LEE [16], which gives the convergence in (4.53) under conditions (4.55) and (4.56). □

Remark 4.5.2. Instead of assumption (iv) in Proposition 4.5.2 we can assume that

$$\limsup_{x \to 0} \frac{f_i(\omega,x)}{x_i} > 1 \quad \text{and} \quad \liminf_{\min_j x_j \to \infty} \frac{f_i(\omega,x)}{x_i} < 1$$

for each $i = 1, \ldots, d$ uniformly with respect to $\omega \in \Omega$, where $f_i(\omega,x)$ are the components of the mapping $f = (f_1, \ldots, f_d)$. This observation makes it possible to relax the hypotheses concerning the function $f(\omega,x)$ in BHAT-TACHARYA/LEE [16, Sect.4]. It was assumed there that for each ω the mapping $f(\omega,x)$ is a strictly concave continuously differentiable function with some properties of the derivatives near 0 and infinity.

4.6 Positive Affine RDS

In this section we consider affine and linear order-preserving RDS which leave
a cone to be invariant. The results given below show that the order-preserving
property provides us with an alternative approach to the study of affine RDS
and makes it possible to obtain additional information in contrast with the
more general affine RDS studied in Sect.1.9.

Let V be a real Banach space with closed convex cone V_+. Recall (see
Definition 1.2.3) that RDS (θ, φ) in V is affine if the cocycle φ is of the form

$$\varphi(t, \omega)x = \Phi(t, \omega)x + \psi(t, \omega), \tag{4.57}$$

where $\Phi(t, \omega)$ is a cocycle over θ consisting of bounded linear operators of V.
The function $\psi : \mathbb{T}_+ \times \Omega \to V$ satisfies the relation

$$\psi(t + s, \omega) = \Phi(t, \theta_s\omega)\psi(s, \omega) + \psi(t, \theta_s\omega), \ t, s \geq 0. \tag{4.58}$$

The affine RDS (θ, φ) is said to be positive (with respect to the cone V_+) if
$\varphi(t, \omega)V_+ \subset V_+$ for all $t > 0$ and $\omega \in \Omega$. If $\psi(t, \omega) \equiv 0$ then the affine RDS is
said to be linear. The simplest properties of positive affine RDS are collected
in the following assertion.

Proposition 4.6.1. *The affine RDS (θ, φ) with the cocycle φ of the form
(4.57) is positive with respect to the cone V_+ if and only if $\Phi(t, \omega)$ is positive,
i.e. maps V_+ to itself, and $\psi : \mathbb{T}_+ \times \Omega \to V_+$. Any positive affine RDS is a
sublinear order-preserving system. It is strongly sublinear if $\psi(t, \omega) \gg 0$ for
$t > 0$. Furthermore*

$$\psi(t, \theta_{-t}\omega) \geq \psi(\tau, \theta_{-\tau}\omega) \geq 0 \quad \text{for all} \quad t \geq \tau \geq 0 \tag{4.59}$$

and $a_t(\omega) := \psi(t, \theta_{-t}\omega)$ is a sub-equilibrium for any $t \geq 0$.

Proof. If $\Phi(t, \omega)$ is positive and $\psi(t, \omega) \geq 0$, then the RDS (θ, φ) is obviously
positive and order-preserving. On the other hand, if (θ, φ) is a positive RDS,
then $\psi(t, \omega) = \varphi(t, \omega)0 \geq 0$. Since

$$\Phi(t, \omega)x + \frac{1}{\lambda} \cdot \psi(t, \omega) = \frac{1}{\lambda} \cdot \varphi(t, \omega)[\lambda x] \geq 0$$

for any $x \geq 0$ and $\lambda > 0$, letting $\lambda \to +\infty$ we obtain $\Phi(t, \omega)x \geq 0$ for
$x \geq 0$. Since $w = 0$ is a sub-equilibrium, Proposition 3.4.1 implies that
$a_t(\omega) = \psi(t, \theta_{-t}\omega) = \varphi(t, \theta_{-t}\omega)0$ is also a sub-equilibrium for any $t \geq 0$.
From (4.57) we have

$$\varphi(t, \omega)[\lambda x] - \lambda\varphi(t, \omega)x = (1 - \lambda)\psi(t, \omega), \quad 0 < \lambda < 1, \ x \in V_+.$$

This relation implies the sublinear properties of (θ, φ).   □

Example 4.6.1 (1D Positive Affine RDE). Consider one-dimensional RDE

$$\dot{x} = \alpha(\theta_t\omega)x + \beta(\theta_t\omega) \tag{4.60}$$

over a metric dynamical system θ, where $\alpha(\omega)$ and $\beta(\omega)$ are random variables such that $t \mapsto \alpha(\theta_t\omega)$ and $t \mapsto \beta(\theta_t\omega)$ are locally integrable. Equation (4.60) generates an affine RDS in $\mathbb{R}$. The cocycle φ has the form (4.57) with $\Phi(t,\omega)x = x \exp\left\{\int_0^t \alpha(\theta_\tau\omega)d\tau\right\}$ and

$$\psi(t,\omega) = \int_0^t \beta(\theta_s\omega) \exp\left\{\int_s^t \alpha(\theta_\tau\omega)d\tau\right\} ds.$$

If $\beta(\omega) \geq 0$ for all $\omega \in \Omega$, then (θ,φ) is a positive affine RDS. It is strongly sublinear provided that $\beta(\omega) > 0$.

Theorem 4.6.1. *Let (θ,φ) be a positive affine RDS with the cocycle φ represented in the form (4.57). Assume that there exists $t_0 = t_0(\omega) > 0$ such that $\{\psi(t,\theta_{-t}\omega) : t \geq t_0\}$ is a relatively compact set for each $\omega \in \Omega$. Then*

$$u(\omega) := \lim_{t\to\infty} \psi(t,\theta_{-t}\omega) = \sup_{t>0} \psi(t,\theta_{-t}\omega) \tag{4.61}$$

exists and is an equilibrium for (θ,φ). Furthermore,

(i) if there are no non-zero equilibria for (θ,Φ), i.e. if the equation $w(\theta_t\omega) = \Phi(t,\omega)w(\omega)$ has no non-trivial solution, then the equilibrium is unique;

(ii) if $\psi(t,\omega) \gg 0$ for $t > 0$, then $u(\omega) \gg 0$, and u is the unique (up to indistinguishability) equilibrium in V_+. It is attracting in the sense that there exists a θ-invariant set $\Omega^* \in \mathcal{F}$ of full measure such that

$$\lim_{t\to+\infty} \varphi(t,\theta_{-t}\omega)w(\theta_{-t}\omega) = u(\omega) \quad \text{for all} \quad \omega \in \Omega^* \tag{4.62}$$

for any random variable w possessing the property $0 \leq w(\omega) \leq \lambda u(\omega)$ for all $\omega \in \Omega^*$ with some nonrandom constant $\lambda > 0$.

Proof. By the compactness condition and the monotonicity property (4.59) the limit (4.61) exists. Equation (4.58) implies

$$\psi(t+s,\theta_{-t-s}\omega) = \Phi(t,\theta_{-t}\omega)\psi(s,\theta_{-s}\theta_{-t}\omega) + \psi(t,\theta_{-t}\omega), \quad t,s \geq 0.$$

Letting $s \to \infty$ and using (4.61)

$$u(\omega) = \Phi(t,\theta_{-t}\omega)u(\theta_{-t}\omega) + \psi(t,\theta_{-t}\omega),$$

hence $u(\omega)$ is an equilibrium.

(i) For the uniqueness just note that the difference of two equilibria satisfies $w(\theta_t\omega) = \Phi(t,\omega)w(\omega)$.

(ii) Since $\psi(t, \omega) \gg 0$, equation (4.61) implies $u(\omega) \gg 0$. Assume now that there is a second equilibrium $v(\omega) \geq 0$. Then a simple calculation shows that

$$w_\beta(\omega) = \beta v(\omega) + (1 - \beta)u(\omega)$$

is also an equilibrium for any $0 \leq \beta \leq 1$. It is clear that $w_\beta(\omega)$ is strongly positive for any $0 < \beta < 1$. Therefore Uniqueness Theorem 4.2.1 implies that

$$\frac{1}{2}\left(v(\omega) + u(\omega)\right) \equiv w_{1/2}(\omega) = u(\omega), \quad \omega \in \Omega^* ,$$

where $\Omega^* \in \mathcal{F}$ is a θ-invariant set of full measure. This is only possible if $v(\omega) = u(\omega)$, $\omega \in \Omega^*$.

Since 0 is a sub-equilibrium and $\psi(t, \theta_{-t}\omega) = \varphi(t, \theta_{-t}\omega)0$, (4.62) follows from (4.61) and Theorem 4.2.2. We use the relation

$$\varphi(t, \theta_{-t}\omega)[\lambda u(\theta_{-t}\omega)] = \lambda u(\omega) + (1 - \lambda)\psi(t, \theta_{-t}\omega)$$

to prove that the orbit emanating from λu is relatively compact for any λ. $\square$

Remark 4.6.1. We note that the assumption on the compactness of

$$\overline{\{\psi(t, \theta_{-t}\omega) : t \geq t_0(\omega)\}}$$

can be replaced by the condition: there exists a random element $v(\omega) \in V_+$ such that $\psi(t, \theta_{-t}\omega) \leq v(\omega)$ for all $\omega \in \Omega$ and $t > 0$ provided that the cone V_+ is regular (see Definition 3.1.6 and Remark 4.2.2).

Example 4.6.2 (1D Positive Affine RDE). Consider the RDS (θ, φ) described in Example 4.6.1. We additionally assume that θ is ergodic, $\alpha \in L^1(\Omega, \mathcal{F}, \mathbb{P})$, and $\beta(\omega) \geq 0$ is a tempered random variable. If $\mathbb{E}\alpha < 0$, then

$$\psi(t, \theta_{-t}\omega) = \int_{-t}^0 \beta(\theta_s\omega) \exp\left\{\int_s^0 \alpha(\theta_\tau\omega)d\tau\right\} ds \leq u(\omega)$$

for all $t \geq 0$, where

$$u(\omega) := \int_{-\infty}^0 \beta(\theta_s\omega) \exp\left\{\int_s^0 \alpha(\theta_\tau\omega)d\tau\right\} ds . \tag{4.63}$$

The finiteness of $u(\omega)$ follows from the Birkhoff–Khinchin ergodic theorem (cf. Remark 1.4.1). Thus Theorem 4.6.1 is applicable here. It is clear that $u(\omega)$ given by (4.63) is an equilibrium for (θ, φ) and (4.61) holds.

If $\beta(\omega) \geq \delta > 0$ and $\mathbb{E}\alpha > 0$, then the integral in (4.63) diverges on a set of positive probability and we cannot apply Theorem 4.6.1. Nevertheless in this case the RDS (θ, φ) possesses an equilibrium (see Example 2.1.2).

As an example of an application of the comparison principle (see Sect.3.7) and Theorem 4.6.1 we have the following assertion.

Proposition 4.6.2. *Assume that a system (θ, φ) on the solid normal cone V_+ is dominated from above by a positive affine RDS $(\theta, \varphi_{\mathrm{aff}})$. Suppose that the RDS $(\theta, \varphi_{\mathrm{aff}})$ satisfies the hypotheses of Theorem 4.6.1 with $\psi(t, \omega) \gg 0$ for $t > 0$ and $\omega \in \Omega$. Let $u(\omega)$ be the strongly positive equilibrium for $(\theta, \varphi_{\mathrm{aff}})$. Then for any $\mu > 1$ the random variable $v_\mu(\omega) = \mu u(\omega)$ is an absorbing super-equilibrium for (θ, φ) in the universe $\mathcal{D}$ consisting of all random closed sets $\{B(\omega)\}$ such that $B(\omega) \subset [0, \alpha u(\omega)]$ for some $\alpha > 0$. Moreover if V is a finite-dimensional space, the RDS (θ, φ) possesses a random attractor in the universe $\mathcal{D}$ and the conclusions of Theorem 3.6.2 hold.*

Proof. Since $\varphi(t, \omega)x \leq \varphi_{\mathrm{aff}}(t, \omega)x$ for all $x \in V_+$, we have

$$\varphi(t, \theta_{-t}\omega)B(\theta_{-t}\omega) \subset [0, \varphi_{\mathrm{aff}}(t, \theta_{-t}\omega)[\alpha u(\theta_{-t}\omega)]]$$

for every $B(\omega) \subset [0, \alpha u(\omega)]$. Theorem 4.6.1 implies

$$\varphi_{\mathrm{aff}}(t, \theta_{-t}\omega)[\alpha u(\theta_{-t}\omega)] \leq \mu u(\omega), \quad t \geq t_0(\omega), \quad \mu > 1 \ .$$

Thus $[0, \mu u(\omega)]$ is an absorbing set for (θ, φ) and therefore (θ, φ) is dissipative. If V is finite-dimensional, then Corollary 1.8.1 implies that a random attractor exists in the universe $\mathcal{D}$ and we can apply Theorem 3.6.2. □

The following assertion characterizes the linear part of positive affine RDS with a strongly positive equilibrium.

Corollary 4.6.1. *Let (θ, φ) be a positive affine RDS in V_+. Assume that $\psi(t, \omega) \gg 0$ for $t > 0$ and that there exists a strongly positive equilibrium $u(\omega)$. Then for any random element $w(\omega)$ such that $0 \leq w(\omega) \leq \alpha u(\omega)$ for all $\omega \in \Omega$ with some nonrandom constant $\alpha > 0$ we have*

$$\lim_{t \to +\infty} \Phi(t, \theta_{-t}\omega)w(\theta_{-t}\omega) = 0, \tag{4.64}$$

where $\Phi(t, \omega)$ is linear part of the affine cocycle (θ, φ).

Proof. It is clear from (4.57) that (θ, Φ) is dominated from above by (θ, φ). Therefore Proposition 4.6.2 implies that $\mu u(\omega)$ is an absorbing super-equilibrium for (θ, Φ) in the universe $\mathcal{D}$ consisting of all random closed sets $\{B(\omega)\}$ such that $B(\omega) \subset [0, \alpha u(\omega)]$ for some $\alpha > 0$. Hence (4.64) follows from Proposition 1.9.1. □

5. Cooperative Random Differential Equations

In this chapter we consider cooperative random differential equations. For every fixed ω these equations can be solved as deterministic nonautonomous ODEs and they generate order-preserving random systems under the standard (deterministic) cooperativity condition which appears in the nonautonomous case (see, e.g., KRASNOSELSKII [69] or SMITH [102] and the references therein). We also note that cooperative ODEs with periodic and almost-periodic right-hand sides are naturally included in the class of cooperative random ODEs.

Deterministic cooperative differential equations are one of the main applications of monotone methods and comparison arguments and have been studied by numerous authors, see KRASNOSELSKII [69], HIRSCH [52, 53, 54], SMITH [102] and the references therein. The term cooperative system came from the population biology literature.

Here we restrict ourselves to random equations with phase space $\mathbb{R}_+^d$, where

$$\mathbb{R}_+^d = \{x = (x_1, \ldots, x_d) \in \mathbb{R}^d : x_i \geq 0, \ i = 1, \ldots, d\}$$

is the standard cone in $\mathbb{R}^d$, for the following reasons: (a) this class of equations appears naturally in many applications (see examples below) and (b) most of the results given here can be easily extended to other choices of state space.

5.1 Basic Assumptions and the Existence Theorem

Let $\theta = (\Omega, \mathcal{F}, \mathbb{P}, \{\theta_t, t \in \mathbb{R}\})$ be a metric dynamical system. We consider in $\mathbb{R}_+^d$ the pathwise ordinary differential equation

$$\dot{x}(t) = f(\theta_t \omega, x(t)) . \tag{5.1}$$

We assume that $f = (f_1, \ldots, f_d) : \Omega \times \mathbb{R}_+^d \to \mathbb{R}^d$ is a measurable function such that $f(\omega, \cdot)$ possesses the following properties for all $\omega \in \Omega$:

(R1) $f(\omega, \cdot)$ is continuously differentiable and $f_i(\omega, \cdot)$ and $\partial f_i(\omega, \cdot)/\partial x_j$, $i, j = 1, \ldots, d$, are bounded on compact sets $K \subset \mathbb{R}_+^d$ by $C_K(\omega)$ such that $t \mapsto C_K(\theta_t \omega)$ is locally integrable;

(R2) there exist random variables C_1 and C_2 such that $t \mapsto C_j(\theta_t\omega)$ is locally integrable and

$$\langle x, f(\omega, x)\rangle \leq C_1(\omega)|x|^2 + C_2(\omega) \, ,$$

where $\langle \cdot, \cdot \rangle$ is the standard inner product in $\mathbb{R}^d$ and $|x|^2 = \langle x, x\rangle$;

(R3) $f(\omega, \cdot)$ is *weakly positive*, i.e.

$$f_i(\omega, x) \geq 0, \quad \text{for all} \quad x \in \Gamma_i, \; \omega \in \Omega, \; i = 1, \dots, d,$$

where

$$\Gamma_i = \left\{ x = (x_1, \dots, x_d) \in \mathbb{R}^d_+ \; : \; x_i = 0 \right\} \, .$$

We note that condition (R3) is satisfied if and only if

$$\langle f(\omega, x), y \rangle \geq 0 \quad \text{whenever} \quad x \in \partial\mathbb{R}^d_+, \; y \geq 0, \; \langle x, y \rangle = 0 \, . \tag{5.2}$$

Sometimes instead of (R3) we will assume that

(R3*) $f(\omega, \cdot)$ is *strongly positive*, i.e.

$$f_i(\omega, x) > 0, \quad \text{for all} \quad x \in \Gamma_i, \; x \neq 0, \; \omega \in \Omega, \; i = 1, \dots, d \, .$$

Proposition 5.1.1. *Assume that conditions (R1), (R2) and (R3) hold. Then for any initial data $x_0 \in \mathbb{R}^d_+$ at the moment $t = 0$ problem (5.1) has a unique global solution $x(t, \omega) \equiv x(t, \omega; x_0)$ (see Definition 2.1.1) such that $x(t, \omega) \in \mathbb{R}^d_+$ for all $t \geq 0$ and $\omega \in \Omega$. This solution is continuously differentiable with respect to the initial data x_0 and relations (2.6) and (2.7) concerning the evolution of the Jacobian and its determinant hold.*

Proof. We first extend the function $f(\omega, x)$ from $\mathbb{R}^d_+$ to $\mathbb{R}^d$ such that the extended function $\tilde{f}(\omega, x)$ belongs to $C^1(\mathbb{R}^d)$ for all $\omega \in \Omega$ and possesses properties (2.1) and (2.2), i.e. for any compact set $K \subset \mathbb{R}^d$ there exists a random variable $C_K(\omega) \geq 0$ such that

$$\int_a^{a+1} C_K(\theta_t\omega) \, dt < \infty \quad \text{for all} \quad a \in \mathbb{R}, \; \omega \in \Omega \, , \tag{5.3}$$

and

$$|\tilde{f}(\omega, x)| \leq C_K(\omega), \quad |\tilde{f}(\omega, x) - \tilde{f}(\omega, y)| \leq C_K(\omega) \cdot |x - y|$$

for any $x, y \in K$ and $\omega \in \Omega$. It is clear that this extension exists. Now we can apply Proposition 2.1.1 to prove that the problem

$$\dot{x}(t) = \tilde{f}(\theta_t\omega, x(t)), \quad x(0) = x_0 \, ,$$

has a unique local solution $\tilde{x}(t, \omega; x_0)$ which is continuously differentiable with respect to the initial data x_0 and possesses properties (2.6) and (2.7). The weak positivity condition (R3) in the form (5.2) implies that

$$\langle \tilde{f}(\omega, x), \nu_x \rangle = \langle f(\omega, x), \nu_x \rangle \leq 0, \quad x \in \partial \mathbb{R}_+^d, \ \omega \in \Omega,$$

where ν_x is an outer normal to $\partial \mathbb{R}_+^d$ at the point $x \in \partial \mathbb{R}_+^d$ (see Definition 2.2.1). Hence it follows from Theorem 2.2.1 that for any $x_0 \in \mathbb{R}_+^d$ the solution $\tilde{x}(t, \omega; x_0)$ does not leave $\mathbb{R}_+^d$ and therefore it gives a unique local solution $x(t, \omega; x_0)$ to problem (5.1). Property (R2) and Corollary 2.2.2 imply that the solution $x(t, \omega; x_0)$ can be extended to the whole time semi-axis $\mathbb{R}_+$. $\square$

5.2 Generation of RDS

The following assertion shows that equation (5.1) generates an RDS in $\mathbb{R}_+^d$.

Proposition 5.2.1. *Assume that conditions (R1)–(R3) hold. Then equation (5.1) generates a C^1 RDS (θ, φ) in $\mathbb{R}_+^d$ with the cocycle $\varphi(t, \omega)$ defined by the formula $\varphi(t, \omega)x = x(t)$, where $x(t)$ is an absolutely continuous solution to the equation*

$$x(t) = x + \int_0^t f(\theta_\tau \omega, x(\tau)) \, d\tau . \tag{5.4}$$

Moreover the Jacobian $D_x \varphi(t, \omega, x)$ satisfies equations (2.8) and (2.9). We also have the relations

$$\varphi(t, \omega)(\mathbb{R}_+^d \setminus \{0\}) \subset \mathbb{R}_+^d \setminus \{0\} \quad \text{for all} \quad t > 0 \quad \text{and} \quad \omega \in \Omega \tag{5.5}$$

and

$$\varphi(t, \omega)\text{int}\mathbb{R}_+^d \subset \text{int}\mathbb{R}_+^d \quad \text{for all} \quad t > 0 \quad \text{and} \quad \omega \in \Omega . \tag{5.6}$$

If we additionally assume that (R3*) holds, then (θ, φ) is strongly positive, i.e.

$$\varphi(t, \omega)(\mathbb{R}_+^d \setminus \{0\}) \subset \text{int}\mathbb{R}_+^d \quad \text{for all} \quad t > 0 \quad \text{and} \quad \omega \in \Omega . \tag{5.7}$$

Proof. It follows from Proposition 5.1.1 that (5.1) generates a global C^1 RDS (θ, φ) in $\mathbb{R}_+^d$ with properties (2.8) and (2.9).

To prove (5.5) we assume that for some fixed $\omega \in \Omega$, $x \in \mathbb{R}_+^d$ and $t_0 = t_0(\omega, x) > 0$ we have $x(t_0) = \varphi(t_0, \omega)x = 0$. Since $f(\theta_\tau \omega, 0) \geq 0$ for all $\tau \in \mathbb{R}$, equation (5.4) implies that

$$0 \leq x(t) \leq - \int_t^{t_0} (f(\theta_\tau \omega, x(\tau)) - f(\theta_\tau \omega, 0)) \, d\tau, \quad 0 \leq t \leq t_0 .$$

Since $\sup_{\tau \in [0,t_0]} |x(\tau, \omega)| \leq r(\omega)$ with some $r(\omega) > 0$, property (R1) gives

$$|f(\theta_\tau \omega, x(\tau)) - f(\theta_\tau \omega, 0)| \leq C(\tau, \omega) \cdot |x(\tau)| ,$$

where

$$C(\tau, \omega) \equiv \sup_{|x| \leq r(\omega)} |D_x f(\theta_\tau \omega, x)| \in L^1_{loc}(\mathbb{R}) \quad \text{for each} \quad \omega \in \Omega .$$

Therefore we have

$$|x(t)| \leq \int_t^{t_0} C(\tau, \omega) \cdot |x(\tau)| \, d\tau, \quad 0 \leq t \leq t_0 .$$

This implies that $x(t) \equiv 0$ for all $0 \leq t \leq t_0$ and therefore $x = 0$. Thus we have (5.5).

Let us prove (5.6 and (5.7). Suppose that for some $\omega \in \Omega$ there exist a solution $x(t)$ to equation (5.1) such that $x(0) = x \geq 0$, a time $t_0 > 0$ and an element $z \in \mathbb{R}_+^d \setminus \{0\}$ such that $x(t_0) = z$ and $z \in \Gamma_i$ for some $i \in \{1, \ldots, d\}$. In this case we have

$$x_i(t) + \int_t^{t_0} f_i(\theta_\tau \omega, x(\tau)) \, d\tau = 0, \quad 0 \leq t \leq t_0 . \tag{5.8}$$

Using (R3) we get

$$x_i(t) + \int_t^{t_0} [f_i(\theta_\tau \omega, x(\tau)) - f_i(\theta_\tau \omega, \hat{x}(\tau))] \, d\tau \leq 0$$

for $0 \leq t \leq t_0$, where

$$\hat{x}(\tau) = (x_1(\tau), \ldots, x_{i-1}(\tau), 0, x_{i+1}(\tau), \ldots x_d(\tau))$$

Therefore, as above (R1) implies that

$$0 \leq x_i(t) \leq \int_t^{t_0} C(\tau, \omega) \cdot x_i(\tau) \, d\tau, \quad 0 \leq t \leq t_0 .$$

Consequently $x_i(t) \equiv 0$ for all $0 \leq t \leq t_0$. This is impossible if $x(0) = x \gg 0$ and therefore we obtain (5.6). Further, if $x_i(t) \equiv 0$ for all $0 \leq t \leq t_0$, we have from (5.8) that

$$\int_t^{t_0} f_i(\theta_\tau \omega, \hat{x}(\tau)) \, d\tau = 0, \quad 0 \leq t \leq t_0 ,$$

which is impossible under condition (R3*) provided that $x(0) = x > 0$. Thus $\varphi(t, \omega)x \gg 0$ for all $x > 0$, i.e. we have (5.7). $\qquad \square$

Now we introduce assumptions that guarantee that the RDS generated by (5.1) in $\mathbb{R}_+^d$ is order-preserving. We assume that

(R4) the function $f(\omega, \cdot)$ is *cooperative*, i.e.

$$f_i(\omega, x) \leq f_i(\omega, y), \quad i = 1, \ldots, d, \quad \omega \in \Omega, \qquad (5.9)$$

for all $x, y \in \mathbb{R}_+^d$ such that $x_i = y_i$ and $x_j \leq y_j$ for $j \neq i$.

It is easy to see that a function $f(\omega, x)$ satisfies condition (R4) if and only if

$$\langle f(\omega, y) - f(\omega, x), z \rangle \geq 0 \quad \text{whenever} \quad 0 \leq x \leq y, \ z \geq 0, \ \langle y - x, z \rangle = 0.$$

We note that the cooperativity condition (R4) is also known as quasi-monotonicity (see WALTER [107]) and it can be written (see, e.g., SMITH [102]) in the differential form as

(R4*) for each $\omega \in \Omega$ we have

$$\frac{\partial f_i(\omega, x)}{\partial x_j} \geq 0 \quad \text{when} \quad i \neq j, \ x = (x_1, \ldots, x_d) \in \mathbb{R}_+^d. \qquad (5.10)$$

As in the deterministic case (see, e.g., HIRSCH [52], KRASNOSELSKII [68, 69], SMITH [102] and the references therein) we need the concept of irreducibility. Recall the following definition.

Definition 5.2.1. *A matrix $A = \{a_{ij}\}_{i,j=1}^d$ is called irreducible if for every nonempty, proper subset I of the set $N = \{1, 2, \ldots, d\}$, there is an $i \in I$ and $j \in N \setminus I$ such that $a_{ij} \neq 0$.*

One can show that a matrix A is irreducible if and only if no nonzero proper subspace spanned by a subset of the standard basis in $\mathbb{R}^d$ is mapped by A into itself.

Theorem 5.2.1. *Let (R1)–(R4) hold. Then equation (5.1) generates a strictly order-preserving RDS (θ, φ) in $\mathbb{R}_+^d$ and*

$$\varphi(t, \omega)\text{int}\mathbb{R}_+^d \subset \text{int}\mathbb{R}_+^d \quad \text{for any} \quad t \geq 0, \ \omega \in \Omega. \qquad (5.11)$$

If the matrix

$$D_x f(\omega, x) \equiv \left\{ \frac{\partial f_i(\omega, x)}{\partial x_j} \right\}_{i,j=1}^d \qquad (5.12)$$

is irreducible for all $x \in \text{int}\mathbb{R}_+^d$ and $\omega \in \Omega$, then

$$\varphi(t, \omega)x \ll \varphi(t, \omega)y \quad \text{for all} \quad 0 \ll x < y \quad \text{and} \quad \omega \in \Omega, \qquad (5.13)$$

i.e. equation (5.1) generates a strongly order-preserving RDS in $\text{int}\mathbb{R}_+^d$. If the matrix (5.12) is irreducible for all positive x from $\mathbb{R}_+^d$ and $\omega \in \Omega$, then the RDS (θ, φ) is strongly order-preserving in $\mathbb{R}_+^d$.

The proof of this theorem follows the line of argument for the deterministic case (see, e.g., HIRSCH [52] or KRASNOSELSKII [68]) and relies on the following assertion.

Lemma 5.2.1. *Let (R1)–(R4) be valid. Let $\varphi(t,\omega)$ be the cocycle generated by (5.1). Then for any $x \in \mathbb{R}_+^d$ the linear mapping $\psi_x(t,\omega) \equiv D_x\varphi(t,\omega,x)$ possesses the properties*

$$\psi_x(t,\omega)\mathbb{R}_+^d \subset \mathbb{R}_+^d \quad \text{for all} \quad t > 0 \quad \text{and} \quad \omega \in \Omega ; \tag{5.14}$$

$$\psi_x(t,\omega)(\mathbb{R}_+^d \setminus \{0\}) \subset \mathbb{R}_+^d \setminus \{0\} \quad \text{for all} \quad t > 0 \quad \text{and} \quad \omega \in \Omega ; \tag{5.15}$$

$$\psi_x(t,\omega)\text{int}\mathbb{R}_+^d \subset \text{int}\mathbb{R}_+^d \quad \text{for all} \quad t > 0 \quad \text{and} \quad \omega \in \Omega . \tag{5.16}$$

If additionally the matrix $D_x f(\theta_t\omega, \varphi(t,\omega,x))$ is irreducible for all $t \geq 0$ and $\omega \in \Omega$, then $\psi_x(t,\omega)$ possesses the property

$$\psi_x(t,\omega)(\mathbb{R}_+^d \setminus \{0\}) \subset \text{int}\mathbb{R}_+^d \quad \text{for all} \quad t > 0 \quad \text{and} \quad \omega \in \Omega . \tag{5.17}$$

Proof. Proposition 5.1.1 implies that

$$y(t) = \psi_x(t,\omega)y_0 \equiv D_x\varphi(t,\omega,x)y_0$$

is a solution to the problem

$$\dot{y} = D_x f(\theta_t\omega, x(t))y, \quad y(0) = y_0 , \tag{5.18}$$

where $x(t) = \varphi(t,\omega)x$. Assumption (R4) implies (R4*) and therefore the right-hand side of the equation (5.18) is weakly positive (see (R3)). Consequently (5.14) follows from Proposition 2.2.1. Relation (5.15) can be proved in the same way as (5.5).

To prove (5.16) let us assume that for some ω there exist $t_0 > 0$, $z \gg 0$ and $i \in \{1,\dots,d\}$ such that we have $y_i(t_0) = 0$ for the solution $y(t) = (y_1(t),\dots,y_d(t))$ to problem (5.18) with $y_0 = z$. Since $\psi_x(t,\omega)$ is a linear order-preserving operator, equation (5.14) implies that $y_i(t_0) = 0$ for a solution to problem (5.18) with arbitrary initial data $y_0 \in \mathbb{R}^d$. This implies that $\text{Det}\psi_x(t_0,\omega) = 0$ which is impossible because of Liouville's equation (2.9).

To obtain the last assertion of the lemma we apply the same method as in the proof of property (5.7). Assume that for some $\omega \in \Omega$ there exist a solution $y(t) = (y_1(t),\dots,y_d(t)) \geq 0$ to (5.18) with nonzero initial data and a moment $t_0 > 0$ such that $y_i(t_0) = 0$ for $i \in I$ and $y_i(t_0) > 0$ when $i \notin I$, where I is a proper subset of $\{1,\dots,d\}$. We note that the relation $y_i(t_0) = 0$ for all $i \in \{1,\dots,d\}$ is impossible because of (5.15). Let $a_{ij}(t,\omega)$ be the entries of the matrix $D_x f(\theta_t \omega, \varphi(t, \omega, x))$, i.e.

$$a_{ij}(t, \omega) = \frac{\partial f_i}{\partial x_j}(\theta_t \omega, \varphi(t, \omega, x)), \quad t \geq 0, \ \omega \in \Omega, \ i, j = 1, \dots, d \ .$$

Since $\{a_{ij}(t, \omega)\}$ is irreducible, there exists a pair $\{k, l\}$ such that $k \in I, l \notin I$ and $a_{kl}(t, \omega) > 0$. These k and l can depend on t and ω. It follows from (5.18) that

$$y_i(t) + \sum_{j \neq i} \int_t^{t_0} a_{ij}(s, \omega) y_j(s) \cdot \exp\left\{ -\int_t^s a_{ii}(\tau, \omega) d\tau \right\} ds = 0$$

for $t \in [0, t_0]$ and $i \in I$. Therefore

$$\sum_{i \in I} y_i(t) + \int_t^{t_0} F_I(t, s, \omega) \, ds \leq 0 \tag{5.19}$$

for $t \in [0, t_0]$, where

$$F_I(t, s, \omega) = \sum_{i \in I} \sum_{j \notin I} a_{ij}(s, \omega) y_j(s) \cdot \exp\left\{ -\int_t^s a_{ii}(\tau, \omega) d\tau \right\} \ .$$

Since $y(t)$ is continuous, we have $y_j(t) > 0$ for $j \notin I$ and for all $t \in [t_0 - \delta, t_0]$ with some $\delta = \delta(\omega) > 0$. Therefore the irreducibility of $\{a_{ij}(t, \omega)\}$ implies that $F_I(t, s, \omega) > 0$ for all $s \in [t, t_0]$ with $t \in [t_0 - \delta, t_0]$. Thus from (5.19) we have $\sum_{i \in I} y_i(t) < 0$ for $t \in [t_0 - \delta, t_0)$ which is impossible. □

Proof of Theorem 5.2.1. We make use of the equation

$$\varphi(t, \omega, y) = \varphi(t, \omega, x) + \int_0^1 D_x \varphi(t, \omega, sy + (1 - s)x) ds (y - x) \tag{5.20}$$

valid for all $t \geq 0$, $\omega \in \Omega$ and $x, y \in \mathbb{R}_+^d$. If $0 \leq x < y$, then from (5.15) and (5.20) we have that $\varphi(t, \omega, y) > \varphi(t, \omega, x)$, i.e. (θ, φ) is strictly order-preserving in $\mathbb{R}_+^d$. If $0 \equiv x \ll y$, then from (5.16) and (5.20) we have that $\varphi(t, \omega, y) \gg \varphi(t, \omega, 0) \geq 0$, i.e. (5.11) is valid. Moreover if for all $x \in \text{int}\mathbb{R}_+^d$ and $\omega \in \Omega$ the matrix (5.12) is irreducible, then (5.11) implies that $D_x f(\theta_t \omega, \varphi(t, \omega, x))$ is irreducible for all $t \in \mathbb{R}_+$ and $\omega \in \Omega$. Therefore (5.17) and (5.20) give (5.13). In a similar way we obtain the last assertion of Theorem 5.2.1 and conclude the proof. □

Remark 5.2.1. Let (θ, φ) be the RDS in $\mathbb{R}_+^d$ generated by (5.1). Since $t \mapsto \varphi(t, \theta_{-t}\omega)x$ is a right continuous function (see Remark 2.1.2(i)), it follows from Proposition 1.5.2 that the closure $\overline{\gamma_x(\omega)}$ of any pull back orbit $\gamma_x(\omega)$ emanating from $x \in \mathbb{R}_+^d$ is a random closed set with respect to the σ-algebra $\mathcal{F}$.

5.3 Random Comparison Principle

The following comparison theorem is of importance in what follows. In the deterministic case it is known as the Kamke theorem (see, e.g., SMITH [102], WALTER [107] or the references in KRASNOSELSKII [68]). We also refer to LADDE/LAKSHMIKANTHAM [75] for a random comparison principle for another class of RDE in $\mathbb{R}^d$.

Let us consider in $\mathbb{R}^d_+$ the system of random ordinary differential equations

$$\dot{y}_i(t) = g_i(\theta_t\omega, y_1(t), \dots, y_d(t)), \quad i = 1, \dots, d, \qquad (5.21)$$

with the function $g = (g_1, \dots, g_d) : \Omega \times \mathbb{R}^d_+ \mapsto \mathbb{R}^d$ possessing properties (2.1) and (2.2), i.e. for any compact set $K \subset \mathbb{R}^d_+$ there exists a random variable $C_K(\omega) \geq 0$ such that (5.3) holds and

$$|g(\omega, x)| \leq C_K(\omega), \quad \text{and} \quad |g(\omega, x) - g(\omega, y)| \leq C_K(\omega) \cdot |x - y|$$

for any $x, y \in K$ and $\omega \in \Omega$. We denote by $y(t, \omega; x)$ the local solution to problem (5.21) with initial data $x \in \mathbb{R}^d_+$ at the time $t = 0$ with the property $y(t, \omega; x) \in \mathbb{R}^d_+$ for $t \in [0, t_0(\omega, x))$, where $t_0(\omega, x)$ is a positive number. The existence of this solution follows from Proposition 2.1.1 at least for initial data from $\mathrm{int}\mathbb{R}^d_+$.

Theorem 5.3.1 (Random Comparison Principle). *Assume that (R1)-(R4) hold for the function f. Let $\varphi(t, \omega)$ be the cocycle of the RDS in $\mathbb{R}^d_+$ generated by (5.1). Then the condition*

$$f(\omega, x) \leq g(\omega, x) \quad \text{for all} \quad x \in \mathbb{R}^d_+, \ \omega \in \Omega, \qquad (5.22)$$

implies that

$$\varphi(t, \omega)x \leq y(t, \omega; x) \quad \text{for all} \quad t \in [0, t_0(\omega, x)), \ \omega \in \Omega, \ x \in \mathbb{R}^d_+. \qquad (5.23)$$

If

$$f(\omega, x) \geq g(\omega, x) \quad \text{for all} \quad x \in \mathbb{R}^d_+, \ \omega \in \Omega, \qquad (5.24)$$

then

$$\varphi(t, \omega)x \geq y(t, \omega; x) \quad \text{for all} \quad t \in [0, t_0(\omega, x)), \ \omega \in \Omega, \ x \in \mathbb{R}^d_+. \qquad (5.25)$$

Proof. Assume (5.22). Then the function $z(t) = y(t, \omega; x) - \varphi(t, \omega)x$ is a local solution to problem

$$\dot{z}_i(t) = h_i(t, \omega, z_1(t), \dots, z_d(t)), \quad z_i(0) = 0, \quad i = 1, \dots, d,$$

where

$$h(t, \omega, z) = g(\theta_t\omega, \varphi(t, \omega)x + z) - f(\theta_t\omega, \varphi(t, \omega)x).$$

From (5.22) and (R4) we have

$$h_i(t, \omega, z) \geq f_i(\theta_t \omega, \varphi(t, \omega) x + z) - f_i(\theta_t \omega, \varphi(t, \omega) x) \geq 0$$

for every $z = (z_1, \ldots, z_d) \in \mathbb{R}^d_+$ with $z_i = 0$. This implies that

$$\langle h(t, \omega, z), \nu_z \rangle \leq 0, \quad t > 0, \ z \in \partial \mathbb{R}^d_+, \ \omega \in \Omega,$$

where ν_z is an outer normal to $\partial \mathbb{R}^d_+$ at z. Therefore Proposition 2.2.1 implies that $z(t) \geq 0$ on the interval $[0, t_0(\omega, x))$, and we have (5.23).

Assume now that (5.24) holds. Then the function $z^*(t) = -z(t)$ satisfies the equation

$$\dot{z}^*(t) = h^*(t, \omega, z^*(t)), \quad z(0) = 0,$$

where $h^*(t, \omega, z) = -h(t, \omega, -z)$. Since $y(t, \omega; x) \in \mathbb{R}^d_+$ for $t \in [0, t_0(\omega, x))$, we also have $z^*(t) \leq \varphi(t, \omega) x$. From (5.24) and (R4) we obtain

$$h_i^*(t, \omega, z) \geq f_i(\theta_t \omega, \varphi(t, \omega) x) - f_i(\theta_t \omega, \varphi(t, \omega) x - z) \geq 0$$

for every $z = (z_1, \ldots, z_d) \in \mathbb{R}^d_+$ with $z_i = 0$ such that $z \leq \varphi(t, \omega) x$. Therefore as above we can conclude that $z^*(t) \geq 0$. Thus we obtain (5.25).   □

From Theorem 5.3.1 we easily have the following assertion.

Corollary 5.3.1. *Assume that f satisfies (R1)-(R4) and g satisfies (R1)-(R3). Let $\varphi(t, \omega)$ and $\psi(t, \omega)$ be the cocycles of the RDS in $\mathbb{R}^d_+$ generated by (5.1) and by (5.21). Then*

(i)   condition (5.22) implies that

$$\varphi(t, \omega) x \leq \psi(t, \omega) x \quad \text{for all} \ \ t \in \mathbb{R}_+, \ \omega \in \Omega, \ x \in \mathbb{R}^d_+ ;$$

(ii) if we have strict inequality in (5.22) for every $x \in \mathrm{int} \mathbb{R}^d_+$ and $\omega \in \Omega$, then

$$\varphi(t, \omega) x < \psi(t, \omega) x \quad \text{for all} \ \ t > 0, \ \omega \in \Omega, \ x \in \mathrm{int} \mathbb{R}^d_+ ;$$

(iii) condition (5.24) implies that

$$\varphi(t, \omega) x \geq \psi(t, \omega) x \quad \text{for all} \ \ t \in \mathbb{R}_+, \ \omega \in \Omega, \ x \in \mathbb{R}^d_+ ;$$

(iv) if we have strict inequality in (5.24) for every $x \in \mathrm{int} \mathbb{R}^d_+$ and $\omega \in \Omega$, then

$$\varphi(t, \omega) x > \psi(t, \omega) x \quad \text{for all} \ \ t > 0, \ \omega \in \Omega, \ x \in \mathrm{int} \mathbb{R}^d_+ .$$

Proof. It is necessary to prove (ii) and (iv) only. Assume that (5.22) with a strict inequality is valid. Suppose that for some $\omega \in \Omega$, $x \in \text{int}\mathbb{R}_+^d$ and $t_0 > 0$ we have that $\varphi(t_0, \omega)x = \psi(t_0, \omega)x$. Denote $x(t) = \varphi(t, \omega)x$ and $y(t) = \psi(t, \omega)x$. Then from (5.22) and from assertion (i) of the corollary we have

$$0 \leq y(t) - x(t) \leq - \int_t^{t_0} \left(g(\theta_\tau \omega, y(\tau)) - g(\theta_\tau \omega, x(\tau)) \right) d\tau$$

for all $t \in [0, t_0]$. This equality allows us to conclude that $y(t) = x(t)$ for all $t \in [0, t_0]$ (cf. the argument given in the proof of Proposition 5.2.1). Therefore from (5.1) and (5.21) we have the equality

$$\int_t^{t_0} \left(g(\theta_\tau \omega, x(\tau)) - f(\theta_\tau \omega, x(\tau)) \right) d\tau = 0, \quad t \in [0, t_0] \, .$$

By Theorem 5.2.1 $x(\tau) \in \text{int}\mathbb{R}_+^d$ and therefore the last equality contradicts to the strict inequality in (5.22) for $x \in \text{int}\mathbb{R}_+^d$.

The proof of (iv) is similar. □

Below we also need a stronger version of the comparison principle.

Theorem 5.3.2 (Strong Random Comparison Principle). *Assume that f satisfies (R1)–(R4) and g satisfies (R1)–(R3). Let (θ, φ) and (θ, ψ) be the RDS in $\mathbb{R}_+^d$ generated by (5.1) and by (5.21). The following assertions hold.*

(i)   If

$$f_i(\omega, x) < g_i(\omega, x) \quad \text{for all} \quad i = 1, \ldots, d, \ x \in \text{int}\mathbb{R}_+^d, \ \omega \in \Omega \, , \quad (5.26)$$

then

$$\varphi(t, \omega)x \ll \psi(t, \omega)x \quad \text{for all} \quad t > 0, \ x \in \text{int}\mathbb{R}_+^d, \ \omega \in \Omega \, . \quad (5.27)$$

(ii)   If

$$f_i(\omega, x) > g_i(\omega, x) \quad \text{for all} \quad i = 1, \ldots, d, \ x \in \text{int}\mathbb{R}_+^d, \ \omega \in \Omega \, , \quad (5.28)$$

then

$$\varphi(t, \omega)x \gg \psi(t, \omega)x \quad \text{for all} \quad t > 0, \ x \in \text{int}\mathbb{R}_+^d, \ \omega \in \Omega \, . \quad (5.29)$$

(iii) If matrix (5.12) is irreducible for all $x \in \text{int}\mathbb{R}_+^d$ and $\omega \in \Omega$, then
  (a) property (5.22) with strict inequality for every $x \in \text{int}\mathbb{R}_+^d$ implies (5.27);

(b) property (5.24) with strict inequality for every $x \in \text{int}\mathbb{R}^d_+$ implies (5.29).

Proof. Let $x(t) = \varphi(t, \omega)x$ and $y(t) = \psi(t, \omega)x$ with $x \in \text{int}\mathbb{R}^d_+$.
(i) We obviously have

$$y(t) - x(t) = y(s) - x(s) + \int_s^t [g(\theta_\tau\omega, y(\tau)) - f(\theta_\tau\omega, x(\tau))] \, d\tau .$$

Corollary 5.3.1 and Theorem 5.2.1 imply that $y(t) > x(t) \gg 0$ for all $t > 0$. Therefore it follows from (5.26) that

$$y_i(t) - x_i(t) > y_i(s) - x_i(s) + \int_s^t [f_i(\theta_\tau\omega, y(\tau)) - f_i(\theta_\tau\omega, x(\tau))] \, d\tau \quad (5.30)$$

for all $t > s \geq 0$ and $i = 1, \ldots, d$. Hence

$$y_i(t) - x_i(t) \geq y_i(s) - x_i(s) + \sum_{j=1}^d \int_s^t a_{ij}(\tau)(y_j(\tau) - x_j(\tau)) d\tau , \quad (5.31)$$

where

$$a_{ij}(\tau) = \int_0^1 \frac{\partial f_i}{\partial x_j} (\theta_\tau\omega, x(\tau) + \lambda(y(\tau) - x(\tau))) \, d\lambda .$$

Cooperativity condition (R4*) and the relation $y_j(\tau) - x_j(\tau) \geq 0$ imply that

$$y_i(t) - x_i(t) \geq y_i(s) - x_i(s) + \int_s^t a_{ii}(\tau)(y_i(\tau) - x_i(\tau)) d\tau$$

for all $t > s \geq 0$ and $i = 1, \ldots, d$. If we suppose that

$$y_i(t_0) = x_i(t_0) \quad \text{for some} \quad t_0 > 0 \quad \text{and} \quad i \in \{1, \ldots, d\} , \quad (5.32)$$

then we obtain that

$$0 \leq y_i(s) - x_i(s) \leq \int_s^{t_0} |a_{ii}(\tau)|(y_i(\tau) - x_i(\tau)) d\tau$$

for all $0 \leq s \leq t_0$. This implies that $y_i(s) = x_i(s)$ for all $s \in [0, t_0]$. Therefore we can apply (R4) in (5.30) and obtain the inequality

$$y_i(t) - x_i(t) > y_i(s) - x_i(s) \geq 0 \quad \text{for all} \quad 0 \leq s < t \leq t_0$$

which contradicts to (5.32).
(ii) In this case Corollary 5.3.1 implies that $x(t) \geq y(t)$. Since (R1)–(R3) hold for g, it follows from (5.6) that $y(t) \gg 0$. Therefore as above from (5.28) we can obtain the inequality

$$x_i(t) - y_i(t) > x_i(s) - y_i(s) + \int_s^t \left[f_i(\theta_\tau \omega, x(\tau)) - f_i(\theta_\tau \omega, y(\tau)) \right] d\tau$$

and derive (5.29).

(iii-a) We use the idea presented in the proof of Lemma 5.2.1.

By Corollary 5.3.1 and Theorem 5.2.1 we obviously have that $y(t) > x(t) \gg 0$ for all $t > 0$. Assume that for some $\omega \in \Omega$ there exist $x \in \text{int} \mathbb{R}_+^d$ and $t_0 > 0$ such that

$$y_i(t_0) = x_i(t_0), \ i \in I, \quad \text{and} \quad y_i(t_0) > x_i(t_0), \ i \notin I, \tag{5.33}$$

where I is a proper subset of $\{1, \dots, d\}$. We note that the relation $y_i(t_0) = x_i(t_0)$ for all $i = 1, \dots, d$ is impossible because of Corollary 5.3.1(ii). An argument similar to given above makes it possible to obtain (5.31) for this case. Hence as above we can prove that $y_i(s) = x_i(s)$ for all $s \in [0, t_0]$ and $i \in I$. Therefore (5.31) implies that

$$\sum_{i \in I} \int_s^{t_0} F_i(\tau) d\tau \leq 0 \quad \text{for all} \quad s \in [0, t_0], \tag{5.34}$$

where

$$F_i(\tau) = \sum_{j \notin I} a_{ij}(\tau)(y_j(\tau) - x_j(\tau)) \geq 0, \quad i \in I.$$

Since $y_j(t) - x_j(t) > 0$ for $j \notin I$ and for all $t \in [t_0 - \delta, t_0]$ with some $\delta = \delta(\omega) > 0$, from the irreducibility condition we get that $\sum_{i \in I} F_i(t) > 0$ for $t \in [t_0 - \delta, t_0]$. This contradicts to (5.34).

To prove (iii-b) we use arguments similar to ones given in the proofs of (ii) and (iii-a). □

As an application of the random comparison principle we prove the following assertion.

Theorem 5.3.3. *Assume that f satisfies (R1)–(R4) and*

$$f(\omega, x) \leq A(\omega)x + b(\omega), \quad x \in \mathbb{R}_+^d, \ \omega \in \Omega, \tag{5.35}$$

where $A(\omega) = \{a_{ij}(\omega)\}_{i,j=1}^d$ and $b(\omega) = (b_1(\omega), \dots, b_d(\omega)) \geq 0$ are tempered random variables such that $t \mapsto \|A(\theta_t \omega)\|$ and $t \mapsto |b(\theta_t \omega)|$ are locally integrable. Assume also that θ is an ergodic metric dynamical system and the linear RDS (θ, Φ) generated by

$$\dot{x}(t) = A(\theta_t \omega)x(t)$$

has top Lyapunov exponent negative. Then there exist a θ-invariant set Ω^* of full measure and a version $(\theta, \tilde{\varphi})$ of the RDS (θ, φ) generated by (5.1) such that $\tilde{\varphi}(t, \omega) = \varphi(t, \omega)$ for all $\omega \in \Omega^*$ and $t \geq 0$, and the RDS $(\theta, \tilde{\varphi})$ possesses an absorbing super-equilibrium $v(\omega) \gg 0$ in the universe $\mathcal{D}$ of all tempered subsets of $\mathbb{R}_+^d$.

Proof. The comparison principle implies that the RDS (θ, φ_c) generated by the equation

$$\dot{x}(t) = A(\theta_t \omega) x(t) + b(\theta_t \omega) + c \cdot \mathbf{e} ,$$

where $\mathbf{e} := (1, \ldots, 1) \in \mathbb{R}_+^d$, dominates the system (θ, φ) from above for any $c \in \mathbb{R}_+$. Condition (5.35) and the weak positivity property (R3) of f imply that $a_{ij}(\omega) \geq 0$, $i \neq j$. Therefore (θ, φ_c) is a strictly order-preserving RDS in $\mathbb{R}_+^d$ (see Theorem 5.2.1). By (2.15) the cocycle $\varphi_c(t, \omega)$ has the form

$$\varphi_c(t, \omega) x = \Phi(t, \omega) x + \psi_c(t, \omega) ,$$

where

$$\psi_c(t, \omega) := \int_0^t \Phi(t - s, \theta_s \omega) \left(b(\theta_s \omega) + c \cdot \mathbf{e} \right) \, ds .$$

Property (5.11) implies that $\psi_c(t, \omega) \gg 0$ for all $\omega \in \Omega$, $t > 0$ and $c > 0$. By Theorem 2.1.2 and Definition 1.9.1 we also have

$$|\psi_c(t, \theta_{-t} \omega)| \leq \int_{-t}^0 R_\varepsilon(\theta_s \omega) e^{-(\lambda + \varepsilon)s} \left(|b(\theta_s \omega)| + c \cdot \sqrt{d} \right) \, ds, \quad \omega \in \Omega^* ,$$

where $\varepsilon > 0$ is arbitrary and Ω^* is the θ-invariant set of full measure given by Theorem 2.1.2. Therefore we can choose $\varepsilon > 0$ such that $\{|\psi_c(t, \theta_{-t} \omega)| : t \geq 0\}$ is bounded for every $\omega \in \Omega^*$. Consequently by Proposition 1.9.3 for any $c > 0$ the system (θ, φ_c) possesses an equilibrium $w_c(\omega)$ on Ω^* such that

$$\lim_{t \to +\infty} \left\{ e^{\gamma t} \sup_{v \in D(\theta_{-t} \omega)} |\varphi(t, \theta_{-t} \omega) v - w_c(\omega)| \right\} = 0, \quad \omega \in \Omega^* ,$$

for any tempered random closed set $D \subset \mathbb{R}_+^d$ and $\gamma < -\lambda$. It is also clear from Theorem 4.6.1 that $w_c(\omega) \gg w_{c'}(\omega)$ when $c > c'$, $\omega \in \Omega^*$. Therefore, if we redefine f by $f(\omega, x) = -x + \mathbf{e}$ on $\Omega \setminus \Omega^*$, then

$$v_c(\omega) = \begin{cases} w_c(\omega) & \text{if } \omega \in \Omega^* , \\ (1 + c)\mathbf{e} & \text{if } \omega \in \Omega \setminus \Omega^* , \end{cases}$$

is a $\mathcal{D}$-absorbing super-equilibrium for the RDS $(\theta, \tilde{\varphi})$ with the cocycle $\tilde{\varphi}$ defined by the formula

$$\tilde{\varphi}(t, \omega) x = \begin{cases} \varphi(t, \omega) x & \text{if } \omega \in \Omega^* , \\ e^{-t} x + (1 - e^{-t}) \mathbf{e} & \text{if } \omega \in \Omega \setminus \Omega^* . \end{cases}$$

$\square$

Corollary 5.3.2. *Under the hypotheses of Theorem 5.3.3 the RDS $(\theta, \tilde{\varphi})$ generated by (5.1) has a global random attractor in the universe $\mathcal{D}$ of all tempered subsets of $\mathbb{R}_+^d$ and it possesses the properties stated in Theorem 3.6.2.*

Proof. This follows directly from Theorem 5.3.3, Corollary 1.8.1 and Theorem 3.6.2. $\square$

5.4 Equilibria, Semi-Equilibria and Attractors

Now we give several results on the existence of equilibria and attractors for the systems considered. We first note that under assumptions (R1)–(R3) Proposition 5.2.1 implies that the element $x \equiv 0$ is a sub-equilibrium for the RDS (θ, φ) generated by (5.1) in $\mathbb{R}_+^d$. The following assertion gives some properties of this sub-equilibrium.

Proposition 5.4.1. *Let assumptions (R1)–(R4) hold. If $f(\omega, 0) > 0$ for all $\omega \in \Omega$, then $\varphi(t, \omega)0 > 0$ for all $t > 0$ and $\omega \in \Omega$. If*

$$f_i(\omega, 0) > 0, \quad \text{for all} \quad \omega \in \Omega, \ i = 1, \dots, d, \tag{5.36}$$

then $\varphi(t, \omega)0 \gg 0$.

Proof. Let $f(\omega, 0) > 0$. Assume that there exists $t_0 > 0$ such that $\varphi(t_0, \omega)0 = 0$ for some $\omega \in \Omega$. Then the same argument as in the proof of property (5.5) gives us that $\varphi(t, \omega)0 = 0$ for all $t \in [0, t_0]$. Thus $x(t) \equiv 0$ is a stationary solution to (5.1) which is impossible because $f(\omega, 0) > 0$.

Assume now that (5.36) holds. Denote $x(t) = \varphi(t, \omega)0$. Suppose that there exists $t_0 > 0$ such that $x_i(t_0) = 0$ for some i and $\omega \in \Omega$. Then

$$-x_i(t) = \int_t^{t_0} f_i(\theta_\tau \omega, 0, \dots, 0, x_i(\tau), 0, \dots, 0) \, d\tau$$
$$+ \int_t^{t_0} (f_i(\theta_\tau \omega, x(\tau)) - f_i(\theta_\tau \omega, 0, \dots, 0, x_i(\tau), 0, \dots, 0)) \, d\tau$$

for all $t \in [0, t_0]$. The cooperativity condition implies that

$$x_i(t) + \int_t^{t_0} f_i(\theta_\tau \omega, 0, \dots, 0, x_i(\tau), 0, \dots, 0) \, d\tau \leq 0. \tag{5.37}$$

Therefore it follows from (5.36) that

$$x_i(t) + \int_t^{t_0} [f_i(\theta_\tau \omega, 0, \dots, 0, x_i(\tau), 0, \dots, 0) - f_i(\theta_\tau \omega, 0)] \, d\tau \leq 0.$$

Thus as in the proof of Proposition 5.2.1 we find that $x_i(t) \equiv 0$ for $t \in [0, t_0]$. Hence (5.37) implies $\int_t^{t_0} f_i(\theta_\tau \omega, 0) d\tau \leq 0$ which is impossible. $\square$

The following assertion contains a sufficient condition for the existence of an equilibrium.

Proposition 5.4.2. *Let assumptions (R1)–(R4) be valid. Assume that for some $x \in \mathbb{R}_+^d$ and for any $\omega \in \Omega$ there exists $t_0 = t_0(\omega)$ such that the closure $\overline{\gamma_x^{t_0(\omega)}}(\omega)$ of the tail of the orbit emanating from x,*

$$\gamma_c^{t_0(\omega)}(\omega) = \bigcup_{t \geq t_0(\omega)} \varphi(t, \theta_{-t}\omega)x \, ,$$

is a compact set in $\mathbb{R}_+^d$. Then there exists an equilibrium $u(\omega)$ for the RDS (θ, φ) generated by (5.1). This equilibrium is positive when $f(\omega, 0) > 0$.

Proof. Since $t \mapsto \varphi(t, \theta_{-t}\omega)x$ is a right continuous function (see Remark 2.1.2) and therefore a separable process, Remark 1.6.1 and Proposition 1.6.4 imply that the omega-limit set $\Gamma_x(\omega)$ exists and is an invariant random compact set. Therefore we can apply Lemma 3.4.1. Hence $v(\omega) := \inf \Gamma_x(\omega)$ is a super-equilibrium such that $v(\omega) \geq 0$. Since 0 is a sub-equilibrium, the existence of an equilibrium $u(\omega) \in [0, v(\omega)]$ now follows from Theorem 3.5.1. □

In applications below we also use the following assertion on the existence of equilibria.

Theorem 5.4.1. *Let assumptions (R1)–(R4) be valid and assume that there exists a C^1 function $W(x)$ from $\mathrm{int}\mathbb{R}_+^d$ into $\mathbb{R}_+$ such that for some nonrandom R we have*

$$|\nabla W(x)| \neq 0, \quad \langle f(\omega, x), \nabla W(x) \rangle \leq 0 \quad \text{for all} \quad \omega \in \Omega \tag{5.38}$$

provided that $W(x) = R$ and $x \in \mathrm{int}\mathbb{R}_+^d$. If the set

$$B = \{x \in \mathrm{int}\mathbb{R}_+^d : W(x) \leq R\} \tag{5.39}$$

is bounded, then there exists an equilibrium $u(\omega)$ for the RDS (θ, φ) generated by (5.1) in $\mathbb{R}_+^d$ such that $0 \leq u(\omega) \leq \sup B$. In this case there also exists a sub-equilibrium $w(\omega)$ with the properties $w(\omega) \geq u(\omega)$ and $\inf B \leq w(\omega) \leq \sup B$. We also have $u(\omega) > 0$ provided that $f(\omega, 0) > 0$ for all $\omega \in \Omega$ and $u(\omega) \gg 0$ if (5.36) holds.

The proof of this theorem relies on the following lemma.

Lemma 5.4.1. *Let assumptions (R1)–(R3) be valid and assume that there exists a C^1 function $W(x)$ from $\mathrm{int}\mathbb{R}_+^d$ into $\mathbb{R}_+$ such that for some nonrandom R we have (5.38) provided that $W(x) = R$. Then the set B given by (5.39) is a forward invariant set with respect to $\varphi(t, \omega)$, i.e. $\varphi(t, \omega)B \subset B$ for all $\omega \in \Omega$. Here $\varphi(t, \omega)$ is the cocycle generated by (5.1).*

Proof. For every x from the set

$$\mathrm{int}\mathbb{R}_+^d \cap \partial\overline{B} = \{x \in \mathrm{int}\mathbb{R}_+^d : W(x) = R\}$$

an outer normal has the form $\nu_x = \frac{\nabla W(x)}{|\nabla W(x)|}$. Since $\mathrm{int}\mathbb{R}_+^d$ is an open forward invariant set for (θ, φ) (see (5.6)), we can apply Theorem 2.2.2 with $\mathbb{O} = \mathrm{int}\mathbb{R}_+^d$ and $\mathbb{D} = \overline{B}$. It shows that $B = \mathrm{int}\mathbb{R}_+^d \cap \overline{B}$ is a deterministic forward invariant set for the RDS (θ, φ). □

Proof of Theorem 5.4.1. Lemma 5.4.1 shows that $\overline{B}$ is a forward invariant compact set for RDS (θ, φ) in $\mathbb{R}_+^d$. Therefore from Proposition 1.6.3 we have that the omega-limit set $\Gamma_B(\omega)$ is an invariant random compact set. Consequently, as in Proposition 5.4.2, we can apply Lemma 3.4.1. Therefore there exist a sub-equilibrium $\inf B \leq w(\omega) \leq \sup B$ and a super-equlibrium $0 \leq v(\omega) \leq \sup B$ such that $v(\omega) \leq w(\omega)$. Since 0 is a super-equilibrium, the existence of an equilibrium $u(\omega) \in [0, v(\omega)]$ such that

$$u(\omega) \geq \sup_{t>0} \varphi(t, \theta_{-t}\omega)0 = \lim_{t\to\infty} \varphi(t, \theta_{-t}\omega)0$$

now follows from Theorem 3.5.1. This last relation and Proposition 5.4.1 imply the positivity properties of $u(\omega)$.   □

Corollary 5.4.1. *Let (R1)–(R4) hold. Assume that there exist positive numbers R and α such that*

$$\sum_{i=1}^{d} x_i^{\alpha-1} f_i(\omega, x_1, \dots, x_d) \leq 0$$

provided that $\sum_{i=1}^{d} x_i^{\alpha} = R$ and $x \in \text{int}\mathbb{R}_+^d$. Then there exists an equilibrium $u(\omega)$ lying in the interval $[0, R^{1/\alpha}\mathbf{e}]$ for the RDS generated by (5.1) in $\mathbb{R}_+^d$. Here $\mathbf{e}$ is the element from $\text{int}\mathbb{R}_+^d$ given by the formula $\mathbf{e} = (1, \dots, 1)$.

Proof. We apply Theorem 5.4.1 with $W(x) = \sum_{i=1}^{d} x_i^{\alpha}$.   □

Below we make use of the following simple sufficient condition for the existence of super- and sub-equilibria for problem (5.1).

Proposition 5.4.3. *Let (R1)–(R3) be valid. Assume that there exists a nonrandom element $w \in \mathbb{R}_+^d$ such that $f(\omega, x)$ satisfies (R4) for all $x \in [0, w]$ and*

$$f_i(\omega, w) \leq 0, \quad \text{for all} \quad i = 1, \dots, d \quad \text{and} \quad \omega \in \Omega . \tag{5.40}$$

Then $w(\omega) \equiv w$ is a super-equilibrium for the RDS (θ, φ) generated by (5.1). If the inequality in the formula above is reversed and (R4) holds for all $x \geq w$, then $w(\omega) \equiv w$ is a sub-equilibrium.

Proof. If $w = 0$, then the weak positivity (R3) and equation (5.40) imply that $f(\omega, 0) = 0$ and therefore w is an equilibrium.

Assume that there exists $w \in \text{int}\mathbb{R}_+^d$ such that $f_i(\omega, w) \leq 0$ for all $\omega \in \Omega$ and $i = 1, \dots, d$. The cooperativity condition implies that

$$f_i(\omega, w - y) \leq f_i(\omega, w) \leq 0 \quad \text{for all} \quad y \in \Gamma_i \cap [0, w], \ \omega \in \Omega , \tag{5.41}$$

where $i = 1, \dots, d$. We apply Theorem 2.2.2 with $\mathbb{O} = \text{int}\mathbb{R}_+^d$ and $\mathbb{D} = [0, w]$. If $x \in \partial\mathbb{D} \cap \text{int}\mathbb{R}_+^d$, then there exist a subset $I \subset \{1, \dots, d\}$ and an element

$y \in \cap_{i \in I} \Gamma_i$ such that $x = w - y$, $y \ll w$, and $y_i > 0$ for $i \notin I$. Any outer normal ν_x at x has the form

$$\nu_x = \sum_{i \in I} \alpha_i e_i \quad \text{with} \quad \alpha_i \geq 0, \ \sum_{i \in I} \alpha_i^2 = 1 \,,$$

where $\{e_i\}$ is the standard basis in $\mathbb{R}^d$. Therefore from (5.41) we have that

$$\langle f(\omega, x), \nu_x \rangle = \sum_{i \in I} \alpha_i f_i(\omega, x) \leq 0 \,.$$

Thus Theorem 2.2.2 implies that the set $[0, w] \cap \text{int} \mathbb{R}_+^d$ is invariant with respect to (θ, φ). Hence $[0, w]$ is also an invariant set and w is a super-equilibrium (see Remark 3.4.1).

Assume now that $w \in \partial \mathbb{R}_+^d \setminus \{0\}$. For the sake of simplicity we consider the case $w_1 = 0$ and $w_j > 0$, $j = 2, \ldots, d$ (for other cases the proof is similar). The weak positivity condition (R3) and (5.40) imply that

$$f_1(\omega, 0, w_2, \ldots, w_d) = 0 \quad \text{and} \quad f_1(\omega, 0, x_2, \ldots, x_d) \geq 0, \ x_j \geq 0 \,.$$

Therefore using the cooperativity condition (R4) it is easy to see that

$$f_1(\omega, 0, x_2, \ldots, x_d) = 0 \quad \text{for} \quad 0 \leq x_j \leq w_j, \ j = 2, \ldots, d \,. \tag{5.42}$$

Applying the argument given above we can conclude that $(w_2, \ldots, w_d)$ is a super-equilibrium for the RDS generated by the RDE

$$\dot{x}_j(t) = f_j(\theta_t \omega, 0, x_2(t), \ldots, x_d(t)), \quad j = 2, \ldots, d \,.$$

Therefore it follows from (5.42) that $w = (0, w_2, \ldots, w_d)$ is a super-equilibrium for the RDS (θ, φ).

Assume that (5.40) holds with the inequality reversed, i.e

$$f_i(\omega, w) \geq 0, \quad \text{for all} \quad i = 1, \ldots, d, \ \omega \in \Omega \,.$$

The cooperativity condition implies that

$$f_i(\omega, w + y) \geq f_i(\omega, w) \geq 0 \quad \text{for all} \quad y \in \Gamma_i, \ \omega \in \Omega \,,$$

where $i = 1, \ldots, d$. Thus the mapping $x \mapsto f(\omega, w + x)$ is weakly positive. Therefore as in the proof of Proposition 5.2.1 we can conclude that the set $w + \mathbb{R}_+^d$ is invariant with respect to (θ, φ). This implies that w is a sub-equilibrium. $\qquad \square$

Now we prove an assertion on the existence of a random attractor.

Theorem 5.4.2. *Let assumptions (R1)–(R4) be valid and assume that there exists a C^1 function $W(x)$ from $\mathbb{R}_+^d$ into $\mathbb{R}$ such that*

$$a_1|x|^{\alpha_1} - b_1 \leq W(x) \leq a_2|x|^{\alpha_2} + b_2 , \tag{5.43}$$

where a_j, α_j, b_j are positive constants, and

$$\langle f(\omega, x), \nabla W(x) \rangle + (\beta + \varrho(\omega)) \cdot W(x) \leq C(\omega) \quad \text{for all} \quad \omega \in \Omega , \tag{5.44}$$

where $C(\omega) \geq 0$ is a tempered random variable, $\beta > 0$ is a nonrandom constant and $\varrho(\omega)$ is a random variable such that $\varrho(\theta_t \omega)$ lies in $L^1_{loc}(\mathbb{R})$ for every $\omega \in \Omega$ and

$$\lim_{t \to +\infty} \frac{1}{t} \int_0^t \varrho(\theta_\tau \omega) \, d\tau = \lim_{t \to +\infty} \frac{1}{t} \int_{-t}^0 \varrho(\theta_\tau \omega) \, d\tau = 0$$

for all $\omega \in \Omega$. Then the RDS (θ, φ) possesses a random attractor $A(\omega)$ in the universe $\mathcal{D}$ of all tempered random closed subsets of $\mathbb{R}_+^d$. This attractor is bounded from above and from below and there exist maximal and minimal equilibria $\bar{u}$ and $\underline{u}$ such that the random interval $[\underline{u}, \bar{u}]$ contains the attractor as well as all other possible tempered equilibria. In particular, if the equilibrium u is unique, then $A = \{u\}$.

Proof. From (5.1) and (5.44) we have

$$\frac{d}{dt} W(\varphi(t, \omega)x) + (\beta + \varrho(\theta_t \omega)) \cdot W(\varphi(t, \omega)x) \leq C(\theta_t \omega) .$$

Therefore we can apply Proposition 1.4.1 and Corollary 1.8.2 to conclude that the RDS (θ, φ) generated in the space $\mathbb{R}_+^d$ by problem (5.1) possesses a random global attractor $A(\omega)$ in the universe $\mathcal{D}$. The existence of the maximal and minimal equilibria $\bar{u}$ and $\underline{u}$ and their properties follow from Theorem 3.6.2. □

Remark 5.4.1. The hypotheses of Theorem 5.4.2 hold with $W(x) = |x|^2$, if $f(\omega, x)$ satisfies (R1), (R3), (R4) and also the inequality

$$\langle x, f(\omega, x) \rangle + (\beta + \varrho(\omega)) \cdot |x|^2 \leq C(\omega)$$

with β, $\varrho(\omega)$ and $C(\omega)$ possessing the properties listed in Theorem 5.4.2.

5.5 Random Equations with Concavity Properties

Here we study the qualitative behavior of random cooperative differential equations possessing some concavity properties. We rely on general results presented in Chap. 4 for random sublinear systems. We start with the following assertion.

Lemma 5.5.1. *Assume that conditions (R1)–(R4) hold and for any $\omega \in \Omega$ the function $f(\omega, \cdot)$ is a sublinear mapping from $\mathbb{R}_+^d$ into $\mathbb{R}^d$, i.e.*

$$\lambda f(\omega, x) \leq f(\omega, \lambda x) \tag{5.45}$$

for $0 < \lambda < 1$ and for all $x \in \mathbb{R}_+^d$ and $\omega \in \Omega$. Then the RDS (θ, φ) generated by (5.1) is sublinear. Moreover (θ, φ) is strongly sublinear if one of the following conditions is satisfied:

(i) $\lambda f_i(\omega, x) < f_i(\omega, \lambda x)$ for all $i = 1, \ldots, d$, $0 < \lambda < 1$, $x \in \mathrm{int}\mathbb{R}_+^d$ and $\omega \in \Omega$;

(ii) the matrix (5.12) is irreducible for all $x \in \mathrm{int}\mathbb{R}_+^d$ and $\omega \in \Omega$ and property (5.45) holds with strict inequality for every $x \in \mathrm{int}\mathbb{R}_+^d$.

Proof. The function $x_\lambda(t) = \lambda \cdot \varphi(t, \omega)x$ is the solution to the problem

$$\dot{x}_\lambda(t) = f_\lambda(\theta_t \omega, x_\lambda), \quad x_\lambda(0) = \lambda x,$$

where $f_\lambda(\omega, x) = \lambda f(\omega, \lambda^{-1}x)$. From (5.45) we have $f_\lambda(\omega, x) \leq f(\omega, x)$. Therefore the comparison principle (see Corollary 5.3.1) gives

$$\lambda \cdot \varphi(t, \omega)x \equiv x_\lambda(t) \leq x(t) \equiv \varphi(t, \omega)[\lambda x].$$

Thus (θ, φ) is sublinear. The strong sublinearity of (θ, φ) under condition either (i) or (ii) follows from Theorem 5.3.2. □

The simplest examples of sublinear mappings are $f(\omega, x) = a(\omega) \cdot \sqrt{x}$ and $f(\omega, x) = a(\omega) \cdot (1 + x)^{-1}$, where $x \in \mathbb{R}_+$ and $a(\omega) \geq 0$. They are strictly (and strongly) sublinear if $a(\omega) > 0$.

Thus we can apply here the results presented in Chap.4 for sublinear systems. For instance an application of Corollary 4.3.1 leads to the following result.

Theorem 5.5.1. *Assume that conditions (R1)–(R4) and relation (5.36) hold. Assume also that the function f satisfies the condition either (i) or (ii) of Lemma 5.5.1. Let (θ, φ) be the RDS generated by (5.1) over the ergodic metric dynamical system θ. Then either*

(i) for any $x \in \mathbb{R}_+^d$ we have $|\varphi(t, \theta_{-t}\omega)x| \to \infty$ almost surely as $t \to \infty$ or

(ii) there exists a unique almost equilibrium $u(\omega) \gg 0$ defined on a θ-invariant set Ω^* of full measure such that

$$\lim_{t \to +\infty} \varphi(t, \theta_{-t}\omega)v(\theta_{-t}\omega) = u(\omega), \quad \omega \in \Omega^*,$$

for any random variable $v(\omega)$ possessing the property $0 \leq v(\omega) \leq \lambda u(\omega)$ for all $\omega \in \Omega^*$ and for some nonrandom $\lambda > 0$.

Proof. Proposition 5.4.1 implies that $\varphi(t, \omega)0 \gg 0$ for all $t > 0$ and $\omega \in \Omega$. It is also clear that any finite-dimensional RDS is conditionally compact. Therefore we can apply Corollary 4.3.1. □

We can also apply the trichotomy theorem (see Sect. 4.4) in our situation.

Theorem 5.5.2 (Limit Set Trichotomy). *Let conditions (R1), (R3), (R4) and either (i) or (ii) of Lemma 5.5.1 be valid. Instead of (R2) we assume that there exist positive nonrandom constants a and b such that*

$$-a \cdot |x|_1 \leq f_j(\omega, x) \leq b \cdot (1 + |x|_1) \quad for \quad x \in \mathbb{R}^d_+, \ \omega \in \Omega, \ j = 1, \ldots, d, \tag{5.46}$$

where $|\cdot|_1$ is the l^1-norm in $\mathbb{R}^d$, i.e. $|x|_1 = \sum_{j=1}^d |x_i|$ for $x = (x_1 \ldots, x_d) \in \mathbb{R}^d$. Let $\mathbf{e} := (1, \ldots, 1) \in \mathbb{R}^d_+$ and let $C_{\mathbf{e}}$ be the collection of random variables $w : \Omega \to \mathbb{R}^d_+$ possessing the property

$$\alpha^{-1} \cdot \mathbf{e} \leq w(\omega) \leq \alpha \cdot \mathbf{e} \quad for \ all \quad \omega \in \Omega$$

for some nonrandom number $\alpha \geq 1$. Then any orbit emanating from $a \in C_{\mathbf{e}}$ does not leave $C_{\mathbf{e}}$, i.e.

$$\varphi(t, \omega)a(\omega) \in C_{\mathbf{e}} \quad for \ all \quad a \in C_{\mathbf{e}}, \ t \geq 0 \tag{5.47}$$

and precisely one of the following three cases applies:

(i) for all $b \in C_{\mathbf{e}}$, the orbit γ_b emanating from b is unbounded;
(ii) for all $b \in C_{\mathbf{e}}$, the orbit γ_b emanating from b is bounded, but the closure of γ_b does not belong to $C_{\mathbf{e}}$;
(iii) there exists a unique $\mathcal{F}$-measurable almost equilibrium $u \in C_{\mathbf{e}}$, and for all $b \in C_{\mathbf{e}}$ the orbit emanating from b converges to u, i.e.

$$\lim_{t \to +\infty} \varphi(t, \theta_{-t}\omega)b(\theta_{-t}\omega) = u(\omega) \quad for \ almost \ all \quad \omega \in \Omega. \tag{5.48}$$

Proof. It follows from (5.46) that

$$-a \cdot |x|_1 \cdot \mathbf{e} \leq f(\omega, x) \leq b \cdot (1 + |x|_1) \cdot \mathbf{e} \quad for \ all \quad x \in \mathbb{R}^d_+, \ \omega \in \Omega.$$

Therefore Comparison Theorem 5.3.1 implies that

$$y_1(t) \leq \varphi(t, \omega)\mathbf{e} \leq y_2(t), \quad t \in [0, t_0),$$

where $y_1(t)$ and $y_2(t)$ are solutions to the nonrandom problems

$$\dot{y}_1(t) = -a \cdot |y_1(t)|_1 \cdot \mathbf{e} \quad and \quad \dot{y}_2(t) = b \cdot (1 + |y_2(t)|_1) \cdot \mathbf{e} \tag{5.49}$$

with initial data $y_{1,2}(0) = \mathbf{e}$ and $t_0 := \sup\{s : y_1(t) \geq 0, t \in [0, s)\}$. It is easy to see that

$$|y_1(t)|_1 = d \cdot \exp\{-adt\} \quad and \quad 1 + |y_2(t)|_1 = (d + 1) \cdot \exp\{bdt\} \tag{5.50}$$

for $t \in [0, t_0)$. Therefore from (5.49) and (5.50) we have

$$y_1(t) = \exp\{-adt\} \cdot \mathbf{e} \quad \text{and} \quad y_2(t) = \mathbf{e} + (1 + d^{-1}) \cdot (\exp\{bdt\} - 1) \cdot \mathbf{e}$$

for $t \in [0, t_0)$. However these relations give solutions to (5.49) for each $t \geq 0$. Consequently $t_0 = \infty$ and

$$\exp\{-adt\} \cdot \mathbf{e} \leq \varphi(t, \omega)\mathbf{e} \leq (1 + d^{-1}) \cdot \exp\{bdt\} \cdot \mathbf{e} . \tag{5.51}$$

Thus the orbit emanating from $\mathbf{e}$ does not leave $C_\mathbf{e}$ and therefore we can apply Theorem 4.4.1. By Remark 5.2.1 the trajectory emanating from $\mathbf{e}$ is a random set with respect to $\mathcal{F}$. Therefore by Remark 4.4.1(i) the equilibrium $u(\omega)$ is $\mathcal{F}$-measurable. This completes the proof. $\qquad \square$

Remark 5.5.1. Relation (5.51) shows that under the conditions of Theorem 5.5.2 property (4.39) holds with $a(\omega) \equiv \mathbf{e}$. Therefore by Corollary 4.4.1 statement (ii) in Theorem 5.5.2 is valid in the form: for all $b \in C_\mathbf{e}$, the orbit γ_b emanating from b is bounded, but

$$\limsup_{t \to \infty} \left\{ \sup_{\omega \in \Omega} p(\varphi(t, \theta_{-t}\omega)b(\theta_{-t}\omega), v(\omega)) \right\} = \infty ,$$

where p is the part metric in $\text{int}\mathbb{R}_+^d$ (see (3.4)).

Under additional assumptions we can obtain a more detailed description of the behaviour of trajectories than that given by Theorem 5.5.2. For example using Theorem 4.2.3 we can prove the following assertion.

Proposition 5.5.1. *Let the conditions of Theorem 5.5.2 hold.*
(i) If there exists a sub-equilibrium $w(\omega)$ such that $w(\omega) \in C_\mathbf{e}$ then either (a) for all $v \in C_\mathbf{e}$, the orbit γ_b emanating from v is unbounded; or (b) there exists a unique equilibrium $u \geq w$ such for all $v \in C_\mathbf{e}$ the orbit emanating from v converges to u, i.e.

$$\lim_{t \to +\infty} \varphi(t, \theta_{-t}\omega)v(\theta_{-t}\omega) = u(\omega) \quad \text{for all} \quad \omega \in \Omega^* , \tag{5.52}$$

where Ω^* is a θ-invariant set of full measure.
(ii) Assume that θ is ergodic and there exists a super-equilibrium $w(\omega) \in C_\mathbf{e}$. If

$$\varphi(t, \omega)(\mathbb{R}_+^d \setminus \{0\}) \subset \text{int}\mathbb{R}_+^d \quad \text{for all} \quad \omega \in \Omega , \tag{5.53}$$

then there exists a θ-invariant set Ω^* of full measure such that either (a) for all $v \in C_\mathbf{e}$, the orbit γ_v emanating from v converges to zero for $\omega \in \Omega^*$, i.e.

$$\lim_{t \to +\infty} \varphi(t, \theta_{-t}\omega)v(\theta_{-t}\omega) = 0, \quad \omega \in \Omega^* , \tag{5.54}$$

or (b) there exists a unique equilibrium $u(\omega)$ such that $0 \ll u(\omega) \leq \alpha \cdot \mathbf{e}$ for $\omega \in \Omega^*$ and we have (5.52) for all $v \in C_\mathbf{e}$.

Proof. If under the condition in (i) option (a) is not valid, then the orbit emanating from w is bounded. Therefore (b) follows from the first part of Theorem 4.2.3 (see also Remark 4.2.2).

As for assertion (ii), the second part of Theorem 4.2.3 implies the existence of an equilibrium $0 \le u(\omega) \le w(\omega) \le \alpha \cdot \mathbf{e}$. Lemma 3.5.1 and (5.53) give that either $u(\omega) = 0$ or $u(\omega) \gg 0$ on a θ-invariant set of full measure. In the first case we obtain (5.54). In the second case we obtain (b). $\qquad \square$

The construction of sub- or super-equilibria in Proposition 5.5.1 usually relies on the comparison principle (see examples below). We also note that by Proposition 5.2.1 property (5.53) holds under assumption (R3*). The result given below provides other conditions that guarantee (5.53).

In the next proposition we use the ordering relation between $d \times d$ matrices which arises from viewing them as vectors from the space $\mathbb{R}^{d^2}$ with the standard cone $\mathbb{R}^{d^2}_+$.

Proposition 5.5.2. *Assume that assumptions (R1)–(R4) are met and that the matrix $D_x f(\omega, x)$ is irreducible for $x \in \text{int}\mathbb{R}^d_+$ and $\omega \in \Omega$. Let the function $f(\omega, x)$ be s-concave, i.e.*

$$D_x f(\omega, x) < D_x f(\omega, y) \quad for \quad x \gg y \gg 0, \quad and \quad \omega \in \Omega . \qquad (5.55)$$

Then (θ, φ) is s-concave (see Definition 4.1.3) and strongly order-preserving in $\mathbb{R}^d_+$. In particular (5.53) holds.

Proof. We first note that $y(t) = D_x\varphi(t, \omega, x)z$ is the solution to the problem

$$\dot{y}(t) = D_x f(\theta_t\omega, \varphi(t, \omega, x))y(t), \quad y(0) = z .$$

Since $D_x f(\omega, x)$ is irreducible in $\text{int}\mathbb{R}^d_+$, we have $\varphi(t, \omega, x) \gg \varphi(t, \omega, y) \gg 0$ for all $x > y \gg 0$ by Theorem 5.2.1. Therefore property (5.55) gives that

$$D_x f(\theta_t\omega, \varphi(t, \omega, x)) < D_x f(\theta_t\omega, \varphi(t, \omega, y))$$

for any $x > y \gg 0$. Consequently the comparison principle implies that

$$D_x\varphi(t, \omega, x)z < D_x\varphi(t, \omega, y)z \quad for \quad x \gg y \gg 0, \ z \in \text{int}\mathbb{R}^d_+ \quad and \quad \omega \in \Omega .$$

Therefore (θ, φ) is an s-concave RDS.

In order to apply Theorem 5.2.1 to prove that (θ, φ) is strongly order-preserving in $\mathbb{R}^d_+$ we need to verify relation (5.53) only. For any $x \in \mathbb{R}^d_+ \setminus \{0\}$ and $0 < \lambda < 1$ we have

$$\varphi(t, \omega, x) \ge \varphi(t, \omega, \lambda x) = \varphi(t, \omega, 0) + \int_0^\lambda D_x\varphi(t, \omega, sx) \, ds \, x .$$

Hence

$$\varphi(t, \omega, x) \ge \lambda \cdot D_x\varphi(t, \omega, 0)x + \lambda \cdot b(t, \omega, x, \lambda) , \qquad (5.56)$$

where

$$b(t, \omega, x, \lambda) = \lambda^{-1} \int_0^\lambda \left(D_x \varphi(t, \omega, sx) - D_x \varphi(t, \omega, 0) \right) x \, ds$$

is a random variable from $\mathbb{R}^d$ such that $\lim_{\lambda \to 0} b(t, \omega, x, \lambda) = 0$ for all $\omega \in \Omega$. From the s-concavity of (θ, φ) we obtain

$$D_x \varphi(t, \omega, 0) \geq D_x \varphi(t, \omega, y/2) > D_x \varphi(t, \omega, y) \quad \text{if} \quad y \gg 0 \quad \text{and} \quad \omega \in \Omega \,.$$

Therefore from Lemma 5.2.1 we have

$$D_x \varphi(t, \omega, 0) \gg 0 \quad \text{for all} \quad \omega \in \Omega \,.$$

Hence for every $\omega \in \Omega$ there exists $\lambda_0(\omega) > 0$ such that

$$D_x \varphi(t, \omega, 0) x + b(t, \omega, x, \lambda) \gg 0 \quad \text{for all} \quad \lambda < \lambda_0(\omega) \,.$$

Thus (5.56) implies that $\varphi(t, \omega, x) \gg 0$ for every $x > 0$, whence (5.53).   □

The following theorem describes possible scenarios in s-concave systems.

Theorem 5.5.3. *Let (R1)–(R4) be valid. Assume that the function $f(\omega, x)$ is s-concave and that the matrix $D_x f(\omega, x)$ is irreducible for all $x \in \mathrm{int} \mathbb{R}_+^d$ and $\omega \in \Omega$. If $f(\omega, 0) \in \mathbb{R}_+^d \setminus \{0\}$ for all $\omega \in \Omega$, then either*

(a) the orbit γ_v emanating from v is unbounded for all $v(\omega) \geq 0$, or
(b) there exists a unique equilibrium $u \gg 0$ such for every $v(\omega)$ possessing the property $0 \leq v(\omega) \leq \alpha \cdot u(\omega)$ with some $\alpha > 0$ the orbit emanating from v converges to u, i.e.

$$\lim_{t \to +\infty} \varphi(t, \theta_{-t}\omega) v(\theta_{-t}\omega) = u(\omega) \quad \text{for all} \quad \omega \in \Omega^* \,, \tag{5.57}$$

where Ω^* is a θ-invariant set of full measure.

If we additionally assume that the affine RDS generated by the equation

$$\dot{y} = f(\theta_t \omega, 0) + D_x f(\theta_t \omega, 0) y \tag{5.58}$$

possesses a super-equilibrium $w(\omega) \in \mathrm{int} \mathbb{R}_+^d$, then there exist bounded orbits and (b) holds. If $f(\omega, 0) \equiv 0$, θ is an ergodic metric dynamical system and the top Lyapunov exponent of the linear RDS (5.58) is less than zero, then we have

$$\lim_{t \to \infty} \phi(t, \theta_{-t}\omega) x = 0 \quad \text{for all} \quad x \in \mathbb{R}_+^d$$

on a θ-invariant set of full measure.

Proof. If $f(\omega, 0) \in \mathbb{R}_+^d \setminus \{0\}$ for all $\omega \in \Omega$, then by Proposition 5.4.1 we have that $\varphi(t, \omega, 0) > 0$ for all $t > 0$ and $\omega \in \Omega$. Therefore Proposition 5.5.2 implies that $\varphi(t, \omega, 0) \gg 0$ for $t > 0$. Consequently $v_s(\omega) = \varphi(s, \theta_{-s}\omega, 0) \gg 0$ is a sub-equilibrium for every $s > 0$ (see Proposition 3.4.1). By Proposition 5.5.2 (θ, φ) is s-concave. Therefore Proposition 4.1.1 implies that (θ, φ) is concave, i.e. it satisfies (4.5). In particular

$$(1 - \lambda)\varphi(t, \omega, 0) + \lambda\varphi(t, \omega, y) \le \varphi(t, \omega, \lambda y), \quad y > 0 .$$

Hence $\lambda\varphi(t, \omega, y) \ll \varphi(t, \omega, \lambda y)$ for all $t > 0$ and $y > 0$. Thus (θ, φ) is strongly sublinear. Therefore we can apply the same argument as in the proof of Proposition 5.5.1 to obtain (a)/(b) dichotomy.

Since

$$f(\omega, x) = f(\omega, 0) + \int_0^1 D_x f(\omega, sx)ds\, x ,$$

under s-concavity condition (5.55) we have

$$f(\omega, x) \le f(\omega, 0) + D_x f(\omega, 0)x .$$

Therefore the affine RDS generated by (5.58) dominates (θ, φ) from above. Hence $w(\omega) \in \operatorname{int}\mathbb{R}_+^d$ is a super-equilibrium for (θ, φ). Thus the orbit γ_w emanating from w is bounded, whence (b).

The proof of the last assertion follows from Proposition 4.2.2 (cf. Theorem 2.1.2). □

5.6 One-Dimensional Explicitly Solvable Random Equations

In this section we consider a class of RDS generated by one-dimensional RDE of the form

$$\dot{x} = \alpha(\theta_t\omega) \cdot f(x) + \beta(\theta_t\omega) \cdot g(x), \quad x \in \mathbb{R} , \tag{5.59}$$

over an ergodic metric dynamical system θ. We assume that $\alpha(\omega)$ and $\beta(\omega)$ are tempered random variables from $L^1(\Omega, \mathcal{F}, \mathbb{P})$. In this case the Birkhoff-Khintchin ergodic theorem (see, e.g., ARNOLD [3, Appendix]) gives the relations

$$\lim_{|t|\to\infty} \frac{1}{t} \int_0^t \alpha(\theta_\tau\omega)d\tau = \mathbb{E}\alpha, \qquad \lim_{|t|\to\infty} \frac{1}{t} \int_0^t \beta(\theta_\tau\omega)d\tau = \mathbb{E}\beta \tag{5.60}$$

on a θ-invariant set $\Omega^* \in \mathcal{F}$ of full measure. Without loss of generality we can suppose that $\Omega^* = \Omega$ (see Remark 1.2.1(ii)). We also assume that $f(x)$ and $g(x)$ are C^1 functions on $\mathbb{R}$ such that

$$\alpha(\omega) \cdot f(0) + \beta(\omega) \cdot g(0) \geq 0, \quad \omega \in \Omega ,$$

and equation (5.59) generates an RDS (θ, φ) in some interval $[0, a] \subseteq \mathbb{R}_+$. This is implied by the relation

$$\alpha(\omega) \cdot f(a) + \beta(\omega) \cdot g(a) \leq 0, \quad \omega \in \Omega ,$$

in the case $0 < a < \infty$ (see Proposition 5.4.3) and by the relation

$$x \cdot [\alpha(\omega) \cdot f(x) + \beta(\omega) \cdot g(x)] \leq C_1(\omega) \cdot x^2 + C_2(\omega), \quad \omega \in \Omega ,$$

in the case $[0, a] = \mathbb{R}_+$, where C_1 and C_2 are random variables such that $t \mapsto C_j(\theta_t \omega)$ is locally integrable (see Proposition 5.2.1).
If

$$g(x) > 0 \quad \text{and} \quad f(x) = g(x) \cdot (\gamma_1 G(x) + \gamma_2) \quad \text{for} \quad 0 < x < a , \tag{5.61}$$

where γ_1 and γ_2 are constants and $G(x)$ is a primitive for $[g(x)]^{-1}$ on the interval $(0, a)$ (i.e. $G'(x) = [g(x)]^{-1}$, $x \in (0, a)$), then the cocycle φ can be represented in the explicit form. Indeed, if $x(t)$ is a solution to (5.59), then $y(t) = G(x(t))$ solves the linear equation

$$\dot{y} = \gamma_1 \alpha(\theta_t \omega) \cdot y + \beta(\theta_t \omega) + \gamma_2 \alpha(\theta_t \omega) .$$

Therefore the cocycle φ has the form

$$\varphi(t, \omega, x) = G^{-1} \left(G(x) \exp\left\{ \gamma_1 \int_0^t \alpha(\theta_\tau \omega) d\tau \right\} \right.$$
$$\left. + \int_0^t (\beta(\theta_s \omega) + \gamma_2 \alpha(\theta_s \omega)) \exp\left\{ \gamma_1 \int_s^t \alpha(\theta_\tau \omega) d\tau \right\} ds \right) . \tag{5.62}$$

This observation is proved to be useful in the study of bifurcation phenomena in one-dimensional RDE (see ARNOLD [3, Chap.9] and also XU [110]).

Example 5.6.1. Consider the RDE

$$\dot{x} = \alpha(\theta_t \omega) \cdot x + \beta(\theta_t \omega) \cdot x^{N+1}, \quad x \in \mathbb{R}_+ , \tag{5.63}$$

where $N > 0$. If we assume that $\beta(\omega) \leq 0$, then equation (5.63) generates an RDS (θ, φ) in $\mathbb{R}_+$. It is easy to see that this system is sublinear. We obviously have relation (5.61) with $\gamma_1 = -N$ and $\gamma_2 = 0$ and therefore

$$\varphi(t, \omega, x) = \frac{x \exp\left\{ \int_0^t \alpha(\theta_\tau \omega) d\tau \right\}}{\left(1 + N x^N \int_0^t |\beta(\theta_s \omega)| \exp\left\{ N \int_0^s \alpha(\theta_\tau \omega) d\tau \right\} ds \right)^{1/N}} \tag{5.64}$$

for $x > 0$ and $\varphi(t, \omega, 0) = 0$. Since $\beta(\omega) \leq 0$, the RDS (θ, φ) is dominated from above by the system generated by the linear equation $\dot{x} = \alpha(\theta_t \omega) \cdot x$.

By the ergodic theorem (see (5.60)) the Lyapunov exponent for this system is $\lambda = \mathbb{E}\alpha$. Therefore if $\mathbb{E}\alpha < 0$, then $\lim_{t\to\infty}\varphi(t, \theta_{-t}\omega, u(\theta_{-t}\omega)) = 0$ for any tempered $u(\omega) \geq 0$. Thus $A(\omega) = \{0\}$ is a random attractor for (θ, φ) in the universe $\mathcal{D}$ of all tempered subsets of $\mathbb{R}_+$.

Assume that $\mathbb{E}\alpha > 0$ and $\beta(\omega) \leq -\beta_0 < 0$ for all $\omega \in \Omega$. Then using Theorem 5.4.2 (see Remark 5.4.1) we can prove that (θ, φ) possesses a random attractor $A(\omega) = [0, u(\omega)]$ in the universe $\mathcal{D}$. Here $u(\omega) \geq 0$ is an equilibrium. Using (5.64) it is easy to find that

$$u(\omega) = \left(N \int_{-\infty}^0 |\beta(\theta_s\omega)| \exp\left\{ -N \int_s^0 \alpha(\theta_\tau\omega)d\tau \right\} ds \right)^{-1/N} .$$

Moreover, there exists $\gamma > 0$ such that

$$\lim_{t\to\infty} e^{\gamma t} |\varphi(t, \theta_{-t}\omega, x) - u(\omega)| = 0 \quad \text{for all } x > 0 \text{ and } \omega \in \Omega . \qquad (5.65)$$

If for some $b > 0$ we have $\alpha(\omega) + \beta(\omega) \cdot b^N \leq 0$ for all $\omega \in \Omega$, then the equilibrium $u(\omega)$ belongs to the interval $(0, b]$. Indeed, in this case by Proposition 5.4.3 b is a super-equilibrium, i.e. $\varphi(t, \theta_{-t}\omega, b) \leq b$. Therefore (5.65) implies that $u(\omega) \leq b$.

Example 5.6.2. We consider the RDE

$$\dot{x} = \beta(\theta_t\omega) \cdot g(x), \quad x \in [0,1] , \qquad (5.66)$$

where $g \in C^1(\mathbb{R})$, $g(x) > 0$ for $x \in (0,1)$ and $g(0) = g(1) = 0$. This equation generates an RDS (θ, φ) in $[0,1]$ with the cocycle

$$\varphi(t, \omega, x) = G^{-1}\left(G(x) + \int_0^t \beta(\theta_\tau\omega)d\tau \right) , \qquad (5.67)$$

where $G(x)$ is a primitive for $[g(x)]^{-1}$ on $(0,1)$ (relation (5.61) holds with $\gamma_1 = \gamma_2 = 0$). It is clear that $G(x)$ is an increasing function such that $G(x) \to +\infty$ as $x \to 1$ and $G(x) \to -\infty$ as $x \to 0$. Therefore using (5.60) and (5.67) we observe the following behaviour of trajectories:

(i)   if $\mathbb{E}\beta < 0$, then

$$\lim_{t\to\infty} \varphi(t, \omega, x) = \lim_{t\to\infty} \varphi(t, \theta_{-t}\omega, x) = 0, \quad x \in [0,1) ;$$

(ii) if $\mathbb{E}\beta > 0$, then

$$\lim_{t\to\infty} \varphi(t, \omega, x) = \lim_{t\to\infty} \varphi(t, \theta_{-t}\omega, x) = 1, \quad x \in (0,1] .$$

In the case $\mathbb{E}\beta = 0$ the dynamics is more complicated. For instance, if $\beta(\theta_t\omega) = \frac{d}{dt}B(\theta_t\omega)$ is a derivative of a stationary process $B(\theta_t\omega)$ with absolutely continuous trajectories, then from (5.67) we have

$$\varphi(t,\omega,x) = G^{-1}\left(G(x) + B(\theta_t\omega) - B(\omega)\right) .$$

Therefore a random variable $u_c(\omega) = G^{-1}(c + B(\omega))$ satisfies the relation $\varphi(t,\omega,u_c(\omega)) = u_c(\theta_t\omega)$ for any $c \in \mathbb{R}$. Thus we have a continuous family $\{u_c(\omega) : c \in \mathbb{R}\}$ of equilibria such that $u_c(\omega) \to 0$ as $c \to -\infty$ and $u_c(\omega) \to 1$ as $c \to \infty$. If $B_0 := \sup_{\omega \in \Omega} |B(\omega)| < \infty$, then

$$0 < G^{-1}\left(G(x) - 2B_0\right) \leq \varphi(t,\omega,x) \leq G^{-1}\left(G(x) + 2B_0\right) < 1$$

for all $t > 0$, $\omega \in \Omega$ and $x \in (0,1)$, i.e. all trajectories emanating from points $x \in (0,1)$ are separated from the equilibria 0 and 1 uniformly with respect to t.

On the other hand, if we assume that $\beta \in L^2(\Omega,\mathcal{F},\mathbb{P})$, $\mathbb{E}\beta = 0$, and the process $\beta(\theta_t\omega)$ satisfies the central limit theorem, i.e. the limit

$$\lim_{t\to\infty} \mathbb{E}\left(t^{-1/2}\int_0^t \beta(\theta_\tau\cdot)d\tau\right)^2 = \sigma > 0$$

exists and

$$\lim_{t\to\infty} \mathbb{P}\left\{\omega : t^{-1/2}\int_0^t \beta(\theta_\tau\omega)d\tau < a\right\} = \frac{1}{\sqrt{2\pi\sigma}}\int_{-\infty}^a e^{-\frac{\xi^2}{2\sigma}}d\xi$$

for any $a \in [-\infty,+\infty]$, then it follows from (5.67) that under the conditions $g'(0) > 0$ and $g'(1) < 0$ we have

$$\lim_{t\to\infty} \mathbb{P}\left\{\omega : e^{-\sqrt{t}\log t} \leq \varphi(t,\omega,x) \leq e^{-\sqrt{t}/\log t}\right\} = \frac{1}{2}$$

and

$$\lim_{t\to\infty} \mathbb{P}\left\{\omega : e^{-\sqrt{t}\log t} \leq 1 - \varphi(t,\omega,x) \leq e^{-\sqrt{t}/\log t}\right\} = \frac{1}{2}$$

for all $x \in (0,1)$. We refer to CHUESHOV/VUILLERMOT [24] for the proof of the last two formulas and for other facts on the long-time dynamics of the RDS generated by (5.66). The last two relations imply that

$$\lim_{t\to\infty} \mathbb{P}\left\{\omega : \varphi(t,\omega,x) \in [\delta, 1 - \delta]\right\} = 0$$

for any $\delta > 0$ and $x \in (0,1)$. This means that $\mathrm{dist}\left(\varphi(t,\omega,x),\{0,1\}\right) \to 0$ in probability as $t \to \infty$. Thus the two-point set $A = \{0,1\}$ is a weak point attractor, i.e. an attractor with respect to convergence in probability.

Example 5.6.3. Consider a one-dimensional RDE of the form

$$\dot{x} = \alpha(\theta_t\omega) \cdot \sin x + \beta(\theta_t\omega) \cdot (1 - \cos x) \tag{5.68}$$

over an ergodic metric dynamical system θ, where $\alpha(\omega)$ and $\beta(\omega)$ are random variables such that $t \mapsto \alpha(\theta_t\omega)$ and $t \mapsto \beta(\theta_t\omega)$ are locally integrable. This equation generates an RDS in $\mathbb{R}$. Since any interval $[2\pi k, 2\pi(k+1)]$ is an invariant set and equation (5.68) is invariant with respect to the change $x \mapsto x+2\pi k$, we consider the problem on the unit circle C which is interpreted as the interval $[0, 2\pi]$ with identified end-points.

A simple calculation shows that relation (5.61) holds with $\gamma_1 = -1$, $\gamma_2 = 0$ and $G(x) = -\cot \frac{x}{2}$. Therefore equation (5.68) generates an RDS (θ, φ) in C with the cocycle

$$\varphi(t, \omega)x = 2\text{arccot}\left(-y(t, \omega; -\cot(x/2))\right), \quad x \in (0, 2\pi),$$

where $y(t, \omega; y_0)$ solves the affine equation

$$\dot{y} = -\alpha(\theta_t\omega) \cdot y + \beta(\theta_t\omega), \quad y(0) = y_0.$$

Therefore $\varphi(t, \omega)0 = 0$ and

$$\varphi(t, \omega)x = 2\text{arccot}\left\{\cot\frac{x}{2} \cdot e^{-\int_0^t \alpha(\theta_\tau\omega)d\tau} - \int_0^t \beta(\theta_s\omega)e^{-\int_s^t \alpha(\theta_\tau\omega)d\tau}\,ds\right\}$$

for $x \in C \setminus \{0\}$. Assume that $\alpha \in L^1(\Omega, \mathfrak{F}, \mathbb{P})$. It follows from the considerations presented in Example 2.1.2 that in both cases $\mathbb{E}\alpha > 0$ and $\mathbb{E}\alpha > 0$ the RDS (θ, φ) on the circle C has two equilibria $u_0 \equiv 0$ and either

$$u_+(\omega) = 2\text{arccot}\left\{-\int_{-\infty}^0 \beta(\theta_s\omega)e^{-\int_s^0 \alpha(\theta_\tau\omega)d\tau}\,ds\right\} \quad \text{if } \mathbb{E}\alpha > 0,$$

or

$$u_-(\omega) = 2\text{arccot}\left\{\int_0^\infty \beta(\theta_s\omega)e^{\int_0^s \alpha(\theta_\tau\omega)d\tau}\,ds\right\} \quad \text{if } \mathbb{E}\alpha < 0.$$

If $\mathbb{E}\alpha > 0$, then Proposition 1.9.3 implies that

$$\varphi(t, \theta_{-t}\omega)v(\theta_{-t}\omega) \to u_+(\omega) \quad \text{as} \quad t \to \infty$$

for any random variable $v(\omega)$ from the interval $[\varepsilon, 2\pi - \varepsilon]$, where $\varepsilon > 0$ is arbitrary.

In the case $\mathbb{E}\alpha < 0$ using the representation

$$y(t, \omega; y_0) = -\cot\frac{u_-(\theta_t\omega)}{2} + \left(y_0 + \cot\frac{u_-(\omega)}{2}\right) \cdot e^{-\int_0^t \alpha(\theta_\tau\omega)d\tau}$$

we obtain that

$$\varphi(t, \theta_{-t}\omega)v(\theta_{-t}\omega) = 2\operatorname{arccot} z(t, \omega)$$

for any $v(\omega) \in C \setminus \{0\}$, where

$$z(t, \omega) = -\cot \frac{u_-(\omega)}{2} - \left(\cot \frac{v(\theta_{-t}\omega)}{2} - \cot \frac{u_-(\theta_{-t}\omega)}{2}\right) \cdot e^{-\int_{-t}^{0} \alpha(\theta_\tau \omega) d\tau}.$$

Hence in the circle C we have the relation

$$\varphi(t, \theta_{-t}\omega)v(\theta_{-t}\omega) \to 0 \quad \text{as} \quad t \to \infty$$

provided that

$$\left| \cot \frac{v(\omega)}{2} - \cot \frac{u_-(\omega)}{2} \right| \geq \delta(\omega) > 0 ,$$

where $\delta(\omega)^{-1}$ is a tempered random variable.

Thus we observe the exchange of stability between the equilibria $u_0(\omega)$ and $u_\pm(\omega)$ when the value $\mathbb{E}\alpha$ changes the sign in the RDS (θ, φ) generated by (5.68) in the circle C.

5.7 Applications

As the main example of an application of the theory developed we consider a random model of the control protein synthesis in the cell (for the deterministic case see, e.g., SMITH [102] and the references therein).

5.7.1 Random Biochemical Control Circuit

Consider the system of random differential equations

$$\dot{x}_1(t) = g(\theta_t \omega, x_d(t)) - \alpha_1(\theta_t \omega)x_1(t) , \tag{5.69}$$

$$\dot{x}_j(t) = x_{j-1}(t) - \alpha_j(\theta_t \omega)x_j(t), \quad j = 2, \dots, d . \tag{5.70}$$

Here $\{\alpha_j(\omega)\}$ are random variables for which $t \mapsto \alpha_j(\theta_t \omega)$ is locally integrable. We assume that $g : \Omega \times \mathbb{R} \mapsto \mathbb{R}$ is measurable and $x \mapsto g(\omega, x)$ is continuously differentiable for every $\omega \in \Omega$. Moreover there exist positive random variables $b(\omega)$ and $c(\omega)$ and a deterministic constant $a > 0$ possessing the properties

$$0 \leq g(\omega, x) \leq a \cdot x + b(\omega) \quad \text{and} \quad 0 < g'(\omega, x) \leq c(\omega) \tag{5.71}$$

for every $\omega \in \Omega$ and for every $x > 0$. We assume also that $t \mapsto b(\theta_t \omega)$ and $t \mapsto c(\theta_t \omega)$ are locally integrable. We use g' to denote the derivative with respect to the space variable.

The values x_j represent concentrations of various macro-molecules in the cell and therefore must be nonnegative. It is easy to see that assumptions (R1)–(R4) are valid here. Hence equations (5.69) and (5.70) generate a strictly order-preserving RDS (θ, φ) in the cone $\mathbb{R}^d_+$. We note that the above equations reduce to the standard deterministic equations of a biochemical control circuit (see SMITH [102]), when $g(x)$ and α_j are nonrandom.

Assume now that there exist positive nonrandom constants α_j, $j = 1, \dots, d$, such that $\alpha_j(\omega) \geq \alpha_j > 0$, $j = 1, \dots, d$, and consider the following affine RDE

$$\dot{x}_1(t) = a \cdot x_d(t) - \alpha_1 x_1(t) + b(\theta_t \omega) , \tag{5.72}$$

$$\dot{x}_j(t) = x_{j-1}(t) - \alpha_j x_j(t), \quad j = 2, \dots, d . \tag{5.73}$$

It is clear that equations (5.72) and (5.73) generate a strictly order-preserving RDS (θ, ψ) in the cone $\mathbb{R}^d_+$. Comparison Theorem 5.3.1 implies that this RDS dominates (θ, φ) from above. The cocycle ψ of this system has the form

$$\psi(t, \omega)x = e^{tA}x + \xi(t, \omega), \quad \xi(t, \omega) := \int_0^t e^{(t-\tau)A} B(\theta_\tau \omega) \, d\tau , \tag{5.74}$$

where A is the matrix with all entries equal to zero, except $a_{jj} = -\alpha_j$, $a_{j,j-1} = 1$ and $a_{1d} = a$, $B(\omega)$ is the column whose only nonzero element is $b_1(\omega) = b(\omega)$. Since the eigenvalues λ of the matrix A satisfy the equation $\prod_{j=1}^d (\lambda + \alpha_j) = a$, it is easy to see that $\text{Re}\lambda < 0$ provided that $\prod_{j=1}^d \alpha_j > a$, which we assume to make A stable.

Assume also that $b(\omega)$ is tempered. Then it is easy to see that $t \mapsto \xi(t, \theta_{-t}\omega)$ is bounded for all $\omega \in \Omega$. Therefore by Theorem 4.6.1 and Remark 4.6.1

$$w(\omega) := \lim_{t \to \infty} \xi(t, \theta_{-t}\omega) = \int_{-\infty}^0 e^{-\tau A} B(\theta_\tau \omega) \, d\tau \tag{5.75}$$

exists and is a tempered equilibrium for (θ, ψ). Since

$$\psi(t, \omega)y - w(\theta_t \omega) = e^{tA}(y - w(\omega)) ,$$

this equilibrium uniformly attracts all tempered random sets with exponential speed, i.e. there exists $\gamma > 0$ such that

$$\lim_{t \to +\infty} e^{\gamma t} \sup_{y \in D(\theta_{-t}\omega)} |\psi(t, \theta_{-t}\omega)y - w(\omega)| = 0$$

for any $D(\omega) \in \mathcal{D}$, where $\mathcal{D}$ is the universe of all random tempered sets in $\mathbb{R}^d_+$.

By the comparison principle the random variable $\mu w(\omega)$ is a super-equilibrium for the RDS (θ, φ) generated by (5.69) and (5.70) for any $\mu \geq 1$.

Consequently by Proposition 3.5.2 the RDS (θ, φ) possesses an equilibrium $u(\omega)$ such that $0 \leq u(\omega) \leq w(\omega)$. By Proposition 3.7.1 (θ, φ) is dissipative in $\mathcal{D}$. Therefore it has a random attractor A in $\mathcal{D}$, and since $v(\omega) \equiv 0$ is evidently a sub-equilibrium for (θ, φ), this attractor belongs to the interval $[0, w](\omega) := \{v : 0 \leq v \leq w(\omega)\}$ and the conclusions of Theorem 3.6.2 on the structure of random attractors are valid.

If we assume in addition that $g(\omega, 0) > 0$ for all $\omega \in \Omega$, then the affine system

$$\dot{x}_1(t) = -\alpha_1(\theta_t \omega) x_1(t) + g(\theta_t \omega, 0) ,$$

$$\dot{x}_j(t) = x_{j-1}(t) - \alpha_j(\theta_t \omega) x_j(t), \quad j = 2, \ldots, d .$$

dominates (θ, φ) from below. It possesses a unique globally asymptotically stable equilibrium $v(\omega) = (v_1(\omega), \ldots, v_d(\omega))$ with

$$v_1(\omega) = \int_{-\infty}^{0} g(\theta_\tau \omega, 0) \cdot \exp\left\{ -\int_{\tau}^{0} \alpha_1(\theta_s \omega) ds \right\} d\tau ,$$

and

$$v_j(\omega) = \int_{-\infty}^{0} v_{j-1}(\theta_\tau \omega) \cdot \exp\left\{ -\int_{\tau}^{0} \alpha_j(\theta_s \omega) ds \right\} d\tau, \quad j = 2, \ldots, d .$$

For every ω the equilibrium $v(\omega)$ belongs to $\mathrm{int} \mathbb{R}_+^d$ and $v(\omega) \leq w(\omega)$, where $w(\omega)$ is given by (5.75). The comparison principle gives that $\mu v(\omega)$ is a sub-equilibrium of (θ, φ) for any $0 \leq \mu \leq 1$. Therefore the random attractor of (θ, φ) is contained in the interval $[v(\omega), w(\omega)]$. According to Theorem 3.6.2 this attractor lies between two equilibria $\underline{u}(\omega)$ and $\overline{u}(\omega)$ such that $0 \ll v(\omega) \leq \underline{u}(\omega) \leq \overline{u}(\omega) \leq w(\omega)$. Moreover, $\underline{u}(\omega) = \overline{u}(\omega)$ almost surely provided that the function $g(\omega, x)$ possesses the property

$$\lambda g(\omega, x) < g(\omega, \lambda x) \quad \text{for} \quad 0 < \lambda < 1, \ \omega \in \Omega .$$

In this case the condition (ii) of Lemma 5.5.1 holds. This implies that the system (θ, φ) is strongly sublinear and therefore we can apply Theorem 4.2.1 on the uniqueness of strongly positive equilibria.

We now consider the case when $g(\omega, 0) = 0$ for all $\omega \in \Omega$. In this case the attractor A lies between the two equilibria $\underline{u}(\omega) \equiv 0$ and $\overline{u}(\omega) \geq 0$. We can guarantee that $\overline{u}(\omega) \gg 0$ if we assume, for instance, that there exist nonrandom constants α_j^*, $j = 1, \ldots, d$, and a function $g_0(x)$ such that

$$\alpha_j(\omega) \leq \alpha_j^*, \ j = 1, \ldots, d, \quad g_0(x) \leq g(\omega, x) \quad \text{for all} \ \omega \in \Omega$$

and

$$\prod_{j=1}^{d} \alpha_j^* < \limsup_{x \to +0} \frac{g_0(x)}{x} \leq +\infty . \tag{5.76}$$

Indeed, let

$$v^{(n)} = (v_1^n, \dots, v_d^n) := \varepsilon_n \left(\prod_{j=2}^{d} \alpha_j^*, \prod_{j=3}^{d} \alpha_j^*, \dots, \alpha_d^*, 1 \right) \gg 0 \, ,$$

where ε_n are positive numbers. Since

$$g(\omega, v_d^n) - \alpha_1(\omega)v_1^n \geq g_0(\varepsilon_n) - \varepsilon_n \prod_{j=1}^{d} \alpha_j^* = \varepsilon_n \left(\frac{g_0(\varepsilon_n)}{\varepsilon_n} - \prod_{j=1}^{d} \alpha_j^* \right) \, ,$$

equation (5.76) implies that there exists a sequence $\{\varepsilon_n\}$, $\varepsilon_n > 0$, $\varepsilon_n \to 0$, such that

$$g(\omega, v_d^n) - \alpha_1(\omega)v_1^n > 0 \, .$$

We also have the relation

$$v_{i-1}^n - \alpha_i(\omega)v_i^n \geq v_{i-1}^n - \alpha_i^* v_i^n = 0 \, .$$

Thus by Proposition 5.4.3 $v^{(n)}$ is a sub-equilibrium for any $n = 1, 2, \dots$. This implies the instability of $\underline{u}(\omega) \equiv 0$ and the strong positivity of $\overline{u}(\omega)$. Since $\overline{u}(\omega)$ is the maximal equilibrium in the attractor, Theorem 3.6.2 implies that $\overline{u}(\omega)$ is stable from above. Its stability from below in the strongly sublinear case is equivalent to the property

$$\varphi(t, \theta_{-t}\omega)v^{(n)} \to \overline{u}(\omega) \quad \text{as} \quad t \to \infty \tag{5.77}$$

for every n almost surely. If (5.77) is not true, then there exists another strongly positive equilibrium $w(\omega)$ such that $\varphi(t, \theta_{-t}\omega)v^{(n)} \to w(\omega)$. This contradicts the uniqueness of strongly positive equilibria for strongly sublinear RDS. Thus the equilibrium $\overline{u}(\omega)$ is stable.

If $g'(\omega, x)$ is a strictly decreasing function for every $\omega \in \Omega$, then Proposition 5.5.2 implies that the RDS (θ, φ) is s-concave. Therefore if for some nonrandom g_0^*

$$g'(\omega, 0) \leq g_0^* < \prod_{j=1}^{d} \alpha_j \, , \tag{5.78}$$

the system (θ, φ) is dominated from above by the RDS generated by the linear equations

$$\dot{x}_1(t) = g_0^* \cdot x_d(t) - \alpha_1 x_1(t) \, , \tag{5.79}$$

$$\dot{x}_j(t) = x_{j-1}(t) - \alpha_j x_j(t), \quad j = 2, \dots, d \, . \tag{5.80}$$

However, assumption (5.78) means that all the eigenvalues of problem (5.79) and (5.80) possesses the property $\mathrm{Re}\lambda_j < 0$. Thus the zero equilibrium of (5.79) and (5.80) is exponentially stable. This implies that the random attractor $A(\omega)$ of (θ, φ) is trivial, i.e. $A(\omega) = \{0\}$.

5.7.2 Random Gonorrhea Model

Let us consider a system of random differential equations of the following form

$$\dot{x}_j(t) = f_j(\theta_t\omega, x_1(t), \dots, x_d(t), p_1 - x_1(t), \dots, p_d - x_d(t)), \quad j = 1, \dots, d.$$
$$(5.81)$$

Here $p = (p_1, \dots, p_d)$ is a fixed point from $\mathrm{int}\,\mathbb{R}^d_+$ and $f_j(\omega, x, y)$ are measurable functions on $\Omega \times [0, p] \times [0, p]$, where $[0, p]$ is the interval in $\mathbb{R}^d_+$ with end-points 0 and p. We also assume that for every $\omega \in \Omega$

(a) $f_j(\omega, x, y)$ is a continuously differentiable function on $[0, p] \times [0, p]$ such that $t \mapsto f_j(\theta_t\omega, 0, p)$ is locally integrable and the partial derivatives of $f_j(\omega, \cdot, \cdot)$ are bounded by $C(\omega)$ such that $C(\theta_t\omega) \in L^1_{loc}(\mathbb{R})$, $j = 1, \dots, d$;
(b) we have

$$f_j(\omega, x, p - x) \geq 0, \quad j = 1, \dots, d,$$

for all $x \in [0, p]$ of the form $x = (x_1, \dots, x_{j-1}, 0, x_{j+1}, \dots, x_d)$ and

$$f_j(\omega, x, p - x) \leq 0, \quad j = 1, \dots, d,$$

for all $x \in [0, p]$ of the form $x = (x_1, \dots, x_{j-1}, p_j, x_{j+1}, \dots, x_d)$;
(c) the function

$$f(\omega, x, p - x) = (f_1(\omega, x, p - x), \dots, f_d(\omega, x, p - x))$$

satisfies the cooperativity condition, i.e.

$$\left.\frac{\partial f_i(\omega, x, y)}{\partial x_j}\right|_{(x, p-x)} - \left.\frac{\partial f_i(\omega, x, y)}{\partial y_j}\right|_{(x, p-x)} \geq 0 \quad \text{if} \quad i \neq j, \, 0 < x < p.$$

It is easy to see that equations (5.81) possess a local solution for any initial data from $\{x : 0 \ll x \ll p\}$. Assumption (b) and Proposition 5.4.3 imply that the interval $[0, p]$ is a forward invariant set. This makes it possible to guarantee the global existence of solutions to (5.81) with initial data from $[0, p]$ for every ω and therefore equations (5.81) generate an RDS with state space $X = [0, p] \subset \mathbb{R}^d_+$. Assumption (c) implies that this RDS is strictly order-preserving (see Theorem 5.2.1). It is clear that $w(\omega) \equiv p$ is a super-equilibrium and $v(\omega) \equiv 0$ is a sub-equilibrium. Therefore Theorem 3.5.1 implies the existence of an equilibrium $u(\omega)$ with the property

$$0 \leq u(\omega) \leq p \quad \text{for all} \quad \omega \in \Omega.$$

If $f(\omega, 0, p) = (f_1(\omega, 0, p), \dots, f_d(\omega, 0, p)) > 0$ then $u(\omega) > 0$ by Proposition 5.4.1.

We note that deterministic version of the equations (5.81) first appeared in HIRSCH [53] as a generalization of a gonorrhea transmission models previously considered. The time-periodic version of (5.81) was studied in SMITH [101], TAKAČ [103], see also KRAUSE/RANFT [73].

The assumptions above are met, if we suppose, for instance,

$$f_j(\omega, x, p - x) = -\alpha_j(\omega)x_j + (p_j - x_j) \sum_{i=1}^{d} \beta_{ji}(\omega)x_i \ . \tag{5.82}$$

Here $\alpha_j(\omega)$ and $\beta_{ji}(\omega)$ are random variables such that $t \mapsto \alpha_j(\theta_t\omega)$ and $t \mapsto \beta_{ji}(\theta_t\omega)$ are locally integrable and satisfy the inequalities

$$\alpha_j(\omega) \geq \alpha_j^0 > 0, \quad \beta_{ji}(\omega) \geq \beta_{ji}^0 > 0 \quad \text{for all} \quad \omega \in \Omega \ ,$$

where α_j^0 and β_{ji}^0 are nonrandom constants. Biologically, equations (5.81) with f_j of the type (5.82) correspond to the population divided into d groups, each of constant population size p_j. The variable x_j denotes the number infected with gonorrhea in the j-th group, $p_j - x_j$ is the number of susceptibles in the the j-th group, β_{ji} is the rate at which group i infects the j-th group and α_j is the care rate. The randomness of α_j and β_{ji} can be interpreted as seasonal fluctuations.

The mapping $f = (f_1, \ldots, f_d)$ given by (5.82) is s-concave (cf. (5.55)) Therefore Proposition 5.5.2 implies that the RDS generated by (5.81) with (5.82) is s-concave. Moreover the function (5.82) admits the estimate

$$-\alpha_j(\omega)x_j + \beta_{jj}(\omega)(p_j - x_j)x_j \leq f_j(\omega, x, p - x) \leq -\alpha_j(\omega)x_j + p_j \sum_{i=1}^{d} \beta_{ji}(\omega)x_i$$

for $x \in [0, p]$. Using the Comparison Theorem 5.3.1 we find that the RDS generated by (5.81) and (5.82) is dominated from above by a linear system and from below by the direct product of one-dimensional RDS. These properties make it possible (see Theorem 5.5.3 and Example 5.6.1) to give some condition on $\alpha_j(\omega)$ and $\beta_{ji}(\omega)$ which ensure one of the following cases:

(a) $\underline{u}(\omega) \equiv 0$ is globally stable and $\mathcal{A} = \{0\}$;
(b) there exists a strictly positive equilibrium $\bar{u}(\omega)$ and $\mathcal{A} \subset [0, \bar{u}]$, where $\underline{u}(\omega) \equiv 0$ is unstable and $\bar{u}(\omega)$ is stable.

5.7.3 Random Model of Symbiotic Interaction

We consider the RDE

$$\dot{x}_j = x_j h_j(\theta_t\omega, x_1, \ldots, x_d), \quad j = 1, \ldots, d \ . \tag{5.83}$$

These equations arise in the model of symbiosis between d populations, for instance. Here x_j is the size of the j-th population. For a deterministic version of this model see, e.g., SMITH [102] and the references therein and also KRAUSE/RANFT [73], where the periodic case is discussed.

We assume that the function $h = (h_1, \ldots, h_d)$ satisfies (R1), (R3), (R4) and is bounded for every $\omega \in \Omega$ (hence (R2) holds). Under these conditions equations (5.83) generate an order-preserving RDS (θ, φ) in $\mathbb{R}_+^d$ possessing the following invariance property: for every subset I of $N = \{1, \ldots, d\}$ the set

$$K_I = \{x = (x_1, \ldots, x_d) \in \mathbb{R}_+^d \; : \; x_j = 0, \, j \in N \setminus I\}$$

is invariant with respect to (θ, φ). The restriction of (θ, φ) to K_I is described by the RDE

$$\dot{x}_j = x_j h_j(\theta_t \omega, p_I x), \quad j \in I \,.$$

where p_I is a projector in $\mathbb{R}^d$ defined by $(p_I x)_j = x_j$ for $j \in I$ and $(p_I x)_j = 0$ if $j \in N \setminus I$.

This example demonstrates a typical symbiotic interaction, i.e. interaction that can result in benefits for several populations as far as their size is concerned. To see this assume that

$$h_j(\omega, x) = \alpha_j(\omega, x_j) + \sum_{i \neq j} g_i(\omega, x_i) \,, \tag{5.84}$$

where $\alpha_j(\omega, x)$ and $g_i(\omega, x)$ are functions on $\Omega \times \mathbb{R}_+$ such that the conditions on h_j mentioned above are valid. We also assume that

$$0 \leq g_i(\omega, 0) \leq g_i(\omega, x) \leq M, \quad \text{and} \quad g_i'(\omega, x) > 0 \quad \text{for every} \quad x > 0 \,. \tag{5.85}$$

We choose $\alpha_j(\omega, x)$ such that the RDS generated in $\mathbb{R}_+$ by the equation

$$\dot{x} = x \cdot \alpha_j(\theta_t \omega, x) \tag{5.86}$$

has a positive equilibrium for every $j = 1, \ldots, d$. For this we can assume, for example, that

$$\alpha_j(\omega, x) \geq \alpha_j^0 > 0 \quad \text{for} \quad 0 \leq x \leq \delta$$

and

$$\alpha_j(\omega, x) \leq -\beta_j^0 < 0 \quad \text{for} \quad x \geq R \,,$$

where α_j^0, δ, β_j^0 and R are positive nonrandom constants. We note that equation (5.86) describes the evolution of the j-th population independent of the others. Denote the positive equilibrium for (5.86) by $v_j(\omega)$. Assumptions (5.85) ensure that the RDS generated in $\mathbb{R}_+^d$ by

$$\dot{x}_j = x_j \cdot \alpha_j(\theta_t \omega, x_j), \quad j = 1, \ldots, d \,,$$

dominates the RDS (θ, φ) generated by (5.83) with (5.84) from below. Therefore $v(\omega) = (v_1(\omega), \ldots, v_d(\omega))$ is a positive sub-equilibrium for (θ, φ). On the other hand under condition (5.85) the system generated by

$$\dot{x}_j = x_j \cdot (\alpha_j(\theta_t \omega, x_j) + (d - 1)M), \quad j = 1, \ldots, d \,,$$

dominates (θ, φ) from above. If $\beta_j^0 > (d-1)M$ this RDS has a super-equilibrium $w(\omega)$ such that $w(\omega) > v(\omega)$. Therefore Theorem 3.5.1 implies the existence of an equilibrium $u(\omega)$ such that $v(\omega) \leq u(\omega) \leq w(\omega)$. This equilibrium attracts (from below) the collection of equilibria $(v_1(\omega), \dots, v_d(\omega))$ which correspond to the isolated dynamics of each population. We hence observe that the interaction results in a benefit for all populations.

5.7.4 Random Gross-Substitute System

The deterministic gross-substitute system represents the law of supply and demand in economics (see, e.g., NAKAJIMA [84] and SELL/NAKAJIMA [97] and the references therein). Here we consider a random version of this system. Such a generalization seems to be natural because it reflects changes due to random impacts. We consider the RDE

$$\dot{x}_i = f_i(\theta_t \omega, x_1, \dots, x_d), \quad i = 1, \dots, d, \tag{5.87}$$

where the function $f : \Omega \times \mathbb{R}_+^d \mapsto \mathbb{R}^d$ satisfies conditions (R1)–(R4) and also Walras' law

$$\sum_{i=1}^d f_i(\omega, x_1, \dots, x_d) = 0 \tag{5.88}$$

for all $\omega \in \Omega$ and $x \in \mathrm{int}\mathbb{R}_+^d$. The simplest example of a system satisfying (R1)–(R4) and (5.88) gives the following RDE

$$\dot{x}_i = \sum_{j=1}^d a_{ij}(\theta_t \omega) \cdot h_j(x_j), \quad i = j, \dots, d,$$

where the matrix $a_{ij}(\omega)$ satisfies the cooperativity condition, i.e. $a_{ij}(\omega) \geq 0$ for all $i \neq j$ and $\omega \in \Omega$ and

$$\sum_{i=1}^d a_{ij}(\omega) = 0 \quad \text{for all} \quad \omega \in \Omega, \ j = 1, \dots, d,$$

and $h_j : \mathbb{R}_+ \mapsto \mathbb{R}_+$ are nondecreasing functions such that $h_j(0) = 0$.

Theorem 5.2.1 implies that (5.87) generates a strictly order-preserving RDS in $\mathbb{R}_+^d$. It is easy to see (cf. Corollary 5.4.1) that for any $\beta \geq 0$ the set

$$\Sigma_\beta = \left\{ x \in \mathrm{int} \ \mathbb{R}_+^d \ : \ \sum_{i=1}^d x_i = \beta \right\}$$

is a forward invariant set for the RDS (θ, φ) generated by (5.87). Therefore the closure $\overline{\Sigma}_\beta$ of Σ_β is also a forward invariant set.

Let us prove that the RDS (θ, φ) is nonexpansive with respect to the l^1-norm defined by the formula $|x|_1 = \sum_{i=1}^{d} |x_i|$ for $x = (x_1, \ldots, x_d)$, i.e.

$$|\varphi(t, \omega)x - \varphi(t, \omega)y|_1 \leq |x - y|_1, \quad x, y \in \mathbb{R}_+^d, \ t > 0, \ \omega \in \Omega. \qquad (5.89)$$

Indeed, let

$$x(t) = \varphi(t, \omega)x, \quad y(t) = \varphi(t, \omega)y \quad \text{and} \quad z(t) = \varphi(t, \omega)z,$$

where $z = x \vee y \equiv \sup\{x, y\}$. It is clear that $z(t) \geq x(t) \vee y(t)$ for all $t > 0$. Therefore

$$|x(t) - y(t)|_1 = \sum_{i=1}^{d} (2 \max\{x_i(t), y_i(t)\} - x_i(t) - y_i(t))$$

$$(5.90)$$

$$\leq \sum_{i=1}^{d} (2z_i(t) - x_i(t) - y_i(t)).$$

Consequently the invariance of $\overline{\Sigma}_\beta$ and the relation $z_i = \max\{x_i, y_i\}$ imply that

$$|x(t) - y(t)|_1 \leq \sum_{i=1}^{d} (2z_i - x_i - y_i) = \sum_{i=1}^{d} |x_i - y_i| = |x - y|_1.$$

Thus we have (5.89) for any $x, y \in \mathbb{R}_+^d$.

If the matrix $\left\{ \frac{\partial f_i}{\partial x_j}(\omega, x) \right\}$ is irreducible for every $x \gg 0$ and $\omega \in \Omega$, then by Theorem 5.2.1 the RDS (θ, φ) is strongly order-preserving in int $\mathbb{R}_+^d$. In this case the restriction (θ, φ_β) of (θ, φ) to Σ_β is contractive for each $\beta > 0$. Indeed, if $x, y \in \Sigma_\beta$ for some $\beta > 0$ and $x \neq y$, then $z > x$ and $z > y$. Since (θ, φ) is strongly order-preserving, the last relation implies that

$$z(t) = \varphi(t, \omega)z \gg x(t) \quad \text{and} \quad z(t) = \varphi(t, \omega)z \gg y(t).$$

Consequently $z(t) > x(t) \vee y(t)$ for $t > 0$. Therefore we have strict inequality in (5.90) and obtain (5.89) with strict inequality provided that $x, y \in \Sigma_\beta$ and $x \neq y$. Thus (θ, φ_β) is contractive for each $\beta > 0$. Proposition 1.7.1 implies that every set Σ_β can contain a unique (up to indistinguishability) equilibrium.

Let us attempt to prove the existence of these equilibria. Since $\overline{\Sigma}_\beta$ is a compact set, the RDS (θ, φ_β) possesses a random attractor $A_\beta \subseteq \overline{\Sigma}_\beta$. Lemma 3.4.1 implies that

$$w^\beta(\omega) = (w_1^\beta(\omega), \ldots, w_d^\beta(\omega)) := \sup A_\beta(\omega)$$

is a sub-equilibrium such that $\beta \le |w^\beta(\omega)|_1 \le d \cdot \beta$. Moreover $0 \le w^\beta(\omega) \le \beta \cdot \mathbf{e}$, where $\mathbf{e} = (1, \dots, 1)$. It is clear that $\varphi(t, \theta_{-t}\omega)(\beta \mathbf{e}) \le \beta d \cdot \mathbf{e}$. Therefore, since the cone $\mathbb{R}_+^d$ is regular, Proposition 3.5.2 implies that

$$u^\beta(\omega) := \lim_{t \to +\infty} \varphi(t, \theta_{-t}\omega)w^\beta(\theta_{-t}\omega) = \sup_{t>0} \varphi(t, \theta_{-t}\omega)w^\beta(\theta_{-t}\omega)$$

exists for any $\beta > 0$ and is an equilibrium such that

$$u^\beta(\omega) \in \left\{ x \in \mathbb{R}_+^d \ : \ \beta \le \sum_{i=1}^d x_i \le d^2\beta \right\} \bigcap [0, \beta d\mathbf{e}] \ .$$

It follows from the last relation that the RDS (θ, φ) possesses infinitely many equilibria. Since $x(t) = u^\beta(\theta_t\omega)$ solves (5.87), we have that

$$\sum_{i=1}^d u_i^\beta(\theta_t\omega) = \sum_{i=1}^d u_i^\beta(\omega), \quad t > 0, \ \omega \in \Omega \ .$$

Therefore if θ is an ergodic metric dynamical system, then there exist $\beta^* \in [\beta, d^2\beta]$ and a θ-invariant set $\Omega^* \subset \Omega$ of full measure such $u^\beta(\omega) \in \overline{\Sigma}_{\beta^*}$ for all $\omega \in \Omega^*$. Moreover in the ergodic case the invariance of Σ_β implies that any semi-equilibrium is an equilibrium on a θ-invariant set of full measure. We conjecture that *every* set $\overline{\Sigma}_\beta$ contains an equilibrium. However we cannot prove it now.

5.8 Order-Preserving RDE with Non-Standard Cone

In previous sections we have considered RDE with order-preserving properties with respect to the standard cone $\mathbb{R}_+^d$. However by considering other cones besides $\mathbb{R}_+^d$ we can enlarge the area of possible applications of monotone methods to the study of random equations.

Here we restrict our attention to the case that the cone is one of the orthants of $\mathbb{R}^d$. Let $m = (m_1, \dots, m_d)$, where $m_i \in \{0, 1\}$. Consider the cone

$$K_m := \left\{ x = (x_1, \dots, x_d) \in \mathbb{R}^d \ : \ (-1)^{m_i} x_i \ge 0, \ i = 1, \dots, d \right\} \ .$$

This cone generates a partial order $\le_m$ defined by $x \le_m y$ if and only if $y - x \in K_m$. Let P be the diagonal matrix defined by

$$P = \text{diag} \left\{ (-1)^{m_1}, \dots, (-1)^{m_d} \right\} \ .$$

We note that $x \le_m y$ if and only if $Px \le Py$, where $\le$ is the order relation generated by $\mathbb{R}_+^d$.

Assume that (θ, φ) is an RDS in some domain $\mathbb{D} \subseteq \mathbb{R}^d$ generated by the random equation

$$\dot{x}(t) = f(\theta_t \omega, x(t)) \, . \tag{5.91}$$

It is easy to see that the cocycle φ can be represented in the form

$$\varphi(t, \omega)x = P^{-1}\psi(t, \omega)Px, \quad x \in \mathbb{D} \, ,$$

where ψ is the cocycle of the RDS in $P\mathbb{D}$ generated by the RDE

$$\dot{y}(t) = g(\theta_t \omega, y(t)), \quad \text{with} \quad g(\omega, y) = Pf(\omega, P^{-1}y) \, . \tag{5.92}$$

Thus the random systems generated by (5.91) and (5.92) are conjugate. Since $P = P^{-1}$ is an order isomorphism, the RDS (θ, φ) is order-preserving with respect to the cone K_m if and only if (θ, ψ) is order-preserving with respect to the cone $\mathbb{R}^d_+$. Hence after simple calculations we find that the condition

$$(-1)^{m_i + m_j} \frac{\partial f_i(\omega, x)}{\partial x_j} \geq 0, \quad i \neq j, \, x = (x_1, \dots, x_d) \in \mathbb{D}, \, \omega \in \Omega \, , \tag{5.93}$$

implies that (θ, φ) generated by (5.91) is order-preserving with respect to the cone K_m.

As an example we consider the random version of the model of two competing populations occupying an environment consisting of two discrete patches between which they can migrate. This model of competition and migration is described by the following system of random equations

$$\begin{aligned}
\dot{x}_1 &= \varepsilon(x_2 - x_1) + x_1(a_1 - b_1(\theta_t\omega)x_1 - c_1(\theta_t\omega)x_3) \, , \\
\dot{x}_2 &= \varepsilon(x_1 - x_2) + x_2(a_2 - b_2(\theta_t\omega)x_2 - c_2(\theta_t\omega)x_4) \, , \\
\dot{x}_3 &= \delta(x_4 - x_3) + x_3(a_3 - b_3(\theta_t\omega)x_3 - c_3(\theta_t\omega)x_1) \, , \\
\dot{x}_4 &= \delta(x_3 - x_4) + x_4(a_4 - b_4(\theta_t\omega)x_4 - c_4(\theta_t\omega)x_2) \, .
\end{aligned} \tag{5.94}$$

Here x_1 and x_2 denote the population density of one population in patches 1 and 2 respectively and x_3 and x_4 denote the population density of the second population in patches 1 and 2. No mortality is suffered during migration between patches. We assume that ε, δ and a_j are positive deterministic parameters. The parameters ε and δ are the migration coefficients and a_j is the intrinsic growth rate of the corresponding population in patch 1 or 2. The terms $b_j(\theta_t\omega)$ and $c_j(\theta_t\omega)$ describe randomly fluctuating interaction rates between the populations. This randomness can occur under seasonal fluctuations, for instance. For deterministic version of this model we refer to SMITH [102] and to the references therein.

We assume that $b_j(\omega)$ and $c_j(\omega)$ are nonnegative random variables such that $t \mapsto b_j(\theta_t\omega)$ and $t \mapsto c_j(\theta_t\omega)$ are locally integrable for every $\omega \in \Omega$. These assumptions imply conditions (R1)–(R3) and therefore by Proposition 5.2.1

equations (5.94) generate C^1 RDS (θ, φ) in $\mathbb{R}_+^4$. It is clear that (5.93) holds with $m = (0, 0, 1, 1)$. Therefore (θ, φ) is an order-preserving RDS in $\mathbb{R}_+^4$ with respect to the cone

$$K_{(0,0,1,1)} := \{x = (x_1, x_2, x_3, x_4) \in \mathbb{R}^4 \ : \ x_1, x_2 \geq 0, \ x_3, x_4 \leq 0\} \ .$$

Assume additionally that

$$b_j(\omega) \geq b_{j0} > 0 \quad \text{for all} \quad \omega \in \Omega, \ j = 1, 2, 3, 4 \ . \tag{5.95}$$

Then from (5.94) we have that

$$\frac{d}{dt}|x|^2 \leq 2 \sum_{i=1}^{4} (a_i - b_{i0}x_i)x_i^2 \leq -|x|^2 + C$$

with some constant $C > 0$. Consequently

$$|\varphi(t, \omega)x|^2 \leq |x|^2 e^{-t} + C(1 - e^{-t}), \quad x \in \mathbb{R}_+^4, \ \omega \in \Omega \ .$$

Therefore (θ, ϕ) is a dissipative RDS in $\mathbb{R}_+^4$ in the universe $\mathcal{D}$ of all tempered random sets of $\mathbb{R}_+^4$ and

$$B = \{x \in \mathbb{R}_+^4 : |x| \leq R\}, \quad R = (1 + C)^{1/2} \ ,$$

is a forward invariant absorbing set. Hence by Corollary 1.8.1 (θ, φ) possesses a random pull back attractor $A(\omega)$ in $\mathcal{D}$.

To study the attractor $A(\omega)$ we first note that the structure of equations (5.94) implies that the sets

$$V_1 = \{x \in \mathbb{R}_+^4 : x_3 = x_4 = 0\} \quad \text{and} \quad V_2 = \{x \in \mathbb{R}_+^4 : x_1 = x_2 = 0\}$$

are forward invariant for (θ, φ). The restriction (θ, φ_1) of (θ, φ) to V_1 is generated by the equations

$$\dot{x}_1 = \varepsilon(x_2 - x_1) + x_1(a_1 - b_1(\theta_t\omega)x_1) \ ,$$

$$\dot{x}_2 = \varepsilon(x_1 - x_2) + x_2(a_2 - b_2(\theta_t\omega)x_2) \ ,$$

for which the cooperativity condition (R4*) holds. Hence (θ, φ_1) is order-preserving in V_1 with respect to the standard cone. It follows from (5.95) and from Proposition 5.4.3 that there exists a super-equilibrium $\overline{w} = (w_1, w_2, 0, 0)$ with $w_i \geq R$ for (θ, φ_1) in V_1.

Similarly, the restriction (θ, φ_2) of (θ, φ) to V_2 is an order-preserving RDS with respect to the standard cone and there exists a super-equilibrium $\underline{w} = (0, 0, w_3, w_4)$ with $w_i \geq R$ for (θ, φ_2) in V_2. We obviously have

$$\varphi(t, \omega)\overline{w} = (\varphi_1(t, \omega)(w_1, w_2), 0, 0) \leq_m (w_1, w_2, 0, 0) = \overline{w}$$

and

$$\varphi(t,\omega)\underline{w} = (0,0,\varphi_2(t,\omega)(w_3,w_4)) \geq_m (0,0,w_3,w_4) = \underline{w} .$$

Thus $\overline{w}$ is a super-equilibrium and $\underline{w}$ is a sub-equilibrium for (θ,φ) with respect to the cone $K_{(0,0,1,1)}$. Moreover $\overline{w} \geq_m \underline{w}$ and the absorbing set B belongs to the interval $\{x : \underline{w} \leq_m x \leq_m \overline{w}\}$. Consequently by Theorem 3.6.2 the attractor $A(\omega)$ possesses the property

$$A(\omega) \subset \{x \in \mathbb{R}_+^4 : \underline{u}(\omega) \leq_m x \leq_m \overline{u}(\omega)\} ,$$

where $\underline{u}(\omega)$ and $\overline{u}(\omega)$ are equilibria. It is also easy to find that $\overline{u}(\omega) \in V_1$ and $\underline{u}(\omega) \in V_2$. Moreover $\overline{u}$ (resp. $\underline{u}$) is globally asymptotically stable from above in V_1 (resp. V_2) with respect to the standard cone.

6. Cooperative Stochastic Differential Equations

In this chapter we consider order-preserving RDS generated in $\mathbb{R}_+^d$ by stochastic differential equations (SDE). The order-preserving property for stochastic equations requires a special form of the diffusion terms. The corresponding assumptions concerning the drift terms are the same as in the deterministic case. In fact we deal here with some classes of stochastic perturbations of deterministic order-preserving systems.

In this chapter we rely essentially on the Wong–Zakaï type approximation theorem and on the results on conjugacy of random and stochastic equations (see Sect. 2.5). We deal with Stratonovich equations only. However most results given below remain true for the Itô case, at least after obvious minor changes.

We refer to Chap. 2 for a description of the basic definitions and results on stochastic differential equations.

6.1 Main Assumptions

We consider the following system of Stratonovich stochastic differential equations

$$dx_i = f_i(x_1, \dots, x_d)dt + \sum_{j=1}^{m} \sigma_{ij}(x_i) \circ dW_t^j, \quad i = 1, \dots, d, \tag{6.1}$$

where $W_t(\omega) = (W_t^1(\omega), \dots, W_t^m(\omega))$ is a Wiener process with values in $\mathbb{R}^m$ and two-sided time $\mathbb{R}$, $m \geq 1$. Below we denote by θ the metric dynamical system corresponding to this process (see Example 1.1.7).

In this chapter our main assumptions are follows:

(S1) every function $f_i : \mathbb{R}_+^d \to \mathbb{R}$ belongs to the class $C_b^{1,\delta}(\mathbb{R}_+^d)$ (see Definition 2.4.1), i.e. $f_i(x)$ is a continuously differentiable function, with derivatives bounded and globally δ-Hölder continuous:

$$\left| \frac{\partial f_i}{\partial x_j}(x) - \frac{\partial f_i}{\partial x_j}(y) \right| \leq C|x - y|^\delta, \quad 0 < \delta \leq 1, \quad i, j = 1, \dots, d;$$

(S2) for every $i = 1, \dots, d$ and $j = 1, \dots, m$ the functions $\sigma_{ij} : \mathbb{R}_+ \mapsto \mathbb{R}$ are twice continuously differentiable, with first derivative bounded and second derivative bounded and globally δ-Hölder continuous, $0 < \delta \leq 1$, such that $\sigma_{ij} \cdot \sigma'_{ij} \in C_b^{1,\delta}(\mathbb{R}_+^d)$;

(S3) the property of *weak positivity* holds, i.e. $\sigma_{ij}(0) = 0$ for all $i = 1, \dots, d$, $j = 1, \dots, m$ and

$$f_i(x) \geq 0, \quad i = 1, \dots, d \,,$$

for all $x \in \mathbb{R}_+^d$ of the form $x = (x_1, \dots, x_{i-1}, 0, x_{i+1}, \dots, x_d)$;

(S4) the function $f = (f_1, \dots, f_d)$ is *cooperative*, i.e.

$$f_i(x) \leq f_i(y), \quad i = 1, \dots, d \,,$$

for all $x, y \in \mathbb{R}_+^d$ such that $x_i = y_i$ and $x_j \leq y_j$ for $j \neq i$ or, in equivalent differential form,

$$\frac{\partial f_i(x)}{\partial x_j} \geq 0, \quad i \neq j, \ x \in \mathbb{R}_+^d \,.$$

6.2 Generation of Order-Preserving RDS

Proposition 6.2.1. *Assume that conditions (S1)–(S3) hold. Then equation (6.1) generates a C^1 RDS (θ, φ) in $\mathbb{R}_+^d$ such that the conclusions of Theorem 2.4.3 hold in $\mathbb{R}_+^d$.*

Proof. Let $\tilde{f}(x) = (\tilde{f}_1(x), \dots, \tilde{f}_d(x))$ be a function from $\mathbb{R}^d$ into itself such that $\tilde{f}_i(x) \in C^{1,\delta}(\mathbb{R}^d)$ and $\tilde{f}_i(x) = f_i(x)$ for all $x \in \mathbb{R}_+^d$, $i = 1, \dots, d$. Let $\tilde{\sigma}_{ij}(x) \in C^{2,\delta}(\mathbb{R})$ be an extension of $\sigma_{ij}(x)$ from $\mathbb{R}_+$ to $\mathbb{R}$ such that $\tilde{\sigma}_{ij} \cdot \tilde{\sigma}'_{ij}$ belongs to $C_b^{1,\delta}(\mathbb{R}_+^d)$, $i = 1, \dots, d$, $i = 1, \dots, m$. It follows from Theorem 2.4.3 that the stochastic equations

$$dx_i = \tilde{f}_i(x_1, \dots, x_d)dt + \sum_{j=1}^{m} \tilde{\sigma}_{ij}(x_i) \circ dW_t^j, \quad i = 1, \dots, d \,,$$

generate a C^1 RDS in $\mathbb{R}^d$. Property (S3) implies that the set $\mathbb{D} = \mathbb{R}_+^d$ satisfies the assumptions of Corollaries 2.5.1 and 2.5.2. Therefore from Corollary 2.5.2 there exists a unique (up to indistinguishability) continuous C^1 RDS (θ, φ) in $\mathbb{R}_+^d$ generated by the system of Stratonovich SDEs (6.1) in the sense of Theorem 2.4.3. $\qquad \square$

Proposition 6.2.2. *Let (S1)–(S4) be valid. Then equation (6.1) generates a strictly order-preserving C^1 RDS (θ, φ) in $\mathbb{R}_+^d$ and*

$$\varphi(t, \omega)(\mathbb{R}_+^d \setminus \{0\}) \subset \mathbb{R}_+^d \setminus \{0\} \quad \text{for any } t \geq 0, \ \omega \in \Omega \,. \tag{6.2}$$

Proof. We first approximate (6.1) by the system of random differential equations

$$\dot{x}_i = f_i(x_1, \dots, x_d) + \sum_{j=1}^{m} \sigma_{ij}(x_i) \cdot \eta_\varepsilon^j(\theta_t \omega), \quad i = 1, \dots, d, \qquad (6.3)$$

where the random variables $\eta_j^\varepsilon(\omega)$ are defined as in Sect.2.5:

$$\eta_j^\varepsilon(\omega) = -\frac{1}{\varepsilon^2} \int_0^\varepsilon \dot{\phi}(\tau/\varepsilon) W_\tau^j(\omega) \, d\tau$$

with a nonnegative function $\phi(t) \in C^1(\mathbb{R})$ such that

$$\operatorname{supp} \phi(t) \subset [0, 1], \quad \int_0^1 \phi(t) \, dt = 1 .$$

Theorem 5.2.1 implies that for every $\varepsilon > 0$ equations (6.3) generate an order-preserving C^1 RDS $(\theta, \varphi^\varepsilon)$ in $\mathbb{R}_+^d$. Therefore

$$\int_\alpha^\beta l(\varphi^\varepsilon(t, \theta_{-t}\omega)y - \varphi^\varepsilon(t, \theta_{-t}\omega)x) dt \geq 0, \quad 0 \leq x \leq y ,$$

for all $0 \leq \alpha < \beta$, where l is a positive ($l(x) \geq 0$ whenever $x \geq 0$) linear functional on $\mathbb{R}^d$. Consequently from (2.46) as in the proof of Corollary 2.5.1 we can conclude that

$$\int_\alpha^\beta l(\varphi(t, \theta_{-t}\omega)y - \varphi(t, \theta_{-t}\omega)x) dt \geq 0, \quad 0 \leq x \leq y, \ \omega \in \Omega^* ,$$

for all $0 \leq \alpha < \beta$, where Ω^* is the θ-invariant set of full measure defined in Remark 2.5.1. From this relation and from the continuity of the function $t \mapsto \varphi(t, \theta_{-t}\omega)x$ (see Remark 2.4.1) we obtain that the inequality $x \leq y$ implies $\varphi(t, \omega)x \leq \varphi(t, \omega)y$ for all $t \geq 0$ and $\omega \in \Omega^*$. The invertibility of the cocycle $\varphi(t, \omega)$ of the RDS generated by (6.1) (see [3, Theorem 2.3.32]) implies that $\varphi(t, \omega)x < \varphi(t, \omega)y$ for all $0 \leq x < y$, $t \geq 0$ and $\omega \in \Omega^*$. Therefore after redefining the cocycle φ by formula (2.50) (cf. Corollary 2.5.2) we obtain a strictly order-preserving RDS.

If for some $x_0 > 0$, $t_0 > 0$ and $\omega \in \Omega$ we have $\varphi(t_0, \omega)x_0 = 0$, then $\varphi(t_0, \omega)y = 0$ for all $0 \leq y \leq x_0$, which is impossible because of the invertibility of the cocycle $\varphi(t, \omega)$. Thus we have (6.2). □

As for the random case (see Proposition 5.4.3) we have the following simple condition for the existence of semi-equilibria for the RDS generated by (6.1).

Proposition 6.2.3. *Let (S1)–(S3) be valid. Assume that there exists an element $c = (c_1, \dots, c_d)$ in $\mathbb{R}_+^d$ such that $\sigma_{ij}(c_j) = 0$ for each $i = 1, \dots, d$ and $j = 1, \dots, m$. If $f(x)$ satisfies (S4) for all $x \in [0, c]$ and*

$$f_i(c) \leq 0 \quad \text{for all} \quad i = 1, \ldots, d , \tag{6.4}$$

then $v(\omega) \equiv c$ is a super-equilibrium for the RDS (θ, φ) generated by (6.1) and the restriction of (θ, φ) to the interval $[0, c]$ is a strictly order-preserving RDS. If we have the reversed inequality sign in (6.4) and (S4) holds for all $x \geq c$, then $w(\omega) \equiv c$ is a sub-equilibrium and the restriction of (θ, φ) to the set $I_c = \{x \in \mathbb{R}_+^d : x \geq c\}$ is a strictly order-preserving RDS.

Proof. Applying Proposition 5.4.3 to the RDS $(\theta, \varphi^\varepsilon)$ generated by the approximate equation (6.3) we obtain

$$\varphi^\varepsilon(t, \theta_{-t}\omega)c \leq c \quad \text{for all} \quad t \geq 0, \ \omega \in \Omega ,$$

under condition (6.4). Therefore as in the proof of Proposition 6.2.2 transition to the limit gives the relation

$$\varphi(t, \theta_{-t}\omega)c \leq c \quad \text{for all} \quad t \geq 0, \ \omega \in \Omega^* ,$$

where Ω^* is a θ-invariant set of full measure. Therefore redefining the cocycle φ by the formula (2.50) we obtain that $c = (c_1, \ldots, c_d)$ is a super-equilibrium and the interval $[0, c]$ is a forward invariant set for (θ, φ) (see Remark 3.4.1). Consequently by Proposition 6.2.2 (θ, φ) is a strictly order-preserving RDS on the interval $[0, c]$.

The proof of the second part of the proposition is similar. $\square$

6.3 Conjugacy with Random Differential Equations

In this section we describe several situations in which the RDS generated by (6.1) is equivalent to an RDS generated by a random differential equation. In some sense the theorems given below are particular cases of the result by IMKELLER/SCHMALFUSS [59] presented in Theorem 2.5.2. However we do not assume C^∞-smoothness of the coefficients in (6.1).

As in Sect. 2.5 we denote by $z(\omega)$ the random variable in $\mathbb{R}^m$ such that $z(t, \omega) := z(\theta_t\omega) = (z_1(\theta_t\omega), \ldots, z_m(\theta_t\omega))$ is the stationary Ornstein-Uhlenbeck process in $\mathbb{R}^m$ which solves the equations

$$dz_k = -\mu z_k dt + dW_t^k, \quad k = 1, \ldots, m ,$$

for some $\mu > 0$ and possesses the properties described in Lemma 2.5.1.

To present the main idea clearly we start with the simplest case of linear diffusion terms.

Theorem 6.3.1. *Assume that (S1)–(S3) hold. Let (θ, φ) be the RDS generated in $\mathbb{R}_+^d$ by (6.1). If $\sigma_{ij}(x_i) = s_{ij} \cdot x_i$ are linear functions, then (θ, φ) is equivalent to the RDS (θ, ψ) generated in $\mathbb{R}_+^d$ by the RDE:*

$$\dot{y}_i(t) = g_i(\theta_t\omega, y_1(t), \dots, y_d(t)), \quad i = 1, \dots, d, \tag{6.5}$$

with

$$g_i(\omega, y_1, \dots, y_d) = e_i^s(\omega)^{-1} \cdot f_i(y_1 \cdot e_1^s(\omega), \dots, y_d \cdot e_d^s(\omega)) + \mu y_i z_i^s(\omega). \tag{6.6}$$

Here $e_i^s(\omega) = \exp\{z_i^s(\omega)\}$ and

$$z_i^s(\omega) = \sum_{j=1}^{m} s_{ij} z_j(\omega), \quad i = 1, \dots, d, \tag{6.7}$$

where the random variables $z_j(\omega)$ are given by Lemma 2.5.1. Moreover we have the relation

$$\varphi(t, \omega, x) = T(\theta_t\omega, \psi(t, \omega, T^{-1}(\omega, x))), \quad t > 0, \ x \in \mathbb{R}_+^d, \ \omega \in \Omega, \tag{6.8}$$

where the diffeomorphism $T(\omega, \cdot) : \mathbb{R}_+^d \mapsto \mathbb{R}_+^d$ is a linear mapping given by the formula

$$T(\omega, y) = (y_1 \cdot e_1^s(\omega), \dots, y_d \cdot e_d^s(\omega)), \quad \omega \in \Omega.$$

Proof. The functions $g_i(\omega, y)$ given by (6.6) satisfy conditions (R1)-(R3) of Chap.5. Therefore Proposition 5.2.1 implies that the RDE (6.5) generates an RDS (θ, ψ) in $\mathbb{R}_+^d$. If we apply Itô's formula (see Theorem 2.3.1) to the value $x_i(t, \omega) := y_i(t, \omega) \cdot e_i^s(\theta_t\omega)$, then we find that $x(t, \omega) = (x_1(t, \omega), \dots, x_d(t, \omega))$ satisfies (6.1). Therefore using (6.8) we can define the perfect cocycle φ which satisfies in $\mathbb{R}_+^d$ the conclusions of Theorem 2.4.3. $\qquad\square$

Now we consider diffusion coefficients σ_{ij} of a slightly more general form. We assume that

$$\sigma_{ij}(x_i) = \sigma_i(x_i) \cdot s_{ij}, \quad \sigma_i(x) > 0, \ x > 0, \quad \sigma_i'(0) > 0, \tag{6.9}$$

where s_{ij} are constants. To obtain a theorem on conjugacy for this case we need the following results.

Lemma 6.3.1. *Suppose that $H_i(x)$ is a primitive for $\sigma_i(x)^{-1}$ on $\mathbb{R}_+ \setminus \{0\}$ and $z_i^s(\omega)$ is defined by (6.7). Let $T_i(\omega, \cdot) : \mathbb{R}_+ \mapsto \mathbb{R}_+$ be the random mapping given by the formula*

$$T_i(\omega, y) = H_i^{-1}(z_i^s(\omega) + H_i(y)), \ y > 0 \quad and \quad T_i(\omega, 0) = 0.$$

Then $T_i(\omega, y) \in C^3(\mathbb{R}_+ \setminus \{0\}) \cap C^1(\mathbb{R}_+)$ for all $\omega \in \Omega$ and the random mapping $T(\omega, \cdot) : \mathbb{R}^d \mapsto \mathbb{R}_+^d$ defined by the relation

$$T(\omega, y) = (T_1(\omega, y_1), \dots, T_d(\omega, y_d)) \tag{6.10}$$

is a strictly order-preserving diffeomorphism. Moreover the relations

$$\exp\{-a|z_i^s(\omega)|\} \le \frac{T_i(\omega, y)}{y} \le \exp\{a|z_i^s(\omega)|\}, \quad y > 0, \tag{6.11}$$

and

$$\exp\{-a|z_i^s(\omega)|\} \le \frac{\sigma_i(y)}{\sigma_i(T_i(\omega, y))} \le \exp\{a|z_i^s(\omega)|\}, \quad y > 0, \tag{6.12}$$

hold for every $i = 1, \ldots, d$ *and* $\omega \in \Omega$. *Here* $a = \sup_{x \in \mathbb{R}_+} |\sigma'(x)|$.

Proof. It is clear that $H_i(\cdot) : \mathbb{R}_+ \setminus \{0\} \mapsto \mathbb{R}$ is an increasing C^3-function such that

$$H_i(x) = c_i(x) + \frac{1}{\sigma_i'(0)} \log x, \quad x > 0,$$

where $c_i(x)$ belongs to the class $C^3(\mathbb{R}_+ \setminus \{0\}) \cap C^1(\mathbb{R}_+)$. This implies the corresponding smoothness of $T_i(\omega, y)$. It is clear that $T(\omega, \cdot)$ is a diffeomorphism of $\text{int}\mathbb{R}_+^d$. A simple calculation shows that

$$\lim_{y \to 0} T_i'(\omega, y) = \lim_{y \to 0} \frac{\sigma_i(T_i(\omega, y))}{\sigma_i(y)} = \lim_{y \to 0} \frac{T_i(\omega, y)}{y} > 0.$$

Therefore every mapping $y \mapsto T_i(\omega, y)$ is a diffeomorphism on $\mathbb{R}_+$. It is a strictly order-preserving mapping because every function $H_i(x)$ is strictly monotone.

To prove (6.11) we note that $0 < \sigma(x) \le ax$ for $x > 0$. Therefore

$$H_i(y_2) - H_i(y_1) \ge \frac{1}{a} \log \frac{y_2}{y_1}, \quad y_2 > y_1 > 0.$$

This relation implies that

$$H_i(y) \le z + H_i(y) \le H_i(ye^{az}), \quad y > 0, z \ge 0,$$

and

$$H_i(ye^{az}) \le z + H_i(y) \le H_i(y), \quad y > 0, z < 0.$$

From the monotonicity of H_i^{-1} we get (6.11).

Let $\eta_i(x) = \sigma_i(H_i^{-1}(x))$. Since $\frac{d}{dx}\eta_i(x) = \sigma_i'(H_i^{-1}(x)) \cdot \eta_i(x)$, we obtain

$$\frac{\sigma_i(y)}{\sigma_i(T_i(\omega, y))} = \frac{\eta_i(H_i(y))}{\eta_i(z_i^s(\omega) + H_i(y))} = \exp\left\{\int_{z_i^s(\omega) + H_i(y)}^{H_i(y)} \sigma_i'(H_i^{-1}(\xi)) d\xi\right\}.$$

This implies (6.12).   □

Lemma 6.3.2. *If the functions f_i and σ_{ij} satisfy (S1)–(S3) and (6.9) holds, then the functions*

$$g_i(\omega, y) = \sigma_i(y_i) \cdot \frac{f_i\left(T(\omega, y)\right)}{\sigma_i(T_i(\omega, y_i))} + \mu \sigma_i(y_i) \cdot z_i^s(\omega) \qquad (6.13)$$

satisfy conditions (R1)–(R3) of Chap. 5. If $f = (f_1, \ldots, f_d)$ is cooperative, then $g = (g_1, \ldots, g_d)$ satisfies (R4) of Chap. 5.

Proof. This is rather simple and it relies on the properties the functions $T_i(\omega, x)$ described in Lemma 6.3.1. We leave the details to the reader.   □

Theorem 6.3.2. *Assume that the functions f_i and σ_{ij} satisfy (S1)–(S3) and (6.9) holds. Let (θ, φ) be the RDS generated in $\mathbb{R}_+^d$ by (6.1). Then (θ, φ) is equivalent to the RDS (θ, ψ) generated in $\mathbb{R}_+^d$ by the RDE (6.5) with $g_i(\omega, y)$ given by (6.13). Relation (6.8) holds with the diffeomorphism $T(\omega, \cdot) : \mathbb{R}_+^d \mapsto \mathbb{R}_+^d$ given by the formula (6.10). Moreover we have the relations (6.2) and*

$$\varphi(t, \omega)\mathrm{int}\mathbb{R}_+^d \subset \mathrm{int}\mathbb{R}_+^d \quad \text{for any} \quad t \geq 0 \quad \text{and} \quad \omega \in \Omega. \qquad (6.14)$$

If in addition (S4) holds, then (θ, φ) is strictly order preserving.

Proof. Lemma 6.3.2 and Proposition 5.2.1 imply that the RDE (6.5) with $g_i(\omega, y)$ given by (6.13) generates a strictly positive RDS (θ, ψ) in $\mathbb{R}_+^d$ such that

$$\psi(t, \omega)\mathrm{int}\mathbb{R}_+^d \subset \mathrm{int}\mathbb{R}_+^d \quad \text{for any} \quad t \geq 0, \ \omega \in \Omega. \qquad (6.15)$$

Let $y(t, \omega) = \psi(t, \omega)T^{-1}(\omega, x)$ with $x \in \mathrm{int}\mathbb{R}_+^d$. Then we can apply Itô's formula (see Theorem 2.3.1) to $x_i(t, \omega) := T_i(\theta_t\omega, y_i(t, \omega))$ and find that $x(t, \omega) = (x_1(t, \omega), \ldots, x_d(t, \omega))$ satisfies (6.1) with initial data $x(0, \omega) = x$ for every $x \in \mathrm{int}\mathbb{R}_+^d$. Now using the continuous dependence of solutions to (6.1) on initial data we obtain that $T(\theta_t\omega, \psi(t, \omega)T^{-1}(\omega, x))$ is also a solution to this equation with initial data x from $\mathbb{R}_+^d$. Therefore using (6.8) we can define the perfect cocycle φ which satisfies in $\mathbb{R}_+^d$ the conclusions of Theorem 2.4.3.   □

Corollary 6.3.1. *Assume that f_i and σ_{ij} satisfy (S1)–(S3) and (6.9) holds. Let (θ, φ) be the RDS generated in $\mathbb{R}_+^d$ by (6.1). If $f = (f_1, \ldots, f_d)$ is strongly positive, i.e. if*

$$f_i(x) > 0, \quad i = 1, \ldots, d, \qquad (6.16)$$

for all $x \in \mathbb{R}_+^d \setminus \{0\}$ of the form $x = (x_1, \ldots, x_{i-1}, 0, x_{i+1}, \ldots, x_d)$, then (θ, φ) is a strongly positive RDS, i.e. $\varphi(t, \omega)(\mathbb{R}_+^d \setminus \{0\}) \subset \mathrm{int}\mathbb{R}_+^d$ for any $t \geq 0$ and $\omega \in \Omega$.

Proof. In this case the function g given by (6.13) satisfies (R3*). Therefore we can apply Proposition 5.2.1.   □

Theorems 6.3.2 and 5.2.1 imply the following assertion.

Corollary 6.3.2. *Assume that the functions f_i and σ_{ij} satisfy (S1)-(S4) and (6.9) holds. Let (θ, φ) be the RDS generated in $\mathbb{R}_+^d$ by (6.1). If the matrix*

$$D_x f(x) \equiv \left\{ \frac{\partial f_i(x)}{\partial x_j} \right\}_{i,j=1}^d \tag{6.17}$$

is irreducible (see Definition 5.2.1) for all $x \in \mathrm{int}\mathbb{R}_+^d$, then

$$\varphi(t, \omega, x) \ll \varphi(t, \omega, y) \quad \text{for all} \quad 0 \ll x < y \quad \text{and} \quad \omega \in \Omega, \tag{6.18}$$

i.e. the equation (6.1) generates a strongly order-preserving RDS in $\mathrm{int}\mathbb{R}_+^d$. If the matrix (6.17) is irreducible for all positive x from $\mathbb{R}_+^d$ and $\omega \in \Omega$, then RDS (θ, φ) is strongly order-preserving in $\mathbb{R}_+^d$.

Proof. Theorem 5.2.1 is applied to (6.5) with g_i given by (6.13). $\square$

Similar to Theorem 6.3.2 we can also prove the following assertion.

Theorem 6.3.3. *Assume that the functions f_i and σ_{ij} satisfy (S1)-(S4) and $\sigma_{ij}(x_i) = \sigma_i(x_i) \cdot s_{ij}$, where s_{ij} are constants and $\sigma_i(x_i)$ possess the properties (a) there exists $c_i > 0$ such that $\sigma_i(c_i) = 0$, (b) $\sigma_i(x_i) > 0$ for all $x_i \in (0, c_i)$, (c) $\sigma_i'(0) > 0$ and (d) $\sigma_i'(c_i) > 0$. Assume that f_i fulfills (6.4). Let (θ, φ) be the RDS generated in $\mathbb{R}_+^d$ by (6.1). Then the restriction (θ, φ^c) of (θ, φ) to $[0, c]$ is equivalent to the order-preserving RDS (θ, ψ^c) generated in $[0, c]$ by the RDE (6.5) with g_i given by (6.13), where $T(\omega, \cdot) : [0, c] \mapsto [0, c]$ is given by the formula (6.10) with*

$$T_i(\omega, y) = H_i^{-1}(z_i^s(\omega) + H_i(y)), \ y \in (0, c_i)$$

with $T_i(\omega, 0) = 0$ and $T_i(\omega, c_i) = c_i$. Here $H_i(x)$ is a primitive for $\sigma_i(u)^{-1}$ on the interval $(0, c_i)$. Moreover

$$\varphi^c(t, \omega, x) = T(\theta_t\omega, \psi^c(t, \omega, T^{-1}(\omega, x))), \quad t > 0, \ x \in [0, c], \ \omega \in \Omega.$$

6.4 Stochastic Comparison Principle

Analogously to the Random Comparison Theorem 5.3.1 we can prove the corresponding stochastic version (for the one-dimensional case see IKEDA/WATANABE [57] and KARATZAS/SHREVE [62], for the $\mathbb{R}^d$ case see LADDE/LAKSHMIKANTHAM [75] and also GEISS/MANTHEY [46] and the references therein). Here we consider the simplest case assuming (S1)–(S4) for one of the system. We refer to GEISS/MANTHEY [46] for more general comparison theorems.

Theorem 6.4.1 (Stochastic Comparison Principle). *Assume that conditions (S1)–(S4) for (6.1) hold. Consider in $\mathbb{R}^d_+$ the system of Stratonovich stochastic equations*

$$dx_i = g_i(x_1, \dots, x_d)dt + \sum_{j=1}^{m} \sigma_{ij}(x_i) \circ dW^j_t, \quad i = 1, \dots, d. \qquad (6.19)$$

with $g = (g_1, \dots, g_d) : \mathbb{R}^d_+ \mapsto \mathbb{R}^d$ satisfying (S1)–(S3). Let $\psi(t, \omega)$ denote the corresponding cocycle generated by (6.19). Then
(i) the condition

$$f_i(x) \leq g_i(x) \quad \text{for all} \quad x \in \mathbb{R}^d_+, \ i = 1, \dots, d, \qquad (6.20)$$

implies that

$$\varphi(t, \omega)x \leq \psi(t, \omega)x \quad \text{for all} \quad t > 0, \ \omega \in \Omega \quad \text{and} \quad x \in \mathbb{R}^d_+; \qquad (6.21)$$

(ii) if

$$f_i(x) \geq g_i(x) \quad \text{for all} \quad x \in \mathbb{R}^d_+, \qquad (6.22)$$

then

$$\varphi(t, \omega)x \geq \psi(t, \omega)x \quad \text{for all} \quad t > 0, \ \omega \in \Omega \quad \text{and} \quad x \in \mathbb{R}^d_+. \qquad (6.23)$$

Proof. Under the conditions listed above, equations (6.19) generate an RDS in $\mathbb{R}^d_+$ (see Proposition 6.2.1). Let us consider together with (6.3) the system of random differential equations

$$\dot{x}_i = g_i(x_1, \dots, x_d) + \sum_{j=1}^{m} \sigma_{ij}(x_i) \cdot \eta^j_\varepsilon(\theta_t \omega), \quad i = 1, \dots, d, \qquad (6.24)$$

where $\eta^j_\varepsilon(\theta_t \omega)$ is defined as in (6.3). By Proposition 5.2.1 both equations (6.3) and (6.24) generate RDS in $\mathbb{R}^d_+$. Let φ^ε and ψ^ε be the corresponding cocycles.

Let (6.20) hold. The random comparison principle (see Corollary 5.3.1(i)) implies that $\varphi^\varepsilon(t, \omega)x \leq \psi^\varepsilon(t, \omega)x$ for all $t > 0$, $\omega \in \Omega$ and $x \in \mathbb{R}^d_+$. Hence

$$\int_\alpha^\beta l(\psi^\varepsilon(t, \theta_{-t}\omega)x - \varphi^\varepsilon(t, \theta_{-t}\omega)x)dt \geq 0, \quad x \geq 0, \ \omega \in \Omega,$$

for all $0 \leq \alpha < \beta$, where l is a positive ($x \geq 0$ implies $l(x) \geq 0$) linear functional on $\mathbb{R}^d$. Therefore as in the proof of Proposition 6.2.2 after transition to the limit we can obtain (6.21) for all $\tilde{\Omega} = \Omega^*_\varphi \cap \Omega^*_\psi$, where Ω^*_φ (resp. Ω^*_ψ) is the θ-invariant set of full measure defined by the cocycle φ (resp. ψ) as in Remark 2.5.1. Therefore after modifications of the cocycles φ and ψ we obtain (6.21). The same argument proves (ii). $\qquad \square$

Remark 6.4.1. (i) If the diffusion coefficients $\sigma_{ij}(x)$ satisfy (6.9), then using Theorem 6.3.2 we can also prove assertions which are similar to Corollary 5.3.1 and Theorem 5.3.2 for RDS generated by stochastic equations.

(ii) The result of this section remains true if we interpret the equations (6.1) in the Itô sense. The point is that by Theorem 2.4.2 the system of Itô stochastic equations

$$dx_i = f_i(x_1, \ldots, x_d)dt + \sum_{j=1}^{m} \sigma_{ij}(x_i) \cdot dW_t^j, \quad i = 1, \ldots, d . \tag{6.25}$$

is equivalent to the system of Stratonovich equations

$$dx_i = \left(f_i(x_1, \ldots, x_d) - \frac{1}{2} \sum_{j=1}^{m} \sigma'_{ij}(x_i) \cdot \sigma_{ij}(x_i) \right) dt + \sum_{j=1}^{m} \sigma_{ij}(x_i) \circ dW_t^j , \tag{6.26}$$

where $i = 1, \ldots, d$. It is clear that the assumptions on $f_i(x)$ immediately imply the corresponding properties of the functions $f_i(x) - \frac{1}{2} \sum_{j=1}^{m} \sigma'_{ij}(x_i) \cdot \sigma_{ij}(x_i)$ and vice versa. This observation makes it also possible to find the corresponding Itô versions of the results presented in Sect. 6.2 and 6.3.

6.5 Equilibria and Attractors

Now we give a result on the existence of equilibria and attractors for the stochastic systems considered. As in the random case we note that under assumptions (S1)–(S4) Proposition 6.2.1 implies that the element $x \equiv 0$ is a sub-equilibrium for the RDS (θ, φ) generated by (6.1) in $\mathbb{R}_+^d$.

Throughout this section we assume that the diffusion terms in (6.1) have the following particular form:

$$\sigma_{ij}(x_i) = \sigma_i(x_i) \cdot s_j \quad \text{for all} \quad i = 1, \ldots, d, \ j = 1, \ldots, m , \tag{6.27}$$

where s_j are constants.

We start with a Lyapunov function type theorem giving sufficient conditions for the existence of random attractors and equilibria.

Theorem 6.5.1. *Let conditions (S1)–(S3) and (6.27) hold. Assume that there exists a function $V(x) \in C(\mathbb{R}_+^d) \cap C^1(\text{int}\mathbb{R}_+^d)$ possessing the properties*

$$\langle f(x), \nabla V(x) \rangle + \alpha \cdot V(x) \leq \beta \quad \text{for all} \quad x \in \text{int}R_+^d \tag{6.28}$$

and

$$\sum_{i=1}^{d} \frac{\partial V(x)}{\partial x_i} \cdot \sigma_i(x_i) = \gamma \cdot V(x) \quad \text{for all} \quad x \in \text{int} \mathbb{R}_+^d , \qquad (6.29)$$

where $\alpha > 0$, $\beta > 0$ and $\gamma \in \mathbb{R}$ are constants. Let

$$R(\omega) := \beta \cdot \int_{-\infty}^{0} \exp\{\alpha\tau - \gamma W_\tau^{(s)}(\omega)\} \, d\tau ,$$

where $W_t^{(s)} = \sum_{j=1}^{m} s_j W_t^j$. Let (θ, φ) be the RDS in $\mathbb{R}_+^d$ generated by (6.1). Then the random set

$$\mathcal{B}(\omega) := \{x \in \mathbb{R}_+^d \; : \; V(x) \le 2\beta/\alpha + 2R(\omega)\}$$

absorbs every deterministic bounded set, i.e. for any bounded set B from $\mathbb{R}_+^d$ there exists $t_0 = t_0(\omega, B) > 0$ such that $\varphi(t, \omega)B \subset \mathcal{B}(\theta_t\omega)$ for $t \ge t_0$. If we additionally assume that

$$a_1|x|^{\alpha_1} - b_1 \le V(x) \le a_2|x|^{\alpha_2} + b_2, \quad \text{for all} \quad x \in \text{int} \mathbb{R}_+^d , \qquad (6.30)$$

where a_j, α_j, b_j are positive constants, then the RDS (θ, φ) possesses a random attractor $A(\omega)$ in the universe $\mathcal{D}$ of all tempered subsets of $\mathbb{R}_+^d$. This attractor is measurable with respect to the past σ-algebra $\mathcal{F}_-$. If in addition (S4) holds, then the attractor $A(\omega)$ is bounded from above and from below and there exist maximal and minimal equilibria $\bar{u}$ and $\underline{u}$ such that the random interval $[\underline{u}, \bar{u}]$ contains the attractor as well as all other possible tempered equilibria. In particular, if the equilibrium u is unique, then $A = \{u\}$.

Proof. Let us consider the RDE

$$\dot{x}_i = f_i(x_1, \dots, x_d) + \sigma_i(x_i) \cdot \eta_\varepsilon^{(s)}(\theta_t\omega), \quad i = 1, \dots, d , \qquad (6.31)$$

where $\eta_\varepsilon^{(s)}(\omega) = \sum_{j=1}^{m} s_j \eta_\varepsilon^j(\omega)$ and $\eta_\varepsilon^j(\omega)$ is defined as in (6.3). Let $\varphi^\varepsilon(t, \omega)$ be the corresponding cocycle generated by (6.31) and $x^\varepsilon(t) = \varphi^\varepsilon(t, \omega)x$, where $x \in \text{int} \mathbb{R}_+^d$. Using (6.31), (6.28) and (6.29) we find that the function $V_\varepsilon(t) = V(x^\varepsilon(t))$ satisfies the inequality

$$\frac{d}{dt} V_\varepsilon(t) \le \left(-\alpha + \gamma \cdot \eta_\varepsilon^{(s)}(\theta_t\omega) \right) V_\varepsilon(t) + \beta .$$

Therefore we have

$$V_\varepsilon(t) \le V(x) \exp\left\{ -\alpha t + \gamma \int_0^t \eta_\varepsilon^{(s)}(\theta_\tau\omega) d\tau \right\}$$
$$+ \beta \cdot \int_0^t \exp\left\{ -\alpha(t - \xi) + \gamma \int_\xi^t \eta_\varepsilon^{(s)}(\theta_\tau\omega) d\tau \right\} d\xi .$$

Hence

$$V(\varphi^\varepsilon(t, \theta_{-t}\omega)x) \leq V(x) \exp\left\{-\alpha t + \gamma \int_{-t}^{0} \eta_\varepsilon^{(s)}(\theta_\tau\omega)d\tau\right\}$$
$$+ \beta \cdot \int_{-t}^{0} \exp\left\{\alpha\xi + \gamma \int_{\xi}^{0} \eta_\varepsilon^{(s)}(\theta_\tau\omega)d\tau\right\} d\xi . \tag{6.32}$$

Since (see Sect. 2.5)

$$\eta_\varepsilon^{(s)}(\theta_t\omega) = \frac{d}{dt}W_t^{(s),\varepsilon}(\omega) := \sum_{j=1}^{m} s_j \frac{d}{dt}W_t^{j,\varepsilon}(\omega) ,$$

where $W_t^{j,\varepsilon}(\omega)$ is defined by (2.43), using (2.25) we have

$$\int_{\xi}^{0} \eta_\varepsilon^{(s)}(\theta_\tau\omega)d\tau = -\sum_{j=1}^{m} s_j \int_{0}^{\varepsilon} \phi_\varepsilon(\tau)W_\xi^{j}(\theta_\tau\omega)d\tau .$$

Since $\tau \mapsto W_\xi^{j}(\theta_\tau\omega)$ is a continuous function for every ξ and ω (see (2.25)), we have that

$$\lim_{\varepsilon \to 0} \int_{\xi}^{0} \eta_\varepsilon^{(s)}(\theta_\tau\omega)d\tau = -\sum_{j=1}^{m} s_j W_\xi^{j}(\omega) = -W_\xi^{(s)}(\omega) . \tag{6.33}$$

Let Ω^* be a θ-invariant set of full measure such that (2.46) holds. As in the proof of Corollary 2.5.1 (cf. (2.49)) we conclude that for any $\omega \in \Omega^*$ and $x \in \mathbb{R}_+^d$ there exists $\varepsilon_k \to 0$ such that

$$\varphi^{\varepsilon_k}(t, \theta_{-t}\omega)x \to \varphi(t, \theta_{-t}\omega)x \quad \text{for almost all} \quad t > 0 .$$

Therefore from (6.32) and (6.33) we have

$$V(\varphi(t, \theta_{-t}\omega)x) \leq V(x) \exp\left\{-\alpha t - \gamma W_{-t}^{(s)}(\omega)\right\}$$
$$+ \beta \cdot \int_{-t}^{0} \exp\left\{\alpha\xi - \gamma W_\xi^{(s)}(\omega)\right\} d\xi \tag{6.34}$$

for almost all $t > 0$. Since $t \mapsto \varphi(t, \theta_{-t}\omega)x$ is continuous (see Remark 2.4.1), we have inequality (6.34) for all $\omega \in \Omega^*$, $t > 0$ and $x \in \mathbb{R}_+^d$ which implies that

$$V(\varphi(t, \theta_{-t}\omega)x) \leq V(x) \exp\left\{-\alpha t - \gamma W_{-t}^{(s)}(\omega)\right\} + R(\omega) . \tag{6.35}$$

for all $\omega \in \Omega^*$, $t > 0$ and $x \in \mathbb{R}_+^d$.

For $\omega \notin \Omega^*$ we redefine the cocycle $\varphi(t, \omega)$ by the formula $\varphi(t, \omega)x = y(t; x)$, where $y(t) = y(t; x)$ is the solution to the problem

$$\dot{y}_i = f_i(y_1, \ldots, y_d), \quad y(0) = x, \quad i = 1, \ldots, d,$$

It is clear from (6.28) that

$$V(\varphi(t, \theta_{-t}\omega)x) \leq V(x)e^{-\alpha t} + \frac{\beta}{\alpha}\left(1 - e^{-\alpha t}\right), \quad t > 0, \ \omega \notin \Omega^*. \tag{6.36}$$

Inequalities (6.35) and (6.36) imply that the random set $\mathcal{B}(\omega)$ absorbs every deterministic bounded set from $\mathbb{R}_+^d$.

It is clear that $R(\omega)$ is a tempered random variable. Therefore under condition (6.30) we have $\mathcal{B} \in \mathcal{D}$. From (6.35) and (6.36) we also have that $\mathcal{B}$ is $\mathcal{D}$-absorbing for the RDS (θ, φ) (cf. the proof of Proposition 1.4.1). Therefore we can apply Theorem 1.8.1 on the existence of random attractors and assert that the RDS (θ, φ) generated in the space $\mathbb{R}_+^d$ by problem (6.1) possesses a random global $\mathcal{D}$-attractor $A(\omega)$. Since $R(\omega)$ is $\mathcal{F}_-$-measurable, it follows from (1.44) that $A(\omega)$ is $\mathcal{F}_-$-measurable. The existence of the maximal and minimal equilibria $\bar{u}$ and $\underline{u}$ and their properties under condition (S4) follow from Theorem 3.6.2. $\qquad \square$

Corollary 6.5.1. *Let assumptions (S1)–(S3) and (6.27) with $\sigma_i(x_i) \equiv x_i$ be valid. Assume that there exist positive numbers κ, α and β such that*

$$\sum_{i=1}^{d} x_i^{\kappa-1} \cdot f_i(x) \leq -\alpha \cdot \sum_{i=1}^{d} x_i^{\kappa} + \beta, \quad x \in \mathbb{R}_+^d. \tag{6.37}$$

Then the RDS (θ, φ) generated by (6.1) possesses a random attractor $\mathcal{A}(\omega)$ in the universe $\mathcal{D}$ of all tempered subsets of $\mathbb{R}_+^d$ and all the conclusions of Theorem 6.5.1 hold.

Proof. It is easy to see that the function $V(x) = \sum_{i=1}^{d} x_i^{\kappa}$ satisfies all the hypotheses of Theorem 6.5.1. $\qquad \square$

Corollary 6.5.2. *Let assumptions (S1)–(S3) and (6.27) hold. Assume additionally that*

$$\sigma_i(x) > 0 \quad for \quad x > 0, \quad \sigma_i'(0) > 0, \quad i = 1, \ldots, d,$$

and

$$\frac{\sigma_i(x)}{x} = \lambda_i + O(x^{-\gamma_i}) \quad when \quad x \to \infty, \quad i = 1, \ldots, d, \tag{6.38}$$

where $\lambda_i > 0$ and $\gamma_i > 0$ are constants, and there exist positive numbers α, β and $0 \leq \kappa < (\min \lambda_i) \cdot (\max \lambda_i)^{-1}$ such that

$$f_i(x_1 \ldots, x_d) \leq -\alpha \cdot x_i + g_i(x_1 \ldots, x_d), \quad x = (x_1 \ldots, x_d) \in \mathbb{R}_+^d, \quad (6.39)$$

where the function $g_i(x_1 \ldots, x_d)$ *possesses the property*

$$|g_i(x_1 \ldots, x_d)| \leq \beta \cdot \left(1 + \sum_{j=1}^d x_j^\kappa\right), \quad x = (x_1 \ldots, x_d) \in \mathbb{R}_+^d, \quad (6.40)$$

Then the RDS (θ, φ) *generated by (6.1) possesses a random attractor* $A(\omega)$
in the universe $\mathcal{D}$ *of all tempered subsets of* $\mathbb{R}_+^d$ *and all the conclusions of
Theorem 6.5.1 hold.*

Proof. Let

$$V(x) = \sum_{i=1}^d V_i(x_i) \quad \text{with} \quad V_i(x) = \exp\left\{\delta \int_1^x \frac{d\xi}{\sigma_i(\xi)}\right\},$$

where δ is a positive parameter. If we set $V_i(0) = 0$, then $V(x) \in C(\mathbb{R}_+^d) \cap C^3(\text{int}\mathbb{R}_+^d)$. It is clear that $V(x)$ satisfies (6.29) with $\gamma = \delta$ and it follows
from (6.39) that

$$\langle f(x), \nabla V(x) \rangle = \delta \sum_{i=1}^d V_i(x_i) \cdot \frac{f_i(x_1 \ldots, x_d)}{\sigma_i(x_i)}$$

$$\leq -\alpha\delta \sum_{i=1}^d \frac{x_i}{\sigma_i(x_i)} \cdot V_i(x_i) + \delta \sum_{i=1}^d V_i(x_i) \cdot \frac{g_i(x_1 \ldots, x_d)}{\sigma_i(x_i)}. \quad (6.41)$$

It is clear that every function $\frac{x}{\sigma_i(x)} \cdot V_i(x)$ is continuous on $\mathbb{R}_+$. Therefore
(6.38) implies that

$$\frac{x_i}{\sigma_i(x_i)} \cdot V_i(x_i) \geq \frac{1}{2\lambda} V_i(x_i) - C, \quad \lambda := \max \lambda_i,$$

for some constant C. Therefore from (6.41) we obtain

$$\langle f(x), \nabla V(x) \rangle \leq -\frac{\alpha\delta}{2\lambda} \cdot V(x) + \delta \sum_{i=1}^d V_i(x_i) \cdot \frac{g_i(x_1 \ldots, x_d)}{\sigma_i(x_i)} + C. \quad (6.42)$$

A simple calculation shows that

$$C_1 \cdot x_i^{\delta/\lambda_i} \leq V_i(x_i) \leq C_2 \cdot x_i^{\delta/\lambda_i} \quad \text{if} \quad x_i \geq 1, \quad i = 1, \ldots, d,$$

and

$$C_1 \cdot x_i^{\delta/\sigma_i'(0)} \leq V_i(x_i) \leq C_2 \cdot x_i^{\delta/\sigma_i'(0)} \quad \text{if} \quad 0 < x_i < 1, \quad i = 1, \ldots, d,$$

These inequalities imply that under the condition $\delta \geq \max \sigma_i'(0)$ we have

$$\frac{V_i(x_i)}{\sigma_i(x_i)} \leq C \cdot \left[1 + V(x)^{1-\lambda_i/\delta}\right] \quad \text{and} \quad x_i \leq C \cdot \left[1 + V(x)^{\lambda_i/\delta}\right] ,$$

where $x = (x_1, \dots, x_d) \in \mathrm{int}\mathbb{R}_+^d$ and C is a constant. Therefore it is easy to see from (6.40) that

$$\sum_{i=1}^d V_i(x_i) \cdot \frac{g_i(x_1, \dots, x_d)}{\sigma_i(x_i)} \leq C \cdot \left[1 + V(x)^{\kappa^*}\right] , \tag{6.43}$$

where $\kappa^* = 1 - (\min \lambda_i - \kappa \cdot \max \lambda_i) \cdot \delta^{-1}$ and δ is large enough. Since $\kappa^* < 1$ for all $\delta > 0$, relations (6.42) and (6.43) imply (6.28) with appropriate choice of parameters. $\qquad\square$

6.6 One-Dimensional Stochastic Equations

In this section we consider the properties of the RDS generated by a single Stratonovich SDE

$$dx = f(x)dt + \sigma(x) \circ dW_t . \tag{6.44}$$

There are many results concerning this equation (see, e.g., ARNOLD [3], GIHMAN/SKOROHOD [47], IKEDA/WATANABE [57], KARATZAS/SHREVE [62], KHASMINSKII [64] among others). We include this section for the sake of completeness.

Below for any closed interval $I \subseteq \mathbb{R}$ (finite or not) and $k \in \mathbb{Z}_+$, $\delta \in (0,1]$ we denote by $C^{k,\delta}(I)$ the space of k times continuously differentiable functions $f(x)$ on I such that the derivative $f^{(k)}(x)$ satisfies the Hölder condition with the exponent δ in a vicinity of every point from I. We also denote by $C_b^{k,\delta}(I)$ the space of restrictions of functions from $C_b^{k,\delta}(\mathbb{R})$ to the interval I.

6.6.1 Stochastic Equations on $\mathbb{R}_+$

We start with the following assertion on the generation of RDS in $\mathbb{R}_+$ by (6.44). It admits a slightly more general class of drift terms in comparison with the result given by Proposition 6.2.2.

Proposition 6.6.1. *Assume that*

$$f(x) \in C^{1,\delta}(\mathbb{R}_+), \quad f(x) \leq ax + b, \quad f(0) \geq 0 , \tag{6.45}$$

and

$$\sigma(x) \in C_b^{2,\delta}(\mathbb{R}_+), \quad \sigma(x) \cdot \sigma'(x) \in C_b^{1,\delta}(\mathbb{R}_+) ,$$

$$\tag{6.46}$$

$$\sigma(0) = 0, \quad |\sigma'(0)| > 0, \quad |\sigma(x)| > 0 \text{ if } x > 0 .$$

Here above $a, b \in \mathbb{R}_+$ and $\delta \in (0,1]$. Then (6.44) generates a strictly order-preserving RDS in $\mathbb{R}_+$.

Proof. We suppose that $\sigma'(0) > 0$ and $\sigma(x) > 0$ if $x > 0$ for the definiteness.

We first assume that $-\alpha \leq f(x) \leq ax + b$ for some $\alpha > 0$. Denote by $\chi_N(z)$ a function from $C^2(\mathbb{R})$ with the properties (i) $\chi_N(z) = N + 1/2$ for $z \geq N + 1$, (ii) $\chi_N(z) = z$ for $z \in [-\infty, N]$ and (iii) $0 \leq \chi'_N(z) \leq 1$ for all $z \in \mathbb{R}$. Then $f_N(x) := \chi_N(f(x)) \in C_b^{1,\delta}(\mathbb{R}_+)$. Since condition (S4) holds automatically for the one-dimensional case, Proposition 6.2.2 implies that the equation

$$dx = f_N(x)dt + \sigma(x) \circ dW_t \tag{6.47}$$

generates a strictly order-preserving C^1 RDS (θ, φ_N) in $\mathbb{R}_+$. Since $f_N(x) \leq f_{N+1}(x) \leq ax + b$, Comparison Theorem 6.4.1 implies that

$$\varphi_N(t, \omega)x \leq \varphi_{N+1}(t, \omega)x \leq \bar{\varphi}(t, \omega)x, \quad t > 0, \ \omega \in \Omega, \ x \in \mathbb{R}_+ , \tag{6.48}$$

where $(\theta, \bar{\varphi})$ is the RDS generated by (6.44) with $f(x) = ax + b$. Relation (6.48) implies that the limit

$$\varphi(t, \omega)x := \lim_{N \to \infty} \varphi_N(t, \omega)x, \quad t > 0, \ \omega \in \Omega, \ x \in \mathbb{R}_+ , \tag{6.49}$$

exists. By Theorem 6.3.2 the RDS (θ, φ_N) is equivalent to the RDS (θ, ψ_N) generated by the RDE

$$\dot{y} = \sigma(y) \cdot \frac{f_N(T(\theta_t\omega, y))}{\sigma(T(\theta_t\omega, y))} + \mu\sigma(y) \cdot z(\theta_t\omega) ,$$

where $T(\omega, y) = H^{-1}(z(\omega) + H(y))$, $y > 0$ and $T(\omega, 0) = 0$. Here $H(x)$ is a primitive for $\sigma(x)^{-1}$ on $\mathbb{R}_+ \setminus \{0\}$ and $z(\omega) = \int_{-\infty}^0 e^{\mu\tau}dW_\tau$ is a Gaussian random variable which generates a stationary Ornstein-Uhlenbeck process in $\mathbb{R}$. We also have the relation

$$\varphi_N(t, \omega, x) = T(\theta_t\omega, \psi_N(t, \omega, T^{-1}(\omega, x))), \quad t > 0, \ x \in \mathbb{R}_+, \ \omega \in \Omega .$$

Therefore (6.49) implies that

$$y(t, \omega) = \psi(t, \omega, y_0) := T^{-1}(\theta_t\omega, \varphi(t, \omega, T(\omega, y_0))), \quad t > 0, \ \omega \in \Omega , \tag{6.50}$$

is a local solution to the problem

$$\dot{y} = \sigma(y) \cdot \frac{f(T(\theta_t\omega, y))}{\sigma(T(\theta_t\omega, y))} + \mu\sigma(y) \cdot z(\theta_t\omega), \quad y_0 \in \mathbb{R}_+ .$$

Thus $(t, x) \mapsto \varphi(t, \omega, x)$ is a continuous function for every $\omega \in \Omega$. Hence by (6.48) the limit in (6.49) is uniform with respect to (t, x) from compact subsets of $\mathbb{R}_+ \times \mathbb{R}_+$. This implies that (θ, φ) is an order-preserving RDS. It is strictly order-preserving because of (6.50). It is also easy to see that $x(t) = \varphi(t, \omega)x$ solves (6.44).

Now we consider the case of general f satisfying (6.45). Let $\bar{\chi}_N(z)$ be a function from $C^2(\mathbb{R})$ with the properties (i) $\bar{\chi}_N(z) = z$ for $z \geq -N$, (ii) $\bar{\chi}_N(z) = -N - 1/2$ for $z \leq -N - 1$ and (iii) $0 \leq \bar{\chi}'_N(z) \leq 1$ for all $z \in \mathbb{R}$. Let $f_N(x) := \bar{\chi}_N(f(x))$. Since f_N satisfies (6.45) and $-N - 1 \leq f_N(x) \leq ax + b$, equation (6.47) generates a strictly order-preserving RDS (θ, φ_N^*) in $\mathbb{R}_+$ such that

$$\varphi_N^*(t, \omega)x \geq \varphi_{N+1}^*(t, \omega)x \geq 0, \quad t > 0, \ \omega \in \Omega, \ x \in \mathbb{R}_+ .$$

Therefore the limit

$$\varphi(t, \omega)x := \lim_{N \to \infty} \varphi_N^*(t, \omega)x, \quad t > 0, \ \omega \in \Omega, \ x \in \mathbb{R}_+ ,$$

exists. The same argument as above leads to the conclusion. □

Remark 6.6.1. Using Feller's test for non-explosion (see, e.g., KARATZAS/ SHREVE [62, p. 348]) it is also possible to give the sufficient and necessary conditions on the functions $f(x) \in C^{1,\delta}(\mathbb{R}_+)$ and $\sigma(x) \in C_b^{2,\delta}(\mathbb{R}_+)$ for generation of a C^1 RDS by Eq. (6.44) (cf. ARNOLD [3, p. 96] and KUNITA [74, p. 181–184]).

Proposition 6.6.1 and Corollary 6.5.2 imply the following assertion.

Corollary 6.6.1. *Assume in addition to (6.45) and (6.46) that*

$$\limsup_{x \to \infty} \frac{f(x)}{x} < 0 \quad and \quad \frac{|\sigma(x)|}{x} = \lambda + O(x^{-\gamma}), \ x \to \infty , \tag{6.51}$$

where $\lambda > 0$ and $\gamma > 0$ are constants. Then the RDS (θ, φ) generated by (6.44) possesses a random attractor $A(\omega)$ in the universe $\mathcal{D}$ of all tempered subsets of $\mathbb{R}_+$. This attractor is measurable with respect to the past σ-algebra $\mathcal{F}_-$. Moreover $A(\omega) = [\underline{u}(\omega), \bar{u}(\omega)]$, where $\bar{u}$ and $\underline{u}$ are $\mathcal{F}_-$-measurable tempered equilibria such that $0 \leq \underline{u}(\omega) \leq \bar{u}(\omega)$.

Example 6.6.1. We consider an RDS generated in $\mathbb{R}_+$ by the SDE

$$dx = (\alpha x - \beta x^{N+1})dt + \sigma x \circ dW_t , \tag{6.52}$$

where $\beta > 0$, $\alpha, \sigma \in \mathbb{R} \setminus \{0\}$ and $N > 0$. By Proposition 6.6.1 this equation generates a strictly order-preserving RDS (θ, φ) in $\mathbb{R}_+$. By Corollary 6.6.1 this RDS has a random attractor $A(\omega) = [0, u(\omega)]$ in the universe $\mathcal{D}$ of all tempered subsets of $\mathbb{R}_+$, where $u(\omega) \geq 0$ is an $\mathcal{F}_-$-measurable equilibrium.

We note that as in the random case (see Example 5.6.1) the cocycle φ can be represented in the form

$$
\varphi(t,\omega,x) = \frac{x \exp\{\alpha t + \sigma W_t(\omega)\}}{\left(1 + \beta N x^N \int_0^t \exp\{N(\alpha\tau + \sigma W_\tau(\omega))\}\, d\tau\right)^{1/N}}
$$

for $x > 0$ and $\varphi(t,\omega,0) = 0$. Therefore if $\alpha < 0$, then $A(\omega) = \{0\}$. In the case $\alpha > 0$ we have $A(\omega) = [0, u_{\alpha,\beta,N}(\omega)]$, where

$$
u_{\alpha,\beta,N}(\omega) := \left(\beta N \int_{-\infty}^0 \exp\{N(\alpha\tau + \sigma W_\tau(\omega))\}\, d\tau\right)^{-\frac{1}{N}}. \tag{6.53}
$$

Moreover a simple calculation relying on Proposition 1.9.3 shows that there exists $\gamma > 0$ such that

$$
\lim_{t\to\infty} e^{\gamma t} |\varphi(t, \theta_{-t}\omega, x) - u_{\alpha,\beta,N}(\omega)| = 0 \quad \text{for all } x > 0 \text{ and } \omega \in \Omega. \tag{6.54}
$$

If $N = 2m+1$ is odd, $m \geq 1$, then equation (6.52) is invariant with respect to the transformation $x \mapsto -x$. Therefore Proposition 6.6.1 implies that (6.52) generates a strictly order-preserving RDS $(\theta, \bar\varphi)$ in $\mathbb{R}$. This RDS has a random attractor $A(\omega)$ in the universe $\mathcal{D}$ of all tempered subsets of $\mathbb{R}$. We have that $A(\omega) = \{0\}$ if $\alpha < 0$ and $A(\omega) = [-u_{\alpha,\beta,N}(\omega), u_{\alpha,\beta,N}(\omega)]$ when $\alpha > 0$. In the last case the equilibrium $u_{\alpha,\beta,N}(\omega)$ (resp. $-u_{\alpha,\beta,N}(\omega)$) is globally stable in $\mathbb{R}_+ \setminus \{0\}$ (resp. $\mathbb{R}_- \setminus \{0\}$) and $u_0 \equiv 0$ is an unstable equilibrium. Thus we observe here a pitchfork bifurcation as α increases through 0. We refer to ARNOLD [3, Chap.9] for a detailed discussion of the bifurcation phenomena for RDS (see also CRAUEL/FLANDOLI [37], CRAUEL ET AL. [38], ARNOLD [4] and the references therein). We also note that other explicitly solvable SDE can be found in HORSTHEMKE/LEFEVER [55, pp. 139ff] and KLOEDEN/PLATEN [67, pp. 117ff] (see also Example 6.6.3 below).

The next assertion shows that the behaviour of trajectories presented in Example 6.6.1 is typical for a dissipative RDS generated in $\mathbb{R}_+$ by an equation of the form (6.44).

To describe possible scenarios of long-time dynamics we introduce the *speed measure* $m(dx)$ on $\mathbb{R}_+$ by the formula (see, e.g., KARATZAS/SHREVE [62])

$$
m(A) = \int_A \exp\left\{2 \int_1^x \frac{f(\xi)}{\sigma^2(\xi)}\, d\xi\right\} \frac{dx}{|\sigma(x)|}, \quad A \in \mathcal{B}(\mathbb{R}_+). \tag{6.55}
$$

It is easy to see that $m([1, +\infty)) < \infty$ under conditions (6.45), (6.46) and (6.51). Below we also use that, by Theorem 2.4.2, equation (6.44) can be written in Itô's form

$$
dx = \left(f(x) + \frac{1}{2}\sigma'(x)\sigma(x)\right) \cdot dt + \sigma(x) \cdot dW_t. \tag{6.56}
$$

Theorem 6.6.1. *Assume that hypotheses (6.45), (6.46) and (6.51) hold. Let $A(\omega)$ be the random attractor in the universe $\mathcal{D}$ of all tempered subsets of $\mathbb{R}_+$ for the RDS generated by (6.44).*

(i) *If $f(0) = 0$ and $m((0,1]) = \infty$, then $A(\omega) = \{0\}$ almost surely.*
(ii) *If $f(0) = 0$ and $m((0,1]) < \infty$, then $A(\omega) = [0, u(\omega)]$ for some $\mathcal{F}_-$-measurable equilibrium $u(\omega)$ such that $u(\omega) > 0$ almost surely. There are no other (up to indistinguishability) $\mathcal{F}_-$-measurable equilibria in the set $(0, u(\omega)]$ and*

$$\lim_{t\to\infty} \mathbb{P}\{\omega : \varphi(t,\omega)x \in B\} = \mathbb{P}\{\omega : u(\omega) \in B\} \equiv \frac{m(B)}{m(\mathbb{R}_+)} \qquad (6.57)$$

for all $x > 0$ and $B \in \mathcal{B}(\mathbb{R}_+)$.
(iii) *If $f(0) > 0$, then there exists an $\mathcal{F}_-$-measurable equilibrium $u(\omega)$ such that $u(\omega) > 0$ and $A(\omega) = \{u(\omega)\}$ almost surely. Moreover (6.57) holds for any $x \in \mathbb{R}_+$.*

Proof. For definiteness we assume that $\sigma(x) > 0$ for $x > 0$.

(i) As in KARATZAS/SHREVE [62] (see also IKEDA/WATANABE [57]) we define the *scale function* $s : [0, +\infty] \mapsto \mathbb{R} \cup \{\pm\infty\}$ (for equation (6.56)) by the formula

$$s(x) = \int_1^x \exp\left\{-2\int_1^y \frac{f(\xi)}{\sigma^2(\xi)}\,d\xi\right\} \frac{dy}{|\sigma(y)|}\ .$$

Since $m([1, +\infty)) < \infty$, it is easy to prove that $s(+\infty) = +\infty$ (see CRAUEL ET AL. [38, Lemma 2.4]). Hence as in SCHEUTZOW [91] we can conclude that $\varphi(t,\omega)x \to 0$ in probability for every $x \in \mathbb{R}_+$, i.e.

$$\lim_{t\to\infty} \mathbb{P}\{\omega : \varphi(t,\omega)x \geq \varepsilon\} = 0, \quad x \in \mathbb{R}_+, \qquad (6.58)$$

for any $\varepsilon > 0$. Indeed, choose $0 < \gamma < \varepsilon$ such that $\varepsilon \cdot m((\gamma, \varepsilon]) > m([\varepsilon, \infty))$ and consider a function $\tilde{f} \in C^{1,\delta}(\mathbb{R}_+)$ possessing the properties (a) $\tilde{f}(x) \geq f(x)$ for $x \in \mathbb{R}_+$, (b) $\tilde{f}(x) = f(x)$ if $x \geq \gamma$ and (c) $\tilde{m}((0,1]) < \infty$, where $\tilde{m}(dx)$ is defined by (6.55) with $\tilde{f}$ instead of f. It is easy to construct a such function $\tilde{f}$ choosing $\tilde{f}(x) = k \cdot x$ with $k > 0$ at a vicinity of 0. Let $(\theta, \tilde{\varphi})$ be the RDS generated by (6.44) with $\tilde{f}$ instead of f. Comparison Theorem 6.4.1 implies that $\varphi(t,\omega)x \leq \tilde{\varphi}(t,\omega)x$ for all $t \in \mathbb{R}$, $\omega \in \Omega$ and $x > 0$. We also have $\tilde{s}(0) = -\infty$ and $\tilde{s}(+\infty) = \infty$, where $\tilde{s}(x)$ is the scale function for the RDS $(\theta, \tilde{\varphi})$. Therefore by [78, Theorem 7, Chap. 4] we have

$$\lim_{t\to\infty} \mathbb{P}\{\omega : \tilde{\varphi}(t,\omega)x \geq \varepsilon\} = \frac{\tilde{m}([\varepsilon, \infty))}{\tilde{m}([0, \infty))} \leq \frac{\tilde{m}([\varepsilon, \infty))}{\tilde{m}([\gamma, \infty))}\ .$$

Since $\tilde{m}([a, \infty)) = m([a, \infty))$ for any $a \geq \gamma$, we obtain that

$$\tilde{m}([\varepsilon, \infty)) < \varepsilon \cdot m((\gamma, \varepsilon]) \leq \varepsilon \cdot m([\gamma, \infty)) \leq \varepsilon\tilde{m}([\gamma, \infty))\ .$$

Thus

$$\limsup_{t\to\infty} \mathbb{P}\{\omega \, : \, \varphi(t,\omega)x \geq \varepsilon\} \leq \lim_{t\to\infty} \mathbb{P}\{\omega \, : \, \bar{\varphi}(t,\omega)x \geq \varepsilon\} \leq \varepsilon$$

for any $\varepsilon > 0$. This implies (6.58).

Since

$$\{\omega \, : \, \varphi(t,\omega)v(\omega) \geq \varepsilon\} \subset \{\omega \, : \, \varphi(t,\omega)N \geq \varepsilon\} \cup \{\omega \, : \, v(\omega) \geq N\}$$

for any random variable $v(\omega)$ and $N \in \mathbb{N}$, it follows from (6.58) that

$$\limsup_{t\to\infty} \mathbb{P}\{\omega \, : \, \varphi(t,\omega)v(\omega) \geq \varepsilon\} \leq \mathbb{P}\{\omega \, : \, v(\omega) \geq N\}$$

for every $N \in \mathbb{N}$. Therefore

$$\lim_{t\to\infty} \mathbb{P}\{\omega \, : \, \varphi(t,\omega)v(\omega) \geq \varepsilon\} = 0 \tag{6.59}$$

for every $\{v(\omega)\} \in \mathcal{D}$ and $\varepsilon > 0$. If $A(\omega) = \{0\}$ is not true almost surely, then by Corollary 6.6.1 there exists an equilibrium $u(\omega) \geq 0$ such that $\mathbb{P}\{\omega \, : \, u(\omega) \geq \varepsilon\} > 0$ for some $\varepsilon > 0$. However the relation

$$\mathbb{P}\{\omega \, : \, \varphi(t,\omega)u(\omega) \geq \varepsilon\} = \mathbb{P}\{\omega \, : \, u(\theta_t\omega) \geq \varepsilon\} = \mathbb{P}\{\omega \, : \, u(\omega) \geq \varepsilon\} > 0$$

contradicts (6.59). Thus $A(\omega) = \{0\}$ almost surely.

(ii) In this case we have $m(\mathbb{R}_+) < \infty$. Therefore the function

$$\varrho(x) = \frac{N}{\sigma(x)} \exp\left\{2 \int_1^x \frac{f(\xi)}{\sigma^2(\xi)} \, d\xi\right\}, \quad x > 0, \tag{6.60}$$

where $N = [m(\mathbb{R}_+)]^{-1}$, is a stationary solution to the Fokker-Plank equation

$$\frac{\partial \varrho}{\partial t} = \frac{1}{2} \cdot \frac{\partial^2}{\partial x^2}\left(\sigma^2(x)\varrho\right) - \frac{\partial}{\partial x}\left(\left[f(x) + \frac{1}{2}\sigma'(x)\sigma(x)\right]\varrho\right), \quad x > 0,$$

possessing the property $\int_{\mathbb{R}_+} \varrho(x)dx = 1$. Thus $\varrho(x)$ is a density of a stationary measure for the Markov family $\{\varphi(t,\omega)x, \, x > 0\}$. Since in the case considered the stationary measure on $\mathbb{R}_+ \setminus \{0\}$ is unique (see, e.g., HORSTHEMKE/LEFEVER [55]), the measure $\varrho(B) := \int_B \varrho(x)dx$, $B \in \mathcal{B}(\mathbb{R}_+)$, is ergodic. Moreover transition probabilities $P_t(x, \cdot)$ weakly converge to the stationary measure ϱ, i.e.

$$P_t(x, B) = \mathbb{P}\{\omega \, : \, \varphi(t,\omega)x \in B\} \to \varrho(B), \quad \text{as} \quad t \to \infty,$$

for any $x > 0$ and $B \in \mathcal{B}(\mathbb{R}_+)$ (see, e.g., MANDL [78, Theorem 7, Chap.4]). In particular, this implies that $A(\omega) = [0, \bar{u}(\omega)]$, where $\bar{u}(\omega)$ is an $\mathcal{F}_-$-measurable equilibrium such that $\bar{u}(\omega) > 0$ almost surely. Indeed, if $\bar{u}(\omega) = 0$ on a set of positive measure, then by Lemma 3.5.1 $u(\omega) = 0$ almost surely. In this case $A(\omega) = \{0\}$ almost surely and therefore $\mathbb{P}\{\omega : \varphi(t,\omega)x \geq \delta\} \to 0$ for any $x > 0$ and $\delta > 0$, which is impossible because $\varrho([\delta, \infty)) > 0$.

The uniqueness of the stationary measure ϱ and Theorem 1.10.1 imply that

$$\mathbb{P}\{\omega : \tilde{u}(\omega) \in B\} = \mathbb{P}\{\omega : u(\omega) \in B\} = \varrho(B), \quad B \in \mathcal{B}(\mathbb{R}_+),$$

for any $\mathcal{F}_-$-measurable positive equilibrium $\tilde{u}(\omega)$. Therefore the inequality $\tilde{u}(\omega) < \bar{u}(\omega)$ is impossible on a set of positive measure. Thus there are no other positive $\mathcal{F}_-$-measurable equilibria in $A(\omega)$.

(iii) The assumption $f(0) > 0$ implies that $\varphi(t,\omega)0 > 0$ for all $t > 0$ and $\omega \in \Omega$. Thus by Proposition 3.5.2 there exists a positive $\mathcal{F}_-$-measurable equilibrium $\underline{u}(\omega) = \lim_{t\to\infty} \varphi(t, \theta_{-t}\omega)0$ and by Corollary 6.6.1 $A(\omega) = [\underline{u}(\omega), \bar{u}(\omega)]$, where $0 < \underline{u}(\omega) \leq \bar{u}(\omega)$ are $\mathcal{F}_-$-measurable equilibria. Since the property $f(0) > 0$ implies that $m((0,1]) < \infty$, as in case (ii) we can conclude that $\underline{u}(\omega) = \bar{u}(\omega)$ almost surely. □

Remark 6.6.2. Assume that $f(x)$ satisfies a Lipschitz condition on each compact subset of $\mathbb{R}_+$, $f(0) = 0$ and $|f(x)| \leq ax + b$ for some positive a and b. Let $\sigma(x) \in C^2$ with bounded first and second derivatives and $\sigma(0) = 0$. It was proved by SCHEUTZOW [91] that $A(\omega) = \{0\}$ is a random attractor for the RDS generated by (6.44) if and only if $m((0,1]) = \infty$ and $m([1,\infty)) < \infty$.

Example 6.6.2. Consider the following generalization of equation (6.52)

$$dx = (\alpha x - g(x))dt + \sigma x \circ dW_t , \tag{6.61}$$

where $\alpha, \sigma \in \mathbb{R} \setminus \{0\}$ and $g(x) \in C^2(\mathbb{R}_+)$ satisfies

$$\beta_1 x^{N+1} \leq g(x) \leq \beta_2 x^{N+1}, \quad x \geq 0,$$

where $\beta_1, \beta_2 > 0$ and $N > 0$. Proposition 6.6.1 and Corollary 6.6.1 are applied here. Therefore (6.61) generates a strictly order-preserving RDS (θ, φ) in $\mathbb{R}_+$ which has a random attractor $A(\omega) = [0, u(\omega)]$ in the universe $\mathcal{D}$ of all tempered subsets of $\mathbb{R}_+$. Using the comparison principle it is easy to see that

$$\varphi_{\alpha,\beta_2}(t,\omega,x) \leq \varphi(t,\omega,x) \leq \varphi_{\alpha,\beta_1}(t,\omega,x), \quad x \geq 0 , \tag{6.62}$$

where $(\theta, \varphi_{\alpha,\beta})$ is the RDS generated by (6.52). Therefore $A(\omega) = \{0\}$ if $\alpha < 0$ and $A(\omega) = [0, u(\omega)]$ when $\alpha > 0$, where the $\mathcal{F}_-$-measurable equilibrium $u(\omega)$ satisfies the inequality

$$0 < u_{\alpha,\beta_2,N}(\omega) \leq u(\omega) \leq u_{\alpha,\beta_1,N}(\omega), \quad \omega \in \Omega .$$

Here $u_{\alpha,\beta,N}(\omega)$ is given by (6.53). It follows from (6.57) that $u(\omega)$ attracts every trajectory $\varphi(t,\omega)x$ with $x > 0$ with respect to convergence in distribution. However using properties of the RDS generated by (6.52) and relation (6.62) we can prove that

$$\lim_{t\to\infty} \varphi(t, \theta_{-t}\omega, x) = u(\omega) \quad \text{for all} \quad x > 0 \qquad (6.63)$$

almost surely. Indeed, (6.54) and (6.62) imply that the omega-limit set $\Gamma_x(\omega)$ emanating from $x > 0$ (see Definition 1.6.1) possesses the property

$$\Gamma_x(\omega) \subseteq [u_{\alpha,\beta_2,N}(\omega), u_{\alpha,\beta_1,N}(\omega)], \quad \omega \in \Omega .$$

Thus by Lemma 3.4.1 and Remark 3.4.2(ii) $\underline{u}(\omega) := \inf \Gamma_x(\omega)$ and $\overline{u}(\omega) := \sup \Gamma_x(\omega)$ are $\mathcal{F}_-$-measurable equilibria such that

$$0 < u_{\alpha,\beta_2,N}(\omega) \leq \underline{u}(\omega) \leq \overline{u}(\omega) \leq u_{\alpha,\beta_1,N}(\omega), \quad \omega \in \Omega .$$

It is clear that $\underline{u}(\omega), \overline{u}(\omega) \in (0, u(\omega)]$. Hence, applying Theorem 6.6.1(ii) we obtain that $\underline{u}(\omega) = \overline{u}(\omega) = u(\omega)$ almost surely. Thus $\Gamma_x(\omega) = u(\omega)$ almost surely and (6.63) holds. As in ARNOLD [3, Theorem 9.3.3] it is also possible to prove the convergence property (6.63) for a random variable $x = x(\omega) > 0$ such that $x(\omega)$ and $x(\omega)^{-1}$ are tempered.

6.6.2 Stochastic Equations on a Bounded Interval

Now we consider an SDE of the form (6.44) inside a bounded deterministic interval $[l, r]$. We assume that

$$f(x) \in C^{1,\delta}([l, r]), \quad f(l) \geq 0, \quad f(r) \leq 0 , \qquad (6.64)$$

and

$$\sigma(x) \in C_b^{2,\delta}([l, r]), \quad \sigma(l) = \sigma(r) = 0, \quad |\sigma(x)| > 0 \text{ if } l < x < r . \qquad (6.65)$$

Here above $\delta \in (0, 1]$. Under these conditions by Proposition 6.2.3 equation (6.44) generates a strictly order-preserving RDS (θ, φ) in the interval $[l, r]$.

As above we introduce the speed measure $m(dx)$ on $[l, r]$ and the scale function $s : [l, r] \mapsto \mathbb{R} \cup \{\pm\infty\}$ by the formulas (see, e.g., KARATZAS/SHREVE [62])

$$m(A) = \int_A \exp\left\{2 \int_c^x \frac{f(\xi)}{\sigma^2(\xi)} d\xi\right\} \frac{dx}{|\sigma(x)|} , \quad A \in \mathcal{B}([l, r]) , \qquad (6.66)$$

and

$$s(x) = \int_c^x \exp\left\{-2 \int_c^y \frac{f(\xi)}{\sigma^2(\xi)} d\xi\right\} \frac{dy}{|\sigma(y)|} , \quad x \in [l, r] ,$$

where c is a fixed point from (l, r).

Similar to Theorem 6.6.1 we can prove the following result.

Theorem 6.6.2. *Assume that (6.64) and (6.65) hold. Let $A(\omega)$ be the random attractor for the RDS (θ, φ) generated by (6.44) in the interval $[l, r]$.*
(i) If $f(l) = f(r) = 0$, then $A(\omega) = [l, r]$.
(ii) If $f(l) = 0$ and $f(r) < 0$, then

(a) $A(\omega) = \{l\}$ *almost surely provided that* $m((l, c]) = \infty$;

(b) *the property* $m((l, c]) < \infty$ *implies that* $A(\omega) = [l, u(\omega)]$ *for some* $\mathcal{F}_-$-*measurable equilibrium* $u(\omega)$ *such that* $l < u(\omega) < r$ *almost surely. There are no other* $\mathcal{F}_-$-*measurable equilibria in the set* $(l, u(\omega)]$ *and*

$$\lim_{t \to \infty} \mathbb{P}\{\omega \ : \ \varphi(t, \omega)x \in B\} = \mathbb{P}\{\omega \ : u(\omega) \in B\} \equiv \frac{m(B)}{m([l, r])} \tag{6.67}$$

for all $x \in (l, r]$ *and* $B \in \mathcal{B}([l, r])$.

(iii) If $f(l) > 0$ *and* $f(r) = 0$, *then*

(a) $A(\omega) = \{r\}$ *almost surely provided that* $m([c, r)) = \infty$;

(b) $A(\omega) = [u(\omega), r]$ *provided that* $m([c, r)) < \infty$, *where* $u(\omega)$ *is an* $\mathcal{F}_-$-*measurable equilibrium* $u(\omega)$ *such that* $l < u(\omega) < r$ *almost surely. There are no other* $\mathcal{F}_-$-*measurable equilibria in the set* $[u(\omega), r)$ *and (6.67) holds for all* $x \in [l, r)$ *and* $B \in \mathcal{B}([l, r])$.

(iv) If $f(l) > 0$ *and* $f(r) < 0$, *then there exists an* $\mathcal{F}_-$-*measurable equilibrium* $u(\omega)$ *such that* $l < u(\omega) < r$ *and* $A(\omega) = \{u(\omega)\}$ *almost surely. Moreover (6.67) holds for all* $x \in [l, r]$.

Proof. Assertion (i) follows from the property $\varphi(t, \omega)[l, r] = [l, r]$ for all $t > 0$ and $\omega \in \Omega$. To prove the other assertions of the theorem we note that $f(l) > 0$ (resp. $f(r) < 0$) implies that $m((l, c]) < \infty$ (resp. $m([c, r)) < \infty$). Therefore we can apply the same argument as in the proof of Theorem 6.6.1. □

Now we consider the case $f(l) = f(r) = 0$ in details. We are interested in the description of the long-time behaviour of trajectories inside the attractor. The following assertion is an almost direct consequence of the well-known theorems on the boundary behaviour of one-dimensional diffusion processes (see, e.g., IKEDA/WATANABE [57] or KARATZAS/SHREVE [62]).

Theorem 6.6.3. *Assume that (6.64) and (6.65) hold. Let* $f(l) = f(r) = 0$. *Denote by* (θ, φ) *the RDS generated by (6.44) in the interval* $[l, r]$.

(i) *If* $m([l, r]) < \infty$, *then there exists an* $\mathcal{F}_-$-*measurable equilibrium* $u(\omega)$ *such that* $l < u(\omega) < r$ *almost surely and (6.67) holds for all* $x \in (l, r)$ *and* $B \in \mathcal{B}([l, r])$. *Moreover*

$$\mathbb{P}\left\{\omega \ : \ \liminf_{t \to \infty} \varphi(t, \omega)x = l\right\} = \mathbb{P}\left\{\omega \ : \ \limsup_{t \to \infty} \varphi(t, \omega)x = r\right\} = 1 \tag{6.68}$$

for any $x \in (l, r)$ *and the process* $\varphi(t, \omega)x$ *is recurrent, i.e. for any* $y \in (l, r)$ *we have*

$$\mathbb{P}\{\omega \ : \ \varphi(t, \omega)x = y \ \text{for some} \ t \in \mathbb{R}_+\} = 1. \tag{6.69}$$

(ii) *If $m((l,c]) = \infty$ and $m([c,r]) < \infty$, then*

$$\lim_{t \to \infty} \mathbb{P}\{\omega : \varphi(t,\omega)x \geq l + \varepsilon\} = 0, \quad x \in [l,r),$$

for any $\varepsilon > 0$ and

$$\mathbb{P}\left\{\omega : \lim_{t \to \infty} \varphi(t,\omega)x = l\right\} = \mathbb{P}\left\{\omega : \sup_{t \in \mathbb{R}_+} \varphi(t,\omega)x < r\right\} = 1 \quad (6.70)$$

for any $x \in [l,r)$ provided that $s(l) > -\infty$.
(iii) *If $m((l,c]) < \infty$ and $m([c,r]) = \infty$, then*

$$\lim_{t \to \infty} \mathbb{P}\{\omega : \varphi(t,\omega)x \leq r - \varepsilon\} = 0, \quad x \in (l,r],$$

for any $\varepsilon > 0$ and

$$\mathbb{P}\left\{\omega : \lim_{t \to \infty} \varphi(t,\omega)x = r\right\} = \mathbb{P}\left\{\omega : \inf_{t \in \mathbb{R}_+} \varphi(t,\omega)x > l\right\} = 1 \quad (6.71)$$

for any $x \in (l,r]$ provided that $s(r) < \infty$.
(iv) *Let $m((l,c]) = \infty$ and $m([c,r]) = \infty$. Then*

$$\lim_{t \to \infty} \mathbb{P}\{\omega : \varphi(t,\omega)x \in [l + \varepsilon, r - \varepsilon]\} = 0, \quad x \in (l,r), \quad (6.72)$$

for any $\varepsilon > 0$. Moreover
(a) if $s(l) > -\infty$ and $s(r) < \infty$, then

$$\mathbb{P}\left\{\omega : \lim_{t \to \infty} \varphi(t,\omega)x = l\right\} = 1 - \mathbb{P}\left\{\omega : \lim_{t \to \infty} \varphi(t,\omega)x = r\right\}$$
$$= \frac{s(r) - s(x)}{s(r) - s(l)}, \quad x \in [l,r]; \quad (6.73)$$

(b) if $s(l) > -\infty$ and $s(r) = \infty$, then (6.70) hold;
(c) if $s(l) = -\infty$ and $s(r) < \infty$, then (6.71) hold;
(d) if $s(l) = -\infty$ and $s(r) = \infty$, then (6.68) and (6.69) hold.

Proof. (i) As in the proof of Theorem 6.6.1(ii) it is easy to see that $\varrho(x) := \frac{m((l,x])}{m([l,r])}$ solves the stationary Fokker–Plank equation on (l,r), $\int_l^r \varrho(x)dx = 1$, and $\varrho(B) = \int_B \varrho(x)dx$, $B \in \mathcal{B}((l,r))$, is a unique ergodic stationary measure. Therefore using Theorem 2.3.45 ARNOLD [3] we can prove that the limit

$$\int_l^r h(x)\mu_\omega(dx) := \lim_{t \to \infty} \int_l^r h(\varphi(t,\theta_{-t}\omega)x)\varrho(x)dx$$

exists almost surely for all $h \in C_b([l,r])$. Moreover μ_ω is a disintegration of a Markov invariant mesure and $\int_\Omega \mu_\omega(B)\mathbb{P}(d\omega) = \varrho(B)$, $B \in \mathcal{B}(\mathbb{R}_+)$. By Remarks 1.10.1 and 2.4.1 there exists a version of μ_ω such that

$$\int_l^r h(x)\mu_{\theta_t\omega}(dx) = \int_l^r h(\varphi(t,\omega)x)\mu_{\theta_t\omega}(dx) \quad \text{for all} \quad \omega \in \Omega .$$

Therefore by Proposition 3.5.1 there exists an $\mathcal{F}_-$-measurable equilibrium $u(\omega)$ such that $\mu_\omega = \delta_{u(\omega)}$. Since $\varrho((l,r)) = 1$ and

$$\mathbb{P}\{\omega \ : \ l < \alpha \le u(\omega) \le \beta < r\} = \int_\Omega \mu_\omega([\alpha,\beta])\mathbb{P}(d\omega) = \varrho([\alpha,\beta]) ,$$

we have that $l < u(\omega) < r$ almost surely and (6.67) holds for $x \in (l,r)$. Since the property $m([l,r]) < \infty$ implies that $s(l) = -\infty$ and $s(r) = \infty$ (see, e.g., CRAUEL ET AL. [38, Lemma 2.4]), we can apply Proposition 5.5.22 KARATZAS/SHREVE [62] (see also IKEDA/WATANABE [57, Theorem 6.3.1]) to obtain (6.68) and (6.69).

To prove (ii) and (iii) we can repeat with a slight modification the argument given in the proof of Theorem 6.6.1(i) and apply Proposition 5.5.22 KARATZAS/SHREVE [62].

(iv) To prove (6.72) we consider the process

$$y(t,\omega;z) := G\left(\varphi(t,\omega)\left[G^{-1}(z)\right]\right), \quad z \in \mathbb{R} ,$$

where $G(x)$ is a primitive for $[\sigma(x)]^{-1}$ on the interval (l,r). This process solves the SDE

$$dy = \frac{f(G^{-1}(y))}{\sigma(G^{-1}(y))} \cdot dt + dW_t, \quad y(0) = z . \tag{6.74}$$

It follows from FRIEDMAN [45, Chaps.5 and 15] that $\{y(t,\omega;z) \ : \ z \in \mathbb{R}\}$ is a Feller process with transition probability

$$\tilde{P}_t(z,B) := \mathbb{P}\left\{\omega \ : \ y(t,\omega;z) \in B\right\} = \int_B p(t,z,y)dy ,$$

where $p(t,z,y)$ is a continuous strictly positive function on $(\mathbb{R}_+\setminus\{0\})\times\mathbb{R}\times\mathbb{R}$. It is also easy to see that the speed measure $\tilde{m}$ for equation (6.74) possesses the property $\tilde{m}(\mathbb{R}) = \infty$. Therefore it follows from KUNITA [74, Theorem 1.3.10] that

$$\lim_{t\to\infty} \tilde{P}_t(z,[a,b]) = 0 \quad \text{for any} \quad -\infty < a < b < \infty .$$

Now using the relation

$$\mathbb{P}\left\{\omega \ : \ \varphi(t,\omega)x \in C\right\} = \mathbb{P}\left\{\omega \ : \ y(t,\omega;G(x)) \in G(C)\right\} = \tilde{P}_t(G(x),G(C)) ,$$

where $C \in \mathcal{B}(l,r)$, we obtain (6.72).

Assertions (iv-a)–(iv-d) are direct consequences of Proposition 5.5.22 KARATZAS/SHREVE [62].   □

Theorem 6.6.3 implies the following result on a random attractor with respect to the convergence in probability (see OCHS [86] for the theory of attractors based on this type of convergence).

Corollary 6.6.2. *Assume that (6.64) and (6.65) hold. Let $f(l) = f(r) = 0$ and $m((l, r)) = \infty$. Then the two-point set $A := \{l, r\}$ is a weak point random attractor for the RDS (θ, φ) generated by (6.44) in the interval $[l, r]$. This means that (i) $\varphi(t, \omega)A = A$ for all $t \geq 0$ and $\omega \in \Omega$; (ii) $\varphi(t, \omega)x$ converges to A in probability for every $x \in [l, r]$, i.e.*

$$\lim_{t \to \infty} \mathbb{P}\{\omega : \operatorname{dist}(\varphi(t, \omega)x, A) \geq \varepsilon\} = 0, \quad x \in [l, r],$$

for any $\varepsilon > 0$; (iii) A is a minimal set possessing the properties (i) and (ii).

Proof. It follows from Theorem 6.6.3(ii-iv) and the relation

$$\mathbb{P}\{\omega : \varphi(t, \omega)x \in [l + \varepsilon, r - \varepsilon]\} = \mathbb{P}\{\omega : \operatorname{dist}(\varphi(t, \omega)x, A) \geq \varepsilon\}.$$

$\square$

Example 6.6.3. Consider a Stratonovich SDE of the form

$$dx = \alpha\sigma(x)dt + \sigma(x) \circ dW_t, \tag{6.75}$$

where $\alpha \in \mathbb{R}$ is a parameter and

$$\sigma(x) \in C_b^{2,\delta}(\mathbb{R}), \quad \sigma(l) = \sigma(r) = 0, \quad \sigma(x) > 0 \text{ if } x \in (l, r), \tag{6.76}$$

for some bounded interval $[l, r] \subset \mathbb{R}$ and $\delta \in (0, 1]$. In this case the speed measure m has the form

$$m([a, b]) = \int_a^b \frac{dx}{\sigma(x)}, \quad [a, b] \subset [l, r], \text{ if } \alpha = 0,$$

and

$$m([a, b]) = \frac{1}{2\alpha} \cdot \left[\exp\left\{2\alpha \int_c^b \frac{d\xi}{\sigma(\xi)}\right\} - \exp\left\{2\alpha \int_c^a \frac{d\xi}{\sigma(\xi)}\right\}\right], \text{ if } \alpha \neq 0,$$

where c is a fixed point from (l, r). For the scale function s we have the representation

$$s(x) = \int_c^x \frac{d\xi}{\sigma(\xi)}, x \in [l, r], \text{ if } \alpha = 0,$$

and

$$s(x) = \frac{1}{2\alpha} \cdot \left[1 - \exp\left\{-2\alpha \int_c^x \frac{d\xi}{\sigma(\xi)}\right\}\right], \text{ if } \alpha \neq 0.$$

Therefore the application of Theorem 6.6.3 gives the following result.

(i)   If $\alpha < 0$, then (6.70) holds.
(ii) If $\alpha = 0$, then we have (6.68), (6.69) and (6.72).
(iii) If $\alpha > 0$, then (6.71) holds.

From Corollary 6.6.2 we also have that $\{l, r\}$ is a weak point random attractor for all $\alpha \in \mathbb{R}$. On the other hand, relying on the representation

$$\varphi(t, \omega)x = G^{-1}(G(x) + \alpha t + W_t(\omega)), \quad x \in (l, r), \tag{6.77}$$

where $G(x)$ is a primitive for $[\sigma(x)]^{-1}$ on the interval (l, r), and using the law of the iterated logarithm for the one-dimensional Wiener process (see, e.g., FRIEDMAN [45, p.40]) it is easy to prove that in the case $\alpha = 0$ the interval $[l, r]$ is the pull back omega-limit set for the trajectory emanating from any point $x \in (l, r)$.

If in addition we assume that $\sigma'(l) > 0$ and $\sigma'(r) < 0$, then relying on (6.77) after a simple calculation (see CHUESHOV/VUILLERMOT [25]) we find that

$$\lim_{t \to \infty} \frac{1}{t} \log |\varphi(t, \omega)x - l| = \alpha\sigma'(l), \quad \mathbb{P} - \text{a.s.}, \quad x \in (l, r), \quad \alpha < 0,$$

and

$$\lim_{t \to \infty} \frac{1}{t} \log |\varphi(t, \omega)x - r| = \alpha\sigma'(r), \quad \mathbb{P} - \text{a.s.}, \quad x \in (l, r), \quad \alpha > 0.$$

To interpret these results we recall (see, e.g., KHASMINSKII [64]) that (a) an equilibrium u (either l or r) is said to be stable in probability if the relation

$$\lim_{x \to u} \mathbb{P}\left\{\omega \in \Omega : \sup_{t>0} |\varphi(t, \omega)x - u| > \varepsilon\right\} = 0 \tag{6.78}$$

holds for every $\varepsilon > 0$; (b) u is globally asymptotically stable in probability if relation (6.78) holds and if we have

$$\mathbb{P}\left\{\omega \in \Omega : \lim_{t \to \infty} |\varphi(t, \omega)x - u| = 0\right\} = 1$$

for every $x \in (l, r)$; (c) u is unstable in probability if (6.78) does not hold.

Thus in this example, on the one hand, we have the weak point attractor $A = \{l, r\}$ which does not depend on α, on the other hand, we observe the following bifurcation picture (with respect to the parameter α):

(i)   if $\alpha < 0$, l is globally asymptotically stable, whereas r is unstable in probability;
(ii) if $\alpha = 0$, both l and r are unstable in probability;
(iii) if $\alpha > 0$, r is globally asymptotically stable, whereas l is unstable in probability.

We note that a similar character of the exchange of stability between the equilibria l and r can be seen in the random case (see Example 5.6.2) and also in the case of equation (6.44) with σ satisfying (6.76) and with $f \in C^{1,\delta}([l,r])$ possessing the property

$$\beta_1 \sigma(x) \leq f(x) \leq \beta_2 \sigma(x), \quad x \in [l,r] \, ,$$

where β_1 and β_2 are positive constants.

Example 6.6.4. Consider now an Itô SDE on the interval (l,r) of the same form as (6.75):

$$dx = \alpha\sigma(x)dt + \sigma(x)dW_t \, , \tag{6.79}$$

where $\alpha \in \mathbb{R}$ is a parameter and $\sigma(x)$ satisfies (6.76). Assume that $\sigma'(l) > 0$ and $\sigma'(r) < 0$ hold. This equation can be written as a Stratonovich SDE of the form

$$dx = \left(\alpha - \frac{1}{2}\sigma'(x)\right)\sigma(x)dt + \sigma(x) \circ dW_t \, ,$$

The speed measure m and the scale function s for this equation are represented by the formulas

$$m(A) = \sigma(c)\int_A \exp\left\{2\alpha\int_c^x \frac{d\xi}{\sigma(\xi)}\right\}\frac{dx}{\sigma^2(x)}, \quad A \in \mathcal{B}([l,r]) \, ,$$

and

$$s(x) = \frac{1}{\sigma(c)}\int_c^x \exp\left\{-2\alpha\int_c^y \frac{d\xi}{\sigma(\xi)}\right\}dy \, , \quad x \in [l,r] \, ,$$

where c is a fixed point from (l,r). It is easy to see that $m((l,r)) = \infty$ for every $\alpha \in \mathbb{R}$. Therefore, as in Example 6.6.3, Corollary 6.6.2 implies that the two-point set $\{l,r\}$ is a weak point random attractor for the RDS in $[l,r]$ generated by (6.79) for every $\alpha \in \mathbb{R}$. Applying Theorem 6.6.3 we obtain the following more precise information on the dynamics:

(i)   if $\alpha \leq \frac{1}{2}\sigma'(r)$, then (6.70) holds;
(ii) if $\frac{1}{2}\sigma'(r) < \alpha < \frac{1}{2}\sigma'(l)$, then we have (6.73);
(iii) if $\alpha \geq \frac{1}{2}\sigma'(l)$, then (6.71) holds.

Thus we observe that in contrast with the RDS generated by (6.75) the long-time behaviour of $\varphi(t,\omega)x$ is characterized by the absence of any recurrent and oscillatory behavior for *all* values of α. Moreover, it is also possible to prove (see CHUESHOV/VUILLERMOT [26]) that the bifurcation picture in this example is following:

(i)   if $\alpha \leq \frac{1}{2}\sigma'(r)$, l is globally asymptotically stable, whereas r is unstable in probability;

(ii) if $\frac{1}{2}\sigma'(r) < \alpha < \frac{1}{2}\sigma'(l)$, both l and r are stable in probability;

(iii) If $\alpha \geq \frac{1}{2}\sigma'(l)$, r is globally asymptotically stable, whereas l is unstable in probability.

Thus the exchange of stability between the equilibria l and r, when the parameter α varies from minus infinity to plus infinity, is slower (softer) in this example in contrast with the picture that we can see in Example 6.6.3. However in both examples the weak point random attractor $A = \{l, r\}$ is the same for all $\alpha \in \mathbb{R}$.

The following example shows that the idea outlined in Sect.5.6 can be also used in the stochastic case.

Example 6.6.5. Consider the Stratonovich SDE

$$
\begin{aligned}
dx &= (\alpha_1 \sin x + \alpha_2(1 - \cos x)) \ dt \\
&\quad + \sigma_1 \sin x \circ dW_t^1 + \sigma_2(1 - \cos x) \circ dW_t^2 ,
\end{aligned}
\tag{6.80}
$$

where α_i and σ_i are parameters and $W_t = (W_t^1, W_t^2)$ is a Wiener process in $\mathbb{R}^2$. This equation generates an RDS in $\mathbb{R}$. However, as in Example 5.6.3, it is natural to consider equation (6.80) on the unit circle C which is interpreted as the interval $[0, 2\pi]$ with identified end-points. Using Itô's formula for the Stratonovich integrals (see Theorem 2.3.1) it is easy to see that (6.80) generates an RDS (θ, φ) in C with the cocycle

$$
\varphi(t, \omega)x = 2\operatorname{arccot}\left(-\psi(t, \omega)[-\cot(x/2)]\right), \quad 0 < x < 2\pi ,
$$

where $\psi(t, \omega)$ is the cocycle in $\mathbb{R}$ generated by the affine Stratonovich equation

$$
dy = (-\alpha_1 y + \alpha_2) \ dt - \sigma_1 y \circ dW_t^1 + \sigma_2 \circ dW_t^2 .
$$

In the case $\alpha_2 = \sigma_1 = 0$ we obtain the SDE

$$
dy = -\alpha_1 y \ dt + \sigma_2 dW_t^2 .
$$

Therefore the equation

$$
dx = \alpha_1 \sin x \ dt + \sigma_2(1 - \cos x) \circ dW_t^2
$$

generates an RDS $(\theta, \tilde{\varphi})$ with the cocycle

$$
\tilde{\varphi}(t, \omega)x = 2\operatorname{arccot}\left\{ e^{-\alpha_1 t} \cdot \cot \frac{x}{2} - \sigma_2 \int_0^t e^{-\alpha_1(t-s)} \ dW_s^2 \right\}, \quad 0 < x < 2\pi .
$$

This formula implies that for any $\alpha_1 \neq 0$ the RDS $(\theta, \tilde{\varphi})$ has two equilibria $u_0 \equiv 0$ and $u(\omega) \in (0, 2\pi)$ which stability properties can be easily derived from the results presented in Example 2.4.4. The case $\alpha_1 = 0$ is covered by Example 6.6.3 with $\alpha = 0$.

Remark 6.6.3. We also note that it is possible to give examples which show that all the cases listed in Theorem 6.6.3 are really occur. We refer to SCHEUTZOW [91], where the corresponding examples are presented in the case $[l, r) = \mathbb{R}_+$.

6.7 Stochastic Equations with Concavity Properties

We start with conditions that ensure that the RDS generated by the system of stochastic cooperative differential equations (6.1) is sublinear.

Lemma 6.7.1. *Assume that conditions (S1), (S3) and (S4) hold. Let $\sigma_{ij}(x)$ be linear functions, i.e.*

$$\sigma_{ij}(x) = \sigma_{ij} \cdot x \quad \text{for all} \quad x \in \mathbb{R}_+ , \tag{6.81}$$

where σ_{ij} are constants. If the function $f(\cdot)$ is a sublinear mapping from $\mathbb{R}_+^d$ into $\mathbb{R}^d$, i.e. if

$$\lambda f(x) \leq f(\lambda x) \quad \text{for all} \quad 0 < \lambda < 1 \quad \text{and} \quad x \in \mathbb{R}_+^d , \tag{6.82}$$

then the RDS (θ, φ) generated by (6.1) is sublinear. Moreover (θ, φ) is strongly sublinear if one of the following conditions is satisfied:

(i) *f is a strongly sublinear mapping, i.e. $\lambda f(x) \ll f(\lambda x)$ for all $0 < \lambda < 1$ and $x \in \text{int}\mathbb{R}_+^d$;*

(ii) *the matrix $\left\{ \frac{\partial f_i}{\partial x_j} \right\}$ is irreducible for all $x \in \text{int}\mathbb{R}_+^d$ and $\lambda f(x) < f(\lambda x)$ for all $0 < \lambda < 1$ and $x \in \text{int}\mathbb{R}_+^d$.*

Proof. The function $x^\lambda(t) = \lambda \cdot \varphi(t, \omega)x$ is the solution to the problem

$$dx_i^\lambda(t) = f_i^\lambda(x^\lambda(t))dt + \sum_{j=1}^{m} \sigma_{ij} \cdot x_i^\lambda(t) \circ dW_t^j, \quad i = 1, \ldots, d ,$$

with initial data $x^\lambda(0) = \lambda \cdot x$, where $f^\lambda(x) = \lambda f(\lambda^{-1}x)$. From (6.82) we have $f^\lambda(x) \leq f(x)$. Therefore the stochastic comparison principle (see Theorem 6.4.1(i)) gives

$$\lambda \cdot \varphi(t, \omega)x \equiv x^\lambda(t) \leq x(t) \equiv \varphi(t, \omega)[\lambda x] .$$

To prove that (θ, φ) is strongly sublinear under the conditions either (i) or (ii) we note that by Theorem 6.3.1 (θ, φ) is equivalent to the RDS (θ, ψ) generated by (6.5) with g given by (6.6). Therefore we can use Lemma 5.5.1. $\square$

We note that Lemma 6.7.1 remains true if we understand (6.1) in the Itô sense (cf. Remark 6.4.1). It is also easy to see that the RDS generated by equation (6.52) gives an example of a strongly sublinear RDS.

Now we apply the results presented Chap.4 for sublinear systems. The application of Corollary 4.3.1 gives the following assertion.

Theorem 6.7.1. *Assume that conditions (S1), (S3), (S4) and (6.81) hold and $f(0) \gg 0$. Let the assumption either (i) or (ii) of Lemma 6.7.1 be valid. Then either*

(i) for any $x \in \mathbb{R}^d_+$ we have $|\varphi(t, \theta_{-t}\omega)x| \to \infty$ almost surely as $t \to \infty$

or

(ii) there exists a unique almost equilibrium $u(\omega) \gg 0$ defined on a θ-invariant set Ω^* of full measure such that

$$\lim_{t \to +\infty} \varphi(t, \theta_{-t}\omega)v(\theta_{-t}\omega) = u(\omega), \quad \omega \in \Omega^* ,$$

for any random variable $v(\omega)$ possessing the property $0 \le v(\omega) \le \lambda u(\omega)$ for all $\omega \in \Omega^*$ and for some nonrandom $\lambda > 0$.

Proof. The property $f(0) \gg 0$ implies that the function $g(\omega, x)$ given by formula (6.6) satisfies (5.36). Therefore it follows from Proposition 5.4.1 and Theorem 6.3.1 that $\varphi(t, \omega)0 \gg 0$ for all $t > 0$ and $\omega \in \Omega$. Thus (θ, φ) is strongly positive. It is also clear that any finite-dimensional RDS is conditionally compact. Therefore we can apply Corollary 4.3.1.   □

Now we prove a stochastic version of the trichtomy theorem (for the random case see Theorem 5.5.2).

Theorem 6.7.2 (Limit Set Trichotomy). *Let conditions (S1), (S3), (S4), (6.81) and either (i) or (ii) of Lemma 6.7.1 hold. Assume that the coefficients $\sigma_{ij} \equiv \sigma_j$ independent of i and there exist positive constants a and b such that*

$$-a \cdot x_j \le f_j(x) \le b \cdot (1 + |x|_1) \quad \text{for all} \quad x \in \mathbb{R}^d_+, \; j = 1, \dots, d , \quad (6.83)$$

where $|\cdot|_1$ is l^1-norm in $\in \mathbb{R}^d$, i.e. $|x|_1 = \sum_{j=1}^d |x_i|$ for $x \in \mathbb{R}^d$. Let $\mathbf{e} = (1, \dots, 1) \in \mathbb{R}^d_+$ and

$$\eta(\omega) = \int_{-\infty}^{\infty} \exp\left\{-\nu|\tau| - W_\tau^{(\sigma)}(\omega)\right\} d\tau , \quad (6.84)$$

where ν is a positive constant and $W_t^{(\sigma)} = \sum_{j=1}^m \sigma_j W_t^j$. Denote by $C_\eta = C_\eta(\omega)$ the collection of random variables $w : \Omega \to \mathbb{R}^d_+$ possessing the property

$$\alpha^{-1}\eta(\omega)\mathbf{e} \le w(\omega) \le \alpha\eta(\omega)\mathbf{e} \quad \text{for all} \quad \omega \in \Omega$$

for some nonrandom number $\alpha \ge 1$. Let (θ, φ) be the RDS generated by the equation

$$dx_i(t) = f_i(x_1(t), \dots, x_d(t))dt + x_i(t) \circ dW_t^{(\sigma)}, \quad i = 1, \dots, d .$$

Then any orbit of (θ, φ) emanating from $a \in C_\eta$ does not leave C_η, i.e.

$$\varphi(t, \omega)a(\omega) \in C_\eta(\theta_t\omega) \quad \text{for all} \quad a \in C_\eta(\omega), \; t \ge 0 , \quad (6.85)$$

and precisely one of the following three cases applies:

(i) for all $b \in C_\eta(\omega)$, the orbit γ_b emanating from b is unbounded;

(ii) for all $b \in C_\eta(\omega)$, the orbit γ_b emanating from b is bounded, but the closure of γ_b does not belong to $C_\eta(\omega)$ and

$$\limsup_{t \to \infty} \left\{ \sup_{\omega \in \Omega} p(\varphi(t, \theta_{-t}\omega)b(\theta_{-t}\omega), \eta(\omega)) \right\} = \infty \,,$$

where p is the part metric in $\mathbb{R}_+^d$;

(iii) there exists a unique $\mathcal{F}$-measurable almost equilibrium $u \in C_\eta(\omega)$, and for all $b \in C_\eta(\omega)$ the orbit emanating from b converges to u, i.e.

$$\lim_{t \to +\infty} \varphi(t, \theta_{-t}\omega)b(\theta_{-t}\omega) = u(\omega) \quad \text{for almost all} \quad \omega \in \Omega \,. \tag{6.86}$$

Proof. As in the proof of Theorem 5.5.2 it is sufficiently to check the invariance property of $C_\eta(\omega)$. It follows from (6.83) that

$$-a \cdot x \le f(x) \le b \cdot (1 + |x|_1) \cdot \mathbf{e} \quad \text{for all} \quad x \in \mathbb{R}_+^d \,.$$

Therefore Theorem 6.4.1 implies that

$$y^{(1)}(t) \le \varphi(t, \omega)\eta(\omega)\mathbf{e} \le y^{(2)}(t) \,, \tag{6.87}$$

where $y^{(1)}(t)$ and $y^{(2)}(t)$ are solutions to the problems

$$dy^{(1)}(t) = -a \cdot y^{(1)}(t) \cdot dt + y^{(1)}(t) \circ dW_t^{(\sigma)}, \quad y^{(1)}(0) = \eta(\omega)\mathbf{e} \,, \tag{6.88}$$

and

$$dy^{(2)}(t) = b \cdot (1 + |y^{(2)}(t)|_1) \cdot \mathbf{e} \cdot dt + y^{(2)}(t) \circ dW_t^{(\sigma)}, \quad y^{(2)}(0) = \eta(\omega)\mathbf{e} \,. \tag{6.89}$$

Using (6.88) it is easy to find that

$$y_i^{(1)}(t) = \eta(\omega) \cdot \exp\left\{ -at + W_t^{(\sigma)} \right\}, \quad i = 1, \dots, d \,. \tag{6.90}$$

From (6.89) we have that

$$d|y^{(2)}(t)|_1 = bd \cdot (1 + |y^{(2)}(t)|_1) \cdot dt + |y^{(2)}(t)|_1 \circ dW_t^{(\sigma)} \,. \tag{6.91}$$

Therefore

$$\begin{aligned}
y_i^{(2)}(t) \le |y^{(2)}(t)|_1 &= d \cdot \eta(\omega) \cdot \exp\left\{ bd \cdot t + W_t^{(\sigma)} \right\} \\
&+ bd \cdot \int_0^t \exp\left\{ bd \cdot (t - \tau) + W_t^{(\sigma)} - W_\tau^{(\sigma)} \right\} d\tau \,.
\end{aligned} \tag{6.92}$$

Using the equality

$$W_t^{(\sigma)}(\omega) - W_\tau^{(\sigma)}(\omega) = -W_{\tau-t}^{(\sigma)}(\theta_t\omega)$$

and relations (6.84) and (6.87), it is easy to see that

$$e^{-(a+\nu)t}\eta(\theta_t\omega) \cdot \mathbf{e} \le \varphi(t,\omega)\,[\eta(\omega) \cdot \mathbf{e}] \le d(1+b)e^{(\nu+bd)t}\eta(\theta_t\omega) \cdot \mathbf{e}\,.$$

This implies the invariance of $C_\eta(\omega)$. Thus we can apply Theorem 4.4.1 and Corollary 4.4.1. □

In order to show that all three cases of the limit set trichotomy can actually occur we consider the following example.

Example 6.7.1. Consider the Stratonovich stochastic differential equation on $X = \mathbb{R}_+$ given by

$$dx = f(x)dt + \sigma x \circ dW_t\,, \quad \sigma \in \mathbb{R}\,, \tag{6.93}$$

where

$$f(x) = \alpha x + \frac{x}{1+x}\,, \quad \alpha \in \mathbb{R}\,,$$

which is strongly sublinear for any $\alpha \in \mathbb{R}$, hence by Lemma 6.7.1 the RDS (θ, φ) generated by (6.93) is strongly sublinear for any $\alpha, \sigma \in \mathbb{R}$. Theorem 6.7.2 is applicable here for any α and σ.

The point $x = 0$ is always an equilibrium.

Consider first the case $\alpha > 0$. Since $f(x) \ge \alpha x$, the comparison principle yields

$$\varphi(t,\omega)b(\omega) \ge b(\omega)e^{\alpha t + \sigma W_t(\omega)}\,,$$

hence

$$\varphi(t, \theta_{-t}\omega)b(\theta_{-t}\omega) \ge b(\theta_{-t}\omega)e^{\alpha t - \sigma W_{-t}(\omega)}\,.$$

Consequently, for any initial random variable b such that $1/b(\omega)$ is tempered, the orbit γ_b of φ emanating from b is unbounded, in fact converges to infinity with probability one.

Now consider the case $-1 < \alpha < 0$. Then $f(x) < 1 + \alpha x$ and Corollary 6.6.1 implies that (θ, φ) possesses a random attractor $A(\omega) = [0, u(\omega)]$, where $u(\omega) \ge 0$ is an $\mathcal{F}_-$-measurable tempered equilibrium. A simple calculation shows that the speed measure m given by (6.55) satisfies $m((0, 1]) < \infty$ if $\alpha > -1$. Therefore by Theorem 6.6.1 $u(\omega) > 0$ almost surely and it follows from Theorem 4.2.2 that $u(\omega)$ is attractive in the part C_u.

Here we can also clearly see why the initial values have to be restricted somehow, e.g. to those in C_u. Take, for example, a random variable $a(\omega) > 0$

which is not tempered, i.e. for which $\limsup_{t\to\infty} \frac{1}{t} \log a(\theta_{-t}\omega) = +\infty$. Non-tempered random variables exist on any standard probability space and for any ergodic and aperiodic θ ARNOLD/CONG/OSELEDETS [9, Lemma 8.6], which includes to the present situation. Then $a \notin C_u$ and since

$$\varphi(t, \theta_{-t}\omega)a(\theta_{-t}\omega) \geq a(\theta_{-t}\omega)e^{\alpha t - \sigma W_{-t}(\omega)} ,$$

there exists $\alpha < 0$ such that the right-hand side tends to $+\infty$ along some sequence $t_n \to \infty$. Hence the orbit emanating from a does not converge to u.

Finally, if $\alpha < -1$ then φ is dominated by the linear cocycle generated by $dx = (\alpha + 1)xdt + \sigma x \circ dW_t$, hence

$$\varphi(t, \theta_{-t}\omega)b(\theta_{-t}\omega) \leq b(\theta_{-t}\omega)e^{(\alpha+1)t - \sigma W_{-t}(\omega)} .$$

Thus whenever b is tempered

$$\lim_{t\to\infty} \varphi(t, \theta_{-t}\omega)b(\theta_{-t}\omega) = 0$$

almost surely. If b is even ε-slowly varying, i.e. if $b(\theta_t\omega) \leq b(\omega)e^{\varepsilon|t|}$ for some $\varepsilon > 0$ such that $\varepsilon + (\alpha + 1) < 0$, then the orbit γ_b is bounded, but the closure of γ_b contains elements (namely 0) which do not belong to any part $C_v \subset \text{int}\mathbb{R}_+$.

We can combine several of the equations (6.93) to produce more complicated limit behaviour.

The following assertion gives conditions under which the RDS generated by (6.1) is s-concave (see Definition 4.1.3).

Proposition 6.7.1. *Assume that assumptions (S1), (S3), (S4) and (6.81) are met and that the matrix $D_x f(x)$ is irreducible in $\text{int}\mathbb{R}_+^d$. Let the function $f(x)$ be s-concave, i.e.*

$$D_x f(x) < D_x f(y) \quad for \quad x \gg y \gg 0 . \tag{6.94}$$

Then (θ, φ) is s-concave and strongly order-preserving. In particular we have

$$\varphi(t, \omega)(\mathbb{R}_+^d \setminus \{0\}) \subset \text{int}\mathbb{R}_+^d \quad for\ all \quad \omega \in \Omega . \tag{6.95}$$

Proof. This follows from Theorem 6.3.1 and Proposition 5.5.2. $\square$

As in the random case (see Theorem 5.5.3) we can also prove the following assertion.

Theorem 6.7.3. *Let (S1), (S3), (S4) and (6.81) hold. Assume that the fuction $f(x)$ is s-concave and that the matrix $D_x f(x)$ is irreducible for all $x \in \text{int}\mathbb{R}_+^d$. If $f(0) > 0$, then either (a) for all $v(\omega) \geq 0$, the orbit γ_v emanating from v is unbounded; or (b) there exists a unique equilibrium $u \gg 0$ such for every $v(\omega)$ possessing the property $0 \leq v(\omega) \leq \alpha \cdot u(\omega)$ with some*

$\alpha > 0$ *the orbit emanating from* v *converges to* u *on a* θ-*invariant set* Ω^* *of full measure, i.e.*

$$\lim_{t\to+\infty} \varphi(t,\theta_{-t}\omega)v(\theta_{-t}\omega) = u(\omega) \quad \text{for all} \quad \omega \in \Omega^* . \tag{6.96}$$

If we additionally assume that the affine RDS generated by the equation

$$dy = (f(0) + D_x f(0)y) \cdot dt + y \circ d\left(\sum_{j=1}^{m} \sigma_j W_t^j\right) \tag{6.97}$$

possesses a super-equilibrium $w(\omega) \in \text{int}\mathbb{R}_+^d$, *then there exist bounded orbits and assertion (b) holds. If* $f(0) \equiv 0$ *and the top Lyapunov exponent of the linear SDE (6.97) is negative (which is the case if and only if the eigenvalues of* $D_x f(0)$ *have negative real parts), then we have*

$$\lim_{t\to\infty} \varphi(t,\theta_{-t}\omega)x = 0 \quad \text{for all} \quad x \in \mathbb{R}_+^d$$

on a θ-*invariant set* Ω^* *of full measure.*

6.8 Applications

We start with a stochastic version of the model of control protein synthesis in the cell (cf. Subsect.5.7.1).

6.8.1 Stochastic Biochemical Control Circuit

Consider the following system of Stratonovich stochastic equations

$$dx_1(t) = (g(x_d(t)) - \alpha_1 x_1(t))dt + \sigma_1 \cdot x_1(t) \circ dW_t^1 , \tag{6.98}$$

$$dx_j(t) = (x_{j-1}(t) - \alpha_j x_j(t))dt + \sigma_j \cdot x_j(t) \circ dW_t^j, \quad j = 2,\ldots,d . \tag{6.99}$$

Here σ_j are nonnegative and α_j are positive constants, $j = 1,\ldots,d$ and $g : \mathbb{R}_+ \mapsto \mathbb{R}_+$ is a C^1 function such that

$$0 < g(u) \le au + b, \quad \text{and} \quad g'(u) \ge 0 \quad \text{for every} \quad u > 0 \tag{6.100}$$

for some constants a and b. Proposition 6.2.2 implies that equations (6.98) and (6.99) generate a strictly order-preserving RDS (θ, φ) in the cone $\mathbb{R}_+^d$. To construct sub- and super-equilibria for (6.98) and (6.99) we consider the following auxiliary affine problem

$$dy_1(t) = (b + ay_d(t) - \alpha_1 y_1(t))dt + \sigma_1 \cdot y_1(t) \circ dW_t^1 , \tag{6.101}$$

$$dy_j(t) = (y_{j-1}(t) - \alpha_j y_j(t))dt + \sigma_j \cdot y_j(t) \circ dW_t^j, \quad j = 2, \dots, d . \quad (6.102)$$

It is clear (see Proposition 6.2.2 and Corollary 6.3.2) that equations (6.101) and (6.102) generate a strongly order-preserving affine RDS for every $b \geq 0$. We denote the corresponding cocycle by $\varphi_{\mathrm{aff}}(t, \omega)$. It is clear that

$$\varphi_{\mathrm{aff}}(t, \omega)x = \Phi(t, \omega)x + \int_0^t \Phi(t - \tau, \theta_\tau \omega)\mathbf{b}\, d\tau, \quad t > 0, \ \omega \in \Omega ,$$

where $\Phi(t, \omega)$ is the cocycle generated by (6.101) and (6.102) with $b = 0$ and $\mathbf{b} = (b, 0, \dots, 0) \in \mathbb{R}_+^d$. If a is small enough, then the cocycle $\Phi(t, \omega)$ has the negative top Lyapunov exponent, i.e. there exists $\lambda < 0$ such that

$$|\Phi(t, \omega)x| \leq R_\varepsilon(\omega)e^{(\lambda + \varepsilon)t}|x|, \quad \omega \in \Omega^*, \quad t \geq 0, \quad x \in \mathbb{R}^d ,$$

for every $\varepsilon > 0$, where $R_\varepsilon(\omega) > 0$ is a tempered random variable and $\Omega^* \in \mathcal{F}$ is a θ-invariant set of full measure. We can suppose $\Omega^* = \Omega$ (see Remark 1.2.1(iii)). By Theorem 5.6.5 (ARNOLD [3]) the RDS $(\theta, \varphi_{\mathrm{aff}})$ possesses a unique tempered equilibrium

$$v(\omega) \equiv \int_0^\infty \Phi(\tau, \theta_{-\tau}\omega)\mathbf{b}\, d\tau, \quad \omega \in \Omega .$$

It is clear that this equilibrium is strongly positive and measurable with respect to the past σ-algebra $\mathcal{F}_-$. Theorem 6.4.1 implies that the RDS $(\theta, \varphi_{\mathrm{aff}})$ dominates (θ, φ) from above. Therefore $\mu \cdot v(\omega)$ is a super-equilibrium for (θ, φ) for any $\mu \geq 1$.

Let $\mathcal{D}$ be the universe of all tempered random closed sets $D(\omega)$ of $\mathbb{R}_+^d$. Applying Proposition 1.9.3 we have that $v(\omega)$ uniformly attracts all random sets $D(\omega)$ from $\mathcal{D}$ with exponential speed, i.e. there exists $\gamma > 0$ such that

$$\lim_{t \to +\infty} e^{\gamma t} \sup_{y \in D(\theta_{-t}\omega)} |\varphi_{\mathrm{aff}}(t, \theta_{-t}\omega)y - v(\omega)| = 0$$

for any $D \in \mathcal{D}$. This means that for any $\mu > 1$ the super-equilibrium $\mu v(\omega)$ is absorbing for the RDS (θ, φ). On the other hand the affine RDS generated by

$$dy_1(t) = (g(0) - \alpha_1 y_1(t))dt + \sigma_1 \cdot y_1(t) \circ dW_t^1 ,$$

$$dy_j(t) = (y_{j-1}(t) - \alpha_j y_j(t))dt + \sigma_j \cdot y_j(t) \circ dW_t^j, \quad j = 2, \dots, d ,$$

dominates (θ, φ) from below. This system possesses a uniformly attracting equilibrium $w(\omega)$ such that $0 \leq w(\omega) \leq v(\omega)$. It is easy to find for $w(\omega) = (w_1(\omega), \dots, w_d(\omega))$ the recurrence formulae

$$w_1(\omega) = g(0) \int_{-\infty}^0 e^{\alpha_1 t - \sigma_1 W_t^1}\, dt \quad (6.103)$$

and

$$w_j(\omega) = \int_{-\infty}^0 w_{j-1}(\theta_t\omega)e^{\alpha_j t - \sigma_j W_t^j}\, dt, \quad j = 2, \ldots, d\,. \tag{6.104}$$

In particular we have $w(\omega) \gg 0$ when $g(0) > 0$ and $w(\omega) \equiv 0$ if $g(0) = 0$. It is also clear that $\mu^{-1}w(\omega)$ is a sub-equilibrium for (θ, φ) for any $\mu \geq 1$. Thus any interval of the form

$$[\mu^{-1}w(\omega), \mu v(\omega)] = \{u \; : \; \mu^{-1}w(\omega) \leq u \leq \mu v(\omega)\}$$

with $\mu > 1$ is an absorbing invariant set for (θ, φ) and it belongs to $\mathcal{D}$. By Corollary 1.8.1 the RDS (θ, φ) generated by (6.98) and (6.99) in $\mathbb{R}_+^d$ possesses a random attractor $A(\omega) \in \mathcal{D}$. This attractor belongs to the interval $[w, v](\omega)$. Therefore applying Theorem 3.6.2 we obtain the existence of two equilibria $\underline{u}(\omega)$ and $\bar{u}(\omega)$ in $A(\omega)$ such that $\underline{u}(\omega) \leq \bar{u}(\omega)$ and $A(\omega) \subset [\underline{u}, \bar{u}](\omega)$. These equilibria are $\mathcal{F}_-$-measurable and they are globally asymptotically stable from below and from above respectively, i.e.

$$\lim_{t \to +\infty} \varphi(t, \theta_{-t}\omega)w(\theta_{-t}\omega) = \underline{u}(\omega)$$

and

$$\lim_{t \to +\infty} \varphi(t, \theta_{-t}\omega)v(\theta_{-t}\omega) = \bar{u}(\omega)$$

for any tempered $w(\omega)$ and $v(\omega)$ such that $w(\omega) \leq \underline{u}(\omega)$ and $v(\omega) \geq \bar{u}(\omega)$.

Assume in addition to (6.100) that $g(0) > 0$, $g'(x) > 0$ for $x > 0$, and $g(x)$ is strictly sublinear, i.e. $\lambda g(x) < g(\lambda x)$ for any $0 < \lambda < 1$ and $x > 0$. Lemma 6.7.1 implies that (θ, φ) is a strongly sublinear RDS. In this case the sub-equilibrium $w(\omega)$ given by (6.103) and (6.104) is strongly positive. Therefore we can apply Theorem 4.2.1 and Corollary 3.6.1 to prove that the random attractor $A(\omega)$ is a one-point set consisting of a unique globally asymptotically stable equilibrium.

6.8.2 Stochastic Gonorrhea Model

We consider the following stochastic version of the problem (5.81) and (5.82):

$$dx_j = \left(-\alpha_j x_j + (p_j - x_j)\sum_{i=1}^d \beta_{ji}x_i\right) dt + \sigma_j x_j(p_j - x_j) \circ dW_t^j\,. \tag{6.105}$$

Here $j = 1, \ldots, d$ and α_j, β_{ji} and p_j are positive numbers, $\sigma_j \neq 0$. The diffusion term in (6.105) models stochastic fluctuations of the rate β_{jj} at which j-th group infects itself.

By Propositions 6.2.2 and 6.2.3 Eqs. (6.105) generate a global strictly order-preserving RDS with the state space

$$X = [0, p] = \{u \in \mathbb{R}^d \ : \ 0 \leq u \leq p\},$$

where $p = (p_1, \ldots, p_j)$. It is clear that $w(\omega) \equiv p$ is a super-equilibrium and $v(\omega) \equiv 0$ is an equilibrium. Therefore Theorem 3.5.1 implies the existence of an equilibrium $u(\omega)$ with the property

$$0 \leq u(\omega) < p \quad \text{for all} \quad \omega \in \Omega.$$

In order to prove the existence of a strictly positive equilibrium let us consider the following auxiliary linear problem

$$dy_j(t) = h_j(y_j)dt + g_j(y_j) \circ dW_t^j, \quad j = 1, 2, \ldots, d, \tag{6.106}$$

where

$$h_j(x) = -\alpha_j x + \beta_{jj} x(p_j - x) \quad \text{and} \quad g_j(x) = \sigma_j x(p_j - x).$$

The system in (6.106) is decoupled, hence for each j equation (6.106) generates a strictly order-preserving RDS (θ, ψ_j) in the one-dimensional interval $[0, p_j]$. Theorem 6.4.1 implies that the direct product of these systems dominates the system (θ, φ) generated by (6.105) from below.

Under the condition $\alpha_j < p_j \beta_{jj}$ the speed measure (6.66) for problem (6.106) on $[0, p_j]$ possesses the property $m((0, c_j]) < \infty$ for any $c_j \in (0, p_j)$. Therefore by Theorem 6.6.2(ii-b) there exists unique $\mathcal{F}_-$-measurable equilibrium $v_j(\omega) \in (0, p_j)$ having a distribution with the density

$$\varrho_j(x) = \frac{N_j}{g_j(x)} \cdot \exp\left(\int_{c_j}^x \frac{2h_j(v)}{g_j(v)^2}\, dv\right), \quad 0 < x < p_j,$$

where N_j is the normalizing factor and $c_j \in (0, p_j)$. Thus the system generated by (6.105) has a strongly positive sub-equilibrium

$$v(\omega) = (v_1(\omega) \ldots, v_d(\omega)) \in [0, p]$$

provided that $\alpha_j < p_j \beta_{jj}$ for all $j = 1, 2, \ldots, d$. Therefore there exists a strongly positive equilibrium $u(\omega)$ which is globally asymptotically stable from above.

6.8.3 Stochastic Model of Symbiotic Interaction

We consider a stochastic version of the RDE (5.83) with a right-hand side of the form (5.84):

$$dx_j = x_j \left(\alpha_j(x_j) + \sum_{i \neq j} g_i(x_i) \right) \cdot dt + \sigma_j x_j \circ dW_t^j, \quad j = 1, \ldots, d, \quad (6.107)$$

where σ_j are constants and $\alpha_j(x)$ and $g_i(x)$ are smooth functions on $\mathbb{R}_+$ with properties like (S1) and such that

$$0 \leq g_i(0) \leq g_i(x) \leq M, \quad \text{and} \quad g_i'(x) > 0 \quad \text{for every} \quad x > 0 \quad (6.108)$$

Under these conditions equations (6.107) generate an order-preserving RDS (θ, φ) in $\mathbb{R}_+^d$ possessing the following invariance property: for every subset I of integers from $N = \{1, \ldots, d\}$ the set

$$K_I = \{x = (x_1, \ldots, x_d) \in \mathbb{R}_+^d : x_j = 0, j \in N \setminus I\}$$

is invariant with respect to (θ, φ). The restriction of (θ, φ) to K_I is described by the system of stochastic equations

$$dx_j = x_j \left(\alpha_j(x_j) + \sum_{i \neq j, i \in I} g_i(x_i) \right) \cdot dt + \sigma_j x_j \circ dW_t^j, \quad j \in I.$$

Consider the equations

$$dx = x \cdot \alpha_j(x) + \sigma_j x \circ dW_t^j, \quad j = 1, \ldots, d, \quad (6.109)$$

that describe the existence of each population independent of the others. Let $\alpha_j(x)$ be chosen such that the RDS generated by (6.109) in $\mathbb{R}_+$ has a positive equilibrium for every $j = 1, \ldots, d$. This is, for example, implied by

$$\alpha_j(0) > 0, \quad \limsup_{x \to +\infty} \alpha_j(x) \leq -\beta_j < 0, \quad j = 1, \ldots, d,$$

(see Theorem 6.6.1(ii)). Denote by $v_j(\omega)$ the positive equilibrium for (6.109). As in Subsect. 5.7.3 under the condition $\beta_j > (d-1)M$ we can prove the existence of an equilibrium $u(\omega) = (u_1(\omega), \ldots, u_d(\omega))$ for the RDS (θ, φ) generated by (6.107) such that $u_j(\omega) > v_j(\omega)$ and which attracts (from below) the collection $(v_1(\omega), \ldots, v_d(\omega))$ of equilibria that correspond to the isolated dynamics of each population. Thus as in the random case we observe an interaction results in a benefit for all populations.

6.8.4 Lattice Models of Statistical Mechanics

In this subsection we briefly consider an order-preserving RDS generated by an infinite family of coupled stochastic differential equations. We note that formally the results of previous sections are not applied here. However the methods developed allow us to study this RDS.

One of the approaches to the investigation of lattice systems in Statistical Mechanics relies on the study of infinite systems of stochastic differential equations (see, e.g., ALBEVERIO/ET AL. [1] and DA PRATO/ZABCZYK [39, 40] and the references therein). For example in the theory of classical anharmonic crystals the following infinite system of stochastic equations

$$dx_i = \left(\sum_{j \in \mathbb{Z}^d} a_{ij} \cdot x_j + f(x_i) \right) \cdot dt + dW_t^i, \quad i \in \mathbb{Z}^d , \tag{6.110}$$

arises in $\mathbb{R}^{\mathbb{Z}^d}$. Here $\{W_t^i : i \in \mathbb{Z}^d\}$ are independent standard real valued Wiener processes. As in DA PRATO/ZABCZYK [39, 40] we also assume that the coefficients a_{ij} possess the properties (i) $a_{ij} = a_{ji} \geq 0$ for all $i \neq j$; (ii) there exists $r > 0$ such that $a_{ij} = 0$ if $|i - j|_1 \equiv \sum_{k=1}^d |i_k - j_k| > r$; (iii) $|a_{ij}| \leq M$ for all $i, j \in \mathbb{Z}^d$. In this case

$$\beta := \sup_{i \in \mathbb{Z}^d} \sum_{j \in \mathbb{Z}^d} |a_{ij}| < \infty .$$

The discrete Laplace operator on $\mathbb{Z}^d$ corresponds to the choice

$$a_{ii} = -2d, \quad a_{ij} = 1 \text{ if } |i - j|_1 = 1, \quad a_{ij} = 0 \text{ if } |i - j|_1 > 1 ,$$

and the assumptions above are true with $\beta = 4d$.

For the function $f : \mathbb{R} \mapsto \mathbb{R}$ we require that f is a globally Lipschitz function such that

$$\alpha x + \gamma_1 \leq f(x) \leq \alpha x + \gamma_2 , \tag{6.111}$$

where $\beta + \alpha < 0$ and $\gamma_1 \leq \gamma_2$ are constants.

Let $V = l_\varrho^2(\mathbb{Z}^d)$ be the space of sequences $\{x_i : i \in \mathbb{Z}^d\}$ such that

$$\|x\|_{l_\varrho^2(\mathbb{Z}^d)}^2 := \sum_{i \in \mathbb{Z}^d} |x_i|^2 \cdot \varrho(i) < \infty ,$$

where $\varrho(i) = \exp\{-\delta |i|_1\}$ for $i \in \mathbb{Z}^d$ with some $\delta > 0$. Let $V_+ := l_{\varrho,+}^2(\mathbb{Z}^d)$ be the cone of nonnegative elements in $l_\varrho^2(\mathbb{Z}^d)$, i.e. $\{x_i : i \in \mathbb{Z}^d\} \in V_+$ if and only if $x_i \geq 0$ for all $i \in \mathbb{Z}^d$. The matrix $\{a_{ij}\}$ and the function f define mappings of $l_\varrho^2(\mathbb{Z}^d)$ into itself via the formulae

$$(Ax)_i = \sum_{j \in \mathbb{Z}^d} a_{ij} \cdot x_j, \quad (F(x))_i = f(x_i), \quad i \in \mathbb{Z}^d, \ x \in l_\varrho^2(\mathbb{Z}^d) .$$

The mapping A is a bounded linear operator in $l_\varrho^2(\mathbb{Z}^d)$ (see DA PRATO/ZAB-CZYK [39, Proposition 3.2]) and F is a globally Lipschitz mapping in this space. The system of equations (6.110) can be written in operator form as

$$dx = (Ax + F(x)) \cdot dt + dW_t .$$

It follows from DA PRATO/ZABCZYK [39, Theorem 3.4] that equation (6.110) generates an RDS (θ, φ) in the space $l_\varrho^2(\mathbb{Z}^d)$. Moreover the cocycle $\varphi(t, \omega)$ can be represented in the form

$$\varphi(t, \omega)x = z(\theta_t\omega) + \psi(t, \omega)[x - z(\omega)] \ .$$

Here (θ, ψ) is the RDS generated by the random equation

$$\dot{y}_i = \sum_{j \in \mathbb{Z}^d} a_{ij} \cdot y_j + f(y_i + z_i(\theta_t\omega)) + \sum_{j \in \mathbb{Z}^d} a_{ij} \cdot z_j(\theta_t\omega) + \mu z_i(\theta_t\omega) \ , \quad (6.112)$$

where $i \in \mathbb{Z}^d$ and $z(\theta_t\omega) = \{z_i(\theta_t\omega) \ : \ i \in \mathbb{Z}^d\}$ is the stationary Ornstein-Uhlenbeck process in $l_\varrho^2(\mathbb{Z}^d)$ which solves the equations

$$dz_i = -\mu z_i dt + dW_t^i, \quad i \in \mathbb{Z}^d \ ,$$

for some $\mu > 0$ (cf. Sect.2.5). The method used in the proof of Theorem 5.3.1 can be applied here. Therefore the RDS (θ, ψ) is order-preserving. Thus (θ, φ) is also an order-preserving RDS.

Let us consider the affine system

$$dx_i = \left(\sum_{j \in \mathbb{Z}^d} a_{ij} \cdot x_j + \alpha x_i + \gamma \right) \cdot dt + dW_t^i, \quad i \in \mathbb{Z}^d \ . \qquad (6.113)$$

It follows from DA PRATO/ZABCZYK [39, Theorem 3.4] again that for any γ this equation generates an RDS (θ, φ_γ) in the space $l_\varrho^2(\mathbb{Z}^d)$. The comparison principle (cf. Theorem 6.4.1) and property (6.111) give

$$\varphi_{\gamma_1}(t, \omega)x \leq \varphi(t, \omega)x \leq \varphi_{\gamma_2}(t, \omega)x \quad \text{for all} \quad x \in l_\varrho^2(\mathbb{Z}^d) \ , \qquad (6.114)$$

where $t > 0$ and $\omega \in \Omega$. Since $\beta + \alpha < 0$, it is clear (see DA PRATO/ZABCZYK [39, Sect. 3]) that we can choose $\delta > 0$ in the definition of $\varrho(i)$ such that

$$\langle Ax, x \rangle_{l_\varrho^2(\mathbb{Z}^d)} \leq -(\alpha + \varepsilon) \cdot \|x\|_{l_\varrho^2(\mathbb{Z}^d)}^2$$

for some $\varepsilon > 0$. Therefore for any $\gamma \in \mathbb{R}$ the affine RDS (θ, ψ_γ) possesses a $\mathcal{F}_-$-measurable equilibrium $w_\gamma(\omega) \in l_\varrho^2(\mathbb{Z}^d)$. Inequality (6.114) implies that $w_{\gamma_1}(\omega)$ is a sub-equilibrium and $w_{\gamma_2}(\omega)$ is a super-equilibrium for (θ, φ). We obviously have the relation $w_{\gamma_1}(\omega) \leq w_{\gamma_2}(\omega)$. Since the cone $l_{\varrho,+}^2(\mathbb{Z}^d)$ is regular, Theorem 3.5.1 and Remark 3.5.1 imply the existence of an $\mathcal{F}_-$-measurable equilibrium $u(\omega)$ in $l_\varrho^2(\mathbb{Z}^d)$ for (θ, φ). This equilibrium generates a Markov invariant measure for the stochastic equation (6.110) (see Sect. 1.10).

References

1. ALBEVERIO S., DALETSKII A., KONDRATIEV Y., RÖCKNER M. (1999) Fluctuations and their Glauber Dynamics in Lattice Systems. J Funct Anal 166(1):148-167
2. AMANN H. (1983) Gewöhnlicne Differentialgleichungen. Walter de Gruyter, Berlin
3. ARNOLD L. (1998) Random Dynamical Systems. Springer, Berlin Heidelberg New York
4. ARNOLD L. (1999) Recent Progress in Stochastic Bifurcation Theory. Institut für Dynamische Systeme, Universität Bremen. Report 439
5. ARNOLD L., CHUESHOV I. (1998) Order-Preserving Random Dynamical Systems: Equilibria, Attractors, Applications. Dynamics and Stability of Systems 13:265-280
6. ARNOLD L., CHUESHOV I. (2001) A Limit Set Trichotomy for Order-Preserving Random Systems. Positivity 5(2):95-114
7. ARNOLD L., CHUESHOV I. (2001) Cooperative Random and Stochastic Differential Equations. Discrete and Continuous Dynamical Systems 7(1):1-33
8. ARNOLD L., DEMETRIUS L., GUNDLACH M. (1994) Evolutionary Formalism for Products of Positive Random Matrices. Ann of Appl Probab 4:859-901
9. ARNOLD L., NGUYEN DINH CONG, OSELEDETS V. (1999) Jordan Normal Form for Linear Cocycles. Random Operators and Stochastic Equations 7:301-356
10. ARNOLD L., SCHEUTZOW M. (1995) Perfect Cocycles through Stochastic Differential Equations. Probab Theory Relat Fields 101:65-88
11. ARNOLD L., SCHMALFUSS B. (1996) Fixed Points and Attractors for Random Dynamical Systems. In: Naess A., Krenk S. (Eds.) IUTAM Symposium on Advances in Nonlinear Stochastic Mechanics. Kluver, Dordrecht, 19-28
12. ARNOLD L., SCHMALFUSS B. (2001) Lyapunov Second Method for Random Dynamical Systems. J Diff Equations 177:235-265
13. BABIN A., VISHIK M. (1992) Attractors of Evolution Equations. Noth-Holland, Amsterdam
14. BAUER H., BEAR H. S. (1969) The Part Metric in Convex Sets. Pacif J Math 30:15-33
15. BELOPOLSKAYA JA.I., DALECKY YU.L. (1990) Stochastic Equations and Differential Geometry. Kluver Academic Publishers, Dordrecht
16. BHATTACHARYA R., LEE O. (1988, 1997) Asymptotics of a Class of Markov Processes which are not in General Irreducible. Ann of Probab 16:1333-1347; Correction, Ann of Probab 25:1541-1543
17. BONY J.-M. (1969) Principe du Maximum, Inegalite de Harnack et Unicite du Probleme de Cauchy pour les Operateurs Elliptiques Degeneres. Ann Inst Fourier 19:277-304

18. CASTAING C., VALADIER M. (1977) Convex Analysis and Measurable Multi-functions, Lect Notes in Math 580. Springer, Berlin
19. CHICONE C., LATUSHKIN YU. (1999) Evolution Semigroups in Dynamical Systems and Differential Equations. Amer Math Soc, Providence, Rhode Island
20. CHUESHOV I.D. (1999) Introduction to the Theory of Infinite-Dimensional Dissipative Systems. Acta, Kharkov (in Russian)
21. CHUESHOV I.D. (2000) Order-Preserving Random Dynamical Systems Generated by a Class of Coupled Stochastic Semilinear Parabolic Equations. In: Fiedler B, Gröger K., Sprekels J. (Eds.) International Conference on Differential Equations, EQUADIF 99, Berlin, Aug 1–7, 1999, vol 1. World Scientific, Singapore, 711–716
22. CHUESHOV I.D. (2001) Order-Preserving Skew-Product Flows and Nonautonomous Parabolic Systems. Acta Appl Math 65:185–205
23. CHUESHOV I.D., SCHEUTZOW M. (2001) Inertial Manifolds and Forms for Stochastically Perturbed Retarded Semilinear Parabolic Equations. J Dyn Diff Equations 13:355–380
24. CHUESHOV I.D., VUILLERMOT P.-A. (1998) Long-Time Behavior of Solutions to a Class of Quasilinear Parabolic Equations with Random Coefficients. Ann Inst Henri Poincaré, Analyse non Linéaire 15:191–232
25. CHUESHOV I.D., VUILLERMOT P.-A. (1998) Long-Time Behavior of Solutions to a Class of Stochastic Parabolic Equations with White Noise: Stratonovitch's Case. Probab Theory and Relat Fields 112:149–202
26. CHUESHOV I.D., VUILLERMOT P.-A. (2000) Long-Time Behavior of Solutions to a Class of Stochastic Parabolic Equations with White Noise: Ito's Case. Stoch Anal Appl 18:581–615
27. CHUNG K.L., WILLIAMS R.J. (1983) Introduction to Stochastic Integration. Birkhäuser, Boston Basel Stuttgart
28. CODDINGTON E.A., LEVINSON N. (1955) Theory of Ordinary Differential Equations. Springer, New York
29. CORNFELD I.P., FOMIN S.V., SINAI YA. G. (1982) Ergodic Theory. Springer, New York
30. COHN D.L. (1980) Measure Theory. Birkhäuser, Boston Basel Stuttgart
31. CRAUEL H. (1991) Markov Measures for Random Dynamical Systems. Stochastics and Stoch Reports 37:153–173
32. CRAUEL H. (1995) Random Probability Measures on Polish Spaces. Habilitationsschrift, Universität Bremen
33. CRAUEL H. (1999) Global Random Attractors are Uniquely Determined by Attracting Deterministic Compact Sets. Ann Mat Pura Appl, Ser IV 176:57–72
34. CRAUEL H. (2001) Random Point Attractors versus Random Set Attractors. J London Math Society (2) 63:413–427
35. CRAUEL H., DEBUSSCHE A., FLANDOLI F. (1997) Random Attractors. J Dyn Diff Equations 9:307–341
36. CRAUEL H., FLANDOLI F. (1994) Attractors for Random Dynamical Systems. Probab Theory Relat Fields 100:365–393
37. CRAUEL H., FLANDOLI F. (1998) Additive Noise Destroys a Pitchfork Bifurcation. J Dyn Diff Equations 10:259–274
38. CRAUEL H., IMKELLER P., STEINKAMP M. (1999) Bifurcations of One-Dimensional Stochastic Differential Equations. In: Crauel H., Gundlach M. (Eds.) Stochastic Dynamics. Conference on Random Dynamical Systems, Bremen, Germany, April 28 - May 2, 1997. Springer. Berlin Heidelberg New York, 27–47

39. DA PRATO G., ZABCZYK J. (1995) Convergence to Equilibrium for Classical and Quantum Spin Systems. Probab Theory Relat Fields 103:529–552
40. DA PRATO G., ZABCZYK J. (1996) Ergodicity for Infinite Dimensional Systems. Cambridge University Press, Cambridge
41. DEIMLING K. (1977) Ordinary Differential Equations in Banach Spaces, Lect Notes Math 596. Springer, Berlin New York
42. ELLIS R. (1969) Lectures on Topological Dynamics. W.A. Benjamin Inc, New York
43. ELWORTHY K. D. (1982) Stochastic Differential Equations on Manifolds. Cambridge University Press, Cambridge
44. FLANDOLI F., SCHMALFUSS B. (1996) Random Attractors for the $3D$ Stochastic Navier-Stokes Equations with Multiplicative White Noise. Stoch and Stoch Reports 59:21–45
45. FRIEDMAN A. (1975, 1976) Stochastic Differential Equations and Applications, Vol 1, 2. Academic Press, New York
46. GEISS C., MANTHEY R. (1994) Comparison Theorems for Stochastic Differential Equations in Finite and Infinite Dimensions. Stoch Processes Appl 53:23–35
47. GIHMAN I.I., SKOROHOD A.V. (1972) Stochastic Differential Equations. Springer, Berlin Heidelberg New York
48. GIHMAN I.I., SKOROHOD A.V. (1974) The Theory of Stochastic Processes, vol I. Springer, Berlin Heidelberg New York
49. HALE J. K. (1980) Ordinary Differential Equations. Krieger, Malabar Florida
50. HALE J. K. (1988) Asymptotic Behavior of Dissipative Systems. Amer Math Soc, Providence, Rhode Island
51. HARTMAN P. (1982) Ordinary Differential Equations, 2nd edn. Birkhäuser, Boston Basel Stuttgart
52. HIRSCH M. W. (1982, 1985) Systems of Differential Equations that are Competitive or Cooperative, I: Limit Sets. SIAM J Math Anal 13:167–179; II: Convergence Almost Everywhere. SIAM J Math Anal 16:423–439
53. HIRSCH M. W. (1984) The Dynamical System Approach to Differential Equations. Bull Amer Math Soc 11:1–64
54. HIRSCH M. W. (1988) Stability and Convergence in Strongly Monotone Dynamical Systems. J reine Angew Math 383:1–53
55. HORSTHEMKE W., LEFEVER R. (1984) Noise-Induced Transitions. Springer, Berlin Heidelberg New York
56. HU S., PAPAGEORGIOU N. S. (1997) Handbook of Multivalued Analysis, vol 1: Theory. Kluver Academic Publishers, Dordrecht
57. IKEDA N., WATANABE S. (1981) Stochastic Differential Equations and Diffusion Processes. North-Holland, Amsterdam
58. IMKELLER P., LEDERER C. (2001) On the Cohomology of Flows of Stochastic and Random Differential Equations. Probab Theory Relat Fields 120:209–235.
59. IMKELLER P., SCHMALFUSS B. (2001) The Conjugacy of Stochastic and Random Differential Equations and the Existence of Global Attractors. J Dyn Diff Equations 13:215–249
60. IOFFE A.D. (1979) Single-Valued Representation of Set-Valued Mappings. Trans Amer Math Soc 252:133–145
61. KAGER G., SCHEUTZOW M. (1997) Generation of One-sided Random Dynamical Systems by Stochastic differential Equations. Electronic J Probab 2, paper 8
62. KARATZAS I., SHREVE S.E. (1988) Brownian Motion and Stochastic Calculus. Springer, Berlin Heidelberg New York

63. KELLER H., SCHMALFUSS B. (1998) Attractors for Stochastic Differential Equations with Nontrivial Noise. Izvestiya Akad Nauk R Moldova 26(1):43–54
64. KHASMINSKII R. Z. (1980) Stochastic Stability of Differential Equations. Sijthoff and Noorhoff, Alphen
65. KELLERER H. (1995) Order-Preserving Random Dynamical Systems. Universität München. Preprint
66. KIFER Y. (1986) Ergodic Theory of Random Transformations. Birkhäuser, Boston Basel Stuttgart
67. KLOEDEN P.E., PLATEN E. (1992) Numerical Solutions of Stochastic Differential Equations. Springer, Berlin Heidelberg New York
68. KRASNOSELSKII M. A. (1964) Positive Solutions of Operator Equations. Noordhoff, Groningen
69. KRASNOSELSKII M. A. (1968) The Operator of Translation Along Trajectories of Differential Equations. Transl Math Monographs 19. Amer Math Soc, Providence, Rhode Island
70. KRASNOSELSKII M. A., BURD V.S., KOLESOV YU.S. (1973) Nonlinear Almost Periodic Oscillations. John Wiley, New York
71. KRASNOSELSKII M. A., LIFSHITS E.A., SOBOLEV A.V. (1989) Positive Linear Systems – Method of Positive Operators. Sigma Series in Appl Math 5. Heldermann, Berlin
72. KRAUSE U., NUSSBAUM R. G. (1993) A Limit Set Trichotomy for Self-Mappings of Normal Cones in Banach Spaces. Nonlin Anal, Theory, Methods & Appl 20:855–870
73. KRAUSE U., RANFT P. (1992) A Limit Set Trichotomy for Monotone Nonlinear Dynamical Systems. Nonlin Anal, Theory, Methods & Appl 19:375–392
74. KUNITA H. (1990) Stochastic Flows and Stochastic Differential Equations. Cambridge University Press, Cambridge
75. LADDE G.S., LAKSHMIKANTHAM V. (1980) Random Differential Inequalities. Academic Press, New York
76. LADYZHENSKAYA O. (1991) Attractors for Semigroups and Evolution Equations. Cambridge University Press, Cambridge
77. LEVITAN B., ZHIKOV V. (1982) Almost Periodic Functions and Differential Equations. Cambridge University Press, Cambridge
78. MANDL P. (1968) Analytical Treatment of One-Dimensional Markov Processes. Springer, Berlin Heidelberg New York
79. MAÑÉ R. (1987) Ergodic Theory and Differentiable Dynamical Systems. Springer, Berlin Heidelberg New York
80. MAO X. (1994) Exponential Stability of Stochastic Differential Equations. Marcel Dekker, New York Basel Hong Kong
81. MARTIN R.H. (1976) Nonlinear Operators and Differential Equations in Banach Spaces. Wiley, New York
82. MCKEAN H.P. (1969) Stochastic Integrals. Academic Press, New York
83. MEYN S.P., TWEEDIE R.L. (1993) Markov Chains and Stochastic Stability. Springer, London
84. NAKAJIMA F. (1979) Periodic Time Dependent Gross-Substitute Systems. SIAM J Appl Math 36:421–427
85. OCHS G. (1998) Examples of Random Dynamical Systems without Random Fixed Points. Univ Iagellonicae Acta Math 36:133–141
86. OCHS G. (1999) Weak Random Attractors. Institut für Dynamische Systeme, Universität Bremen. Report 449
87. OCHS G., OSELEDETS V.I. (1999) Topological Fixed Point Theorems do not hold for Random Dynamical Systems. J Dyn Diff Equations 11(4):583–593.

88. RUDOLPH D. J., (1990) Fundamentals of Measurable Dynamics. Oxford University Press, Oxford
89. SCHENK-HOPPÉ K. R. (1998) Random Attractors - General Properties, Existence and Applications to Stochastic Bifurcation Theory. Discrete and Continuous Dynamical Systems 4(1):99-130
90. SCHEUTZOW M. (1996) On the Perfection of Crude Cocycles. Random & Comp. Dynamics, 4:235-255
91. SCHEUTZOW M. (2001) Comparison of Various Concepts of a Random Attractor: A Case Study. To be published in Archiv der Mathematik
92. SCHMALFUSS B. (1992) Backward cocycles and attractors for stochastic differential equations. In: Reitmann V., Riedrich T., Koksch N. (Eds.), International Seminar on Applied Mathematics - Nonlinear Dynamics: Attractor Approximation and Global Behaviour. Teubner, Leipzig, 185–192.
93. SCHMALFUSS B. (1997) The Attractor of the Stochastic Lorenz System. Z Angew Math Phys 48:951–975
94. SCHMALFUSS B. (1998) A Random Fixed Point Theorem and the Random Graph Transformation. J Math Anal Appl 225:91–113
95. SCHMALFUSS B. (1999) Measure Attractors and Random Attractors for Stochastic Partial Differential Equations. Stoch Anal Appl 17(6):1075–1101
96. SELGRADE J. (1980) Asymptotic Behavior of Solutions to Single Loop Positive Feedback Systems. J Diff Equations 38:80–103
97. SELL G.R., NAKAJIMA F. (1980) Almost Periodic Time Dependent Gross-Substitute Dynamical Systems. Tôhoku Math J 32:255–263
98. SHARPE M. (1988) General Theory of Markov Processes. Academic Press, Boston
99. SHEN W., YI Y. (1998) Almost Automorphic and Almost Periodic Dynamics in Skew Product Semiflows. Memoirs Amer Math Soc 136(647). Amer Math Soc, Providence Rhode Island
100. SINAI YA. G. (1994) Topics in Ergodic Theory. Princeton University Press, Princeton
101. SMITH H. L. (1986) Cooperative Systems of Differential Equations with Concave Nonlinearities. Nonlin Anal, Theory, Methods & Appl 10:1037–1052
102. SMITH H. L. (1996) Monotone Dynamical Systems. An Introduction to the Theory of Competitive and Cooperative Systems. Amer Math Soc, Providence Rhode Island
103. TAKAČ P. (1990) Asymptotic Behavior of Discrete-Time Semigroups of Sublinear Strongly Increasing Mappings with Applications to Biology. Nonlin Anal, Theory, Methods & Appl 14:35–42
104. TEMAM R. (1988) Infinite–Dimensional Dynamical Systems in Mechanics and Physics. Springer New York
105. TWARDOWSKA K. (1996) Wong – Zakaï Approximations for Stochastic Differential Equations. Acta Appl Math 43:317–359
106. WALTERS P. (1982) An Introduction to Ergodic Theory. Springer, Berlin Heidelberg New York
107. WALTER W. (1970) Differential and Integral Inequalities. Springer, Berlin Heidelberg New York
108. WONG E. (1971) Stochastic Processes in Information and Dynamical Systems. Mc Graw-Hill, New York San Francisco
109. WONG E., ZAKAI M., (1969) Riemann-Stieltjes approximations of stochastic integrals. Z Wahrscheinlichkeitstheorie verw Geb 12:87–97
110. XU KEDAI (1993) Bifurcations of Random Differential Equations in Dimension One. Rand Comput Dynamics 1:277–305

Index

ℙ-complete σ-algebra, 19
ℙ-completion of σ-algebra, 21
φ-ergodic measure, 51
φ-invariant measure, 51
u-norm, 84
u-subordination, 84

almost equilibrium, 122

Bernoulli shifts, 11
binary biochemical model
– random, 17, 30, 58, 62, 94, 97, 102, 110, 114
– stochastic, 72–74, 79, 94
biochemical control circuit, 2
– random, 171
– stochastic, 219
Borel σ-algebra, 9

Chapman-Kolmogorov equation, 53
cocycle, 13
comparison principle, 109
– random, 150
– stochastic, 192
competition and migration model, 181
cone
– minihedral, 87
– normal, 86
– part of, 84, 91
– regular, 86
– solid, 83
conjugacy of RDS, 18
cooperativity condition, 147

disintegration, 51

equilibrium, 38
– stable
–– from above, 107
–– from below, 107
–– in probability, 211
equivalence of RDS, 18

Fokker–Plank equation, 204
function
– cooperative, 147
Furstenberg–Khasminskii formula, 75
future σ-algebra, 52

gonorrhea model, 175, 221
gross-substitute system, 178

interval, 83
– absorbing, 108
invariant measure, 51
irreducible matrix, 147
Itô stochastic equation, 71
Itô stochastic integral, 67
Itô's formula, 69

kick model, 16, 27, 31, 38, 93

lattice model, 223
Liouville's equation, 58, 73
Lyapunov exponent, 60, 75

mapping
– strongly positive, 144
– sublinear, 161, 214
– weakly positive, 144
Markov chain, 15
Markov family, 53
Markov measure, 53
martingale, 67
MDS, see metric dynamical system
– ergodic, 10
measurable selection theorem, 20
metric dynamical system, 10
multifunction, 18

orbit, see trajectory
Ornstein-Uhlenbeck process, 79
outer normal, 61

part (Birkhoff) metric, 84
past σ-algebra, 52
perfection procedure, 14
Polish space, 13
probability space, 9
process
– adapted, 66
– continuous, 66
– predictable, 66
product σ-algebra, 9
projection theorem, 21

radius of dissipativity, 26
random attractor, 41
– weak point, 210
random differential equation, 56
random Dirac measure, 51
random dynamical system, 13
RDE, *see* random differential equation
RDS, *see* random dynamical system
– C^k-smooth, 14
– s-concave, 116
– affine, 14, 46, 138
– asymptotically compact, 31
– compact, 30
– concave, 114
– conditionally compact, 123
– dissipative, 26
– linear, 14, 45
– order-preserving, 93
– strictly sublinear, 113
– strongly positive, 105, 145
– strongly sublinear, 114
– sublinear, 113

scale function, 203, 206
SDE, *see* stochastic differential
    equation
semi-equilibria, 95
semimartingale, 68
separability set, 21
separable collection of random sets, 21
set
– θ-invariant, 10
– absorbing, 26
– bounded from
– – above, 84
– – below, 84
– infimum of, 84
– invariant, 24
– – backward, 24
– – forward, 24
– lower bound, 84
– maximal element of, 84
– minimal element of, 84
– omega-limit, 34
– order-bounded, 84
– random, 18
– – bounded, 19
– – closed, 18
– – compact, 19
– – tempered, 23
– supremum of, 84
– universally measurable, 21
– upper bound, 84
skew-product semiflow, 15
spaces $C^{k,\delta}(I)$, 199
spaces $C_b^{k,\delta}$, 70
spaces $C_b^{k,\delta}(I)$, 199
speed measure, 202, 206
stationary measure, 50, 53
stochastic differential equation, 70
stopping time, 67
Stratonovich stochastic equation, 72
Stratonovich stochastic integral, 68
sub-equilibrium, 95
– absorbing, 108
super-equilibrium, 95
– absorbing, 108
symbiotic interaction, 176, 222

tail of trajectory, 32
tempered random variable, 23
top Lyapunov exponent, 48, 60, 75
trajectory, 32

universal σ-algebra, 21
universe, 25

Walras' law, 178
white noise process, 12
Wiener process, 12, 66
Wiener shift, 12
Wong-Zakaï type theorem, 77

Vol. 1694: A. Braides, Approximation of Free-Discontinuity Problems. XI, 149 pages. 1998.

Vol. 1695: D. J. Hartfiel, Markov Set-Chains. VIII, 131 pages. 1998.

Vol. 1696: E. Bouscaren (Ed.): Model Theory and Algebraic Geometry. XV, 211 pages. 1998.

Vol. 1697: B. Cockburn, C. Johnson, C.-W. Shu, E. Tadmor, Advanced Numerical Approximation of Nonlinear Hyperbolic Equations. Cetraro, Italy, 1997. Editor: A. Quarteroni. VII, 390 pages. 1998.

Vol. 1698: M. Bhattacharjee, D. Macpherson, R. G. Möller, P. Neumann, Notes on Infinite Permutation Groups. XI, 202 pages. 1998.

Vol. 1699: A. Inoue, Tomita-Takesaki Theory in Algebras of Unbounded Operators. VIII, 241 pages. 1998.

Vol. 1700: W. A. Woyczy´ski, Burgers-KPZ Turbulence, XI, 318 pages. 1998.

Vol. 1701: Ti-Jun Xiao, J. Liang, The Cauchy Problem of Higher Order Abstract Differential Equations, XII, 302 pages. 1998.

Vol. 1702: J. Ma, J. Yong, Forward-Backward Stochastic Differential Equations and Their Applications. XIII, 270 pages. 1999.

Vol. 1703: R. M. Dudley, R. Norvaiša, Differentiability of Six Operators on Nonsmooth Functions and p-Variation. VIII, 272 pages. 1999.

Vol. 1704: H. Tamanoi, Elliptic Genera and Vertex Operator Super-Algebras. VI, 390 pages. 1999.

Vol. 1705: I. Nikolaev, E. Zhuzhoma, Flows in 2-dimensional Manifolds. XIX, 294 pages. 1999.

Vol. 1706: S. Yu. Pilyugin, Shadowing in Dynamical Systems. XVII, 271 pages. 1999.

Vol. 1707: R. Pytlak, Numerical Methods for Optimal Control Problems with State Constraints. XV, 215 pages. 1999.

Vol. 1708: K. Zuo, Representations of Fundamental Groups of Algebraic Varieties. VII, 139 pages. 1999.

Vol. 1709: J. Azéma, M. Émery, M. Ledoux, M. Yor (Eds), Séminaire de Probabilités XXXIII. VIII, 418 pages. 1999.

Vol. 1710: M. Koecher, The Minnesota Notes on Jordan Algebras and Their Applications. IX, 173 pages. 1999.

Vol. 1711: W. Ricker, Operator Algebras Generated by Commuting Projections: A Vector Measure Approach. XVII, 159 pages. 1999.

Vol. 1712: N. Schwartz, J. J. Madden, Semi-algebraic Function Rings and Reflectors of Partially Ordered Rings. XI, 279 pages. 1999.

Vol. 1713: F. Bethuel, G. Huisken, S. Müller, K. Steffen, Calculus of Variations and Geometric Evolution Problems. Cetraro, 1996. Editors: S. Hildebrandt, M. Struwe. VII, 293 pages. 1999.

Vol. 1714: O. Diekmann, R. Durrett, K. P. Hadeler, P. K. Maini, H. L. Smith, Mathematics Inspired by Biology. Martina Franca, 1997. Editors: V. Capasso, O. Diekmann. VII, 268 pages. 1999.

Vol. 1715: N. V. Krylov, M. Röckner, J. Zabczyk, Stochastic PDE's and Kolmogorov Equations in Infinite Dimensions. Cetraro, 1998. Editor: G. Da Prato. VIII, 239 pages. 1999.

Vol. 1716: J. Coates, R. Greenberg, K. A. Ribet, K. Rubin, Arithmetic Theory of Elliptic Curves. Cetraro, 1997. Editor: C. Viola. VIII, 260 pages. 1999.

Vol. 1717: J. Bertoin, F. Martinelli, Y. Peres, Lectures on Probability Theory and Statistics. Saint-Flour, 1997. Editor: P. Bernard. IX, 291 pages. 1999.

Vol. 1718: A. Eberle, Uniqueness and Non-Uniqueness of Semigroups Generated by Singular Diffusion Operators. VIII, 262 pages. 1999.

Vol. 1719: K. R. Meyer, Periodic Solutions of the N-Body Problem. IX, 144 pages. 1999.

Vol. 1720: D. Elworthy, Y. Le Jan, X-M. Li, On the Geometry of Diffusion Operators and Stochastic Flows. IV, 118 pages. 1999.

Vol. 1721: A. Iarrobino, V. Kanev, Power Sums, Gorenstein Algebras, and Determinantal Loci. XXVII, 345 pages. 1999.

Vol. 1722: R. McCutcheon, Elemental Methods in Ergodic Ramsey Theory. VI, 160 pages. 1999.

Vol. 1723: J. P. Croisille, C. Lebeau, Diffraction by an Immersed Elastic Wedge. VI, 134 pages. 1999.

Vol. 1724: V. N. Kolokoltsov, Semiclassical Analysis for Diffusions and Stochastic Processes. VIII, 347 pages. 2000.

Vol. 1725: D. A. Wolf-Gladrow, Lattice-Gas Cellular Automata and Lattice Boltzmann Models. IX, 308 pages. 2000.

Vol. 1726: V. Marić, Regular Variation and Differential Equations. X, 127 pages. 2000.

Vol. 1727: P. Kravanja M. Van Barel, Computing the Zeros of Analytic Functions. VII, 111 pages. 2000.

Vol. 1728: K. Gatermann Computer Algebra Methods for Equivariant Dynamical Systems. XV, 153 pages. 2000.

Vol. 1729: J. Azéma, M. Émery, M. Ledoux, M. Yor. Séminaire de Probabilités XXXIV. VI, 431 pages. 2000.

Vol. 1730: S. Graf, H. Luschgy, Foundations of Quantization for Probability Distributions. X, 230 pages. 2000.

Vol. 1731: T. Hsu, Quilts: Central Extensions, Braid Actions, and Finite Groups. XII, 185 pages. 2000.

Vol. 1732: K. Keller, Invariant Factors, Julia Equivalences and the (Abstract) Mandelbrot Set. X, 206 pages. 2000.

Vol. 1733: K. Ritter, Average-Case Analysis of Numerical Problems. IX, 254 pages. 2000.

Vol. 1734: M. Espedal, A. Fasano, A. Mikelić, Filtration in Porous Media and Industrial Applications. Cetraro 1998. Editor: A. Fasano. 2000.

Vol. 1735: D. Yafaev, Scattering Theory: Some Old and New Problems. XVI, 169 pages. 2000.

Vol. 1736: B. O. Turesson, Nonlinear Potential Theory and Weighted Sobolev Spaces. XIV, 173 pages. 2000.

Vol. 1737: S. Wakabayashi, Classical Microlocal Analysis in the Space of Hyperfunctions. VIII, 367 pages. 2000.

Vol. 1738: M. Émery, A. Nemirovski, D. Voiculescu, Lectures on Probability Theory and Statistics. XI, 356 pages. 2000.

Vol. 1739: R. Burkard, P. Deuflhard, A. Jameson, J.-L. Lions, G. Strang, Computational Mathematics Driven by Industrial Problems. Martina Franca, 1999. Editors: V. Capasso, H. Engl, J. Periaux. VII, 418 pages. 2000.

Vol. 1740: B. Kawohl, O. Pironneau, L. Tartar, J.-P. Zolesio, Optimal Shape Design. Tróia, Portugal 1999. Editors: A. Cellina, A. Ornelas. IX, 388 pages. 2000.

Vol. 1741: E. Lombardi, Oscillatory Integrals and Phenomena Beyond all Algebraic Orders. XV, 413 pages. 2000.

Vol. 1742: A. Unterberger, Quantization and Non-holomorphic Modular Forms. VIII, 253 pages. 2000.

Vol. 1743: L. Habermann, Riemannian Metrics of Constant Mass and Moduli Spaces of Conformal Structures. XII, 116 pages. 2000.

Vol. 1744: M. Kunze, Non-Smooth Dynamical Systems. X, 228 pages. 2000.

Vol. 1745: V. D. Milman, G. Schechtman, Geometric Aspects of Functional Analysis. VIII, 289 pages. 2000.

Vol. 1746: A. Degtyarev, I. Itenberg, V. Kharlamov, Real Enriques Surfaces. XVI, 259 pages. 2000.

Vol. 1747: L. W. Christensen, Gorenstein Dimensions. VIII, 204 pages. 2000.

Vol. 1748: M. Ruzicka, Electrorheological Fluids: Modeling and Mathematical Theory. XV, 176 pages. 2001.

Vol. 1749: M. Fuchs, G. Seregin, Variational Methods for Problems from Plasticity Theory and for Generalized Newtonian Fluids. VI, 269 pages. 2001.

Vol. 1750: B. Conrad, Grothendieck Duality and Base Change. X, 296 pages. 2001.

Vol. 1751: N. J. Cutland, Loeb Measures in Practice: Recent Advances. XI, 111 pages. 2001.

Vol. 1752: Y. V. Nesterenko, P. Philippon, Introduction to Algebraic Independence Theory. XIII, 256 pages. 2001.

Vol. 1753: A. I. Bobenko, U. Eitner, Painlevé Equations in the Differential Geometry of Surfaces. VI, 120 pages. 2001.

Vol. 1754: W. Bertram, The Geometry of Jordan and Lie Structures. XVI, 269 pages. 2001.

Vol. 1755: J. Azéma, M. Émery, M. Ledoux, M. Yor, Séminaire de Probabilités XXXV. VI, 427 pages. 2001.

Vol. 1756: P. E. Zhidkov, Korteweg de Vries and Nonlinear Schrödinger Equations: Qualitative Theory. VII, 147 pages. 2001.

Vol. 1757: R. R. Phelps, Lectures on Choquet's Theorem. VII, 124 pages. 2001.

Vol. 1758: N. Monod, Continuous Bounded Cohomology of Locally Compact Groups. X, 214 pages. 2001.

Vol. 1759: Y. Abe, K. Kopfermann, Toroidal Groups. VIII, 133 pages. 2001.

Vol. 1760: D. Filipović, Consistency Problems for Heath-Jarrow-Morton Interest Rate Models. VIII, 134 pages. 2001.

Vol. 1761: C. Adelmann, The Decomposition of Primes in Torsion Point Fields. VI, 142 pages. 2001.

Vol. 1762: S. Cerrai, Second Order PDE's in Finite and Infinite Dimension. IX, 330 pages. 2001.

Vol. 1763: J.-L. Loday, A. Frabetti, F. Chapoton, F. Goichot, Dialgebras and Related Operads. IV, 132 pages. 2001.

Vol. 1764: A. Cannas da Silva, Lectures on Symplectic Geometry. XII, 217 pages. 2001.

Vol. 1765: T. Kerler, V. V. Lyubashenko, Non-Semisimple Topological Quantum Field Theories for 3-Manifolds with Corners. VI, 379 pages. 2001.

Vol. 1766: H. Hennion, L. Hervé, Limit Theorems for Markov Chains and Stochastic Properties of Dynamical Systems by Quasi-Compactness. VIII, 145 pages. 2001.

Vol. 1767: J. Xiao, Holomorphic Q Classes. VIII, 112 pages. 2001.

Vol. 1768: M.J. Pflaum, Analytic and Geometric Study of Stratified Spaces. VIII, 230 pages. 2001.

Vol. 1769: M. Alberich-Carramiñana, Geometry of the Plane Cremona Maps. XVI, 257 pages. 2002.

Vol. 1770: H. Gluesing-Luerssen, Linear Delay-Differential Systems with Commensurate Delays: An Algebraic Approach. VIII, 176 pages. 2002.

Vol. 1771: M. Émery, M. Yor, Séminaire de Probabilités 1967-1980. A Selection in Martingale Theory. IX, 553 pages. 2002.

Vol. 1772: F. Burstall, D. Ferus, K. Leschke, F. Pedit, U. Pinkall, Conformal Geometry of Surfaces in S^4. VII, 89 pages. 2002.

Vol. 1773: Z. Arad, M. Muzychuk, Standard Integral Table Algebras Generated by a Non-real Element of Small Degree. X, 126 pages. 2002.

Vol. 1774: V. Runde, Lectures on Amenability. XIV, 296 pages. 2002.

Vol. 1775: W. H. Meeks, A. Ros, H. Rosenberg, The Global Theory of Minimal Surfaces in Flat Spaces. Martina Franca 1999. Editor: G. P. Pirola. X, 117 pages. 2002.

Vol. 1776: K. Behrend, C. Gomez, V. Tarasov, G. Tian, Quantum Comohology. Cetraro 1997. Editors: P. de Bartolomeis, B. Dubrovin, C. Reina. VIII, 319 pages. 2002.

Vol. 1777: E. García-Río, D. N. Kupeli, R. Vázquez-Lorenzo, Osserman Manifolds in Semi-Riemannian Geometry. XII, 166 pages. 2002.

Vol. 1778: H. Kiechle, Theory of K-Loops. X, 186 pages. 2002.

Vol. 1779: I. Chueshov, Monotone Random Systems. VIII, 234 pages. 2002.

Vol. 1780: J. H. Bruinier, Borcherds Products on O(2,1) and Chern Classes of Heegner Divisors. VIII, 152 pages. 2002.

Vol. 1781: E. Bolthausen, E. Perkins, A. van der Vaart, Lectures on Probability Theory and Statistics. Ecole d' Eté de Probabilités de Saint-Flour XXIX-1999. Editor: P. Bernard. VII, 480 pages. 2002.

Vol. 1783: L. Grüne, Asymptotic Behavior of Dynamical and Control Systems under Perturbation and Discretization. X, 231 pages. 2002.

Vol. 1785: J. Arias de Reyna, Pointwise Convergence of Fourier Series. XVIII, 175 pages. 2002.

Recent Reprints and New Editions

Vol. 1200: V. D. Milman, G. Schechtman, Asymptotic Theory of Finite Dimensional Normed Spaces. 1986. – Corrected Second Printing. X, 156 pages. 2001.

Vol. 1618: G. Pisier, Similarity Problems and Completely Bounded Maps. 1995 – Second, Expanded Edition VII, 198 pages. 2001.

Vol. 1629: J. D. Moore, Lectures on Seiberg-Witten Invariants. 1997 – Second Edition. VIII, 121 pages. 2001.

Vol. 1638: P. Vanhaecke, Integrable Systems in the realm of Algebraic Geometry. 1996 – Second Edition. X, 256 pages. 2001.

Vol. 1702: J. Ma, J. Yong, Forward-Backward Stochastic Differential Equations and Their Applications. 1999. – Corrected Second Printing. XIII, 270 pages. 2000.